Nonlinear Systems and Complexity

Volume 29

Series editor

Albert C. J. Luo
Southern Illinois University
Edwardsville, IL, USA

Nonlinear Systems and Complexity provides a place to systematically summarize recent developments, applications, and overall advance in all aspects of nonlinearity, chaos, and complexity as part of the established research literature, beyond the novel and recent findings published in primary journals. The aims of the book series are to publish theories and techniques in nonlinear systems and complexity; stimulate more research interest on nonlinearity, synchronization, and complexity in nonlinear science; and fast-scatter the new knowledge to scientists, engineers, and students in the corresponding fields. Books in this series will focus on the recent developments, findings and progress on theories, principles, methodology, computational techniques in nonlinear systems and mathematics with engineering applications. The Series establishes highly relevant monographs on wide ranging topics covering fundamental advances and new applications in the field. Topical areas include, but are not limited to: Nonlinear dynamics Complexity, nonlinearity, and chaos Computational methods for nonlinear systems Stability, bifurcation, chaos and fractals in engineering Nonlinear chemical and biological phenomena Fractional dynamics and applications Discontinuity, synchronization and control.

More information about this series at http://www.springer.com/series/11433

Marat Akhmet • Mehmet Onur Fen
Ejaily Milad Alejaily

Dynamics with Chaos and Fractals

Springer

Marat Akhmet
Department of Mathematics
Middle East Technical University
Çankaya, Ankara, Turkey

Mehmet Onur Fen
Department of Mathematics
TED University
Çankaya, Ankara, Turkey

Ejaily Milad Alejaily
Department of Fundamental Sciences
College of Engineering Technology
Houn, Libya

ISSN 2195-9994 ISSN 2196-0003 (electronic)
Nonlinear Systems and Complexity
ISBN 978-3-030-35853-2 ISBN 978-3-030-35854-9 (eBook)
https://doi.org/10.1007/978-3-030-35854-9

This Springer imprint is published by the registered company Springer Nature Switzerland AG.
The registered company address is: Gewerbestrasse 11, 6330 Cham, Switzerland

To our beloved families

Preface

The book concerns the mathematical concepts, which are within the scope of dynamical systems, geometry, measure theory, topology, and numerical analysis during the last several decades. Moreover, they are not only at the focus, but on the frontiers of the research. In our opinion, this is true since there are many questions that have not been clarified and answered yet, and the facts and information that have already been collected in the study are not rigorously embedded to the theories of dynamical systems and differential equations. Additionally, they are the most sophisticated notions in their areas of science. We are talking about *chaos and fractals*.

It so happens that the line of oscillations in the classical theory of dynamical systems, which is founded by H. Poincaré and G. Birkhoff, was broken at Poisson stable motions. The next oscillations were considered as actors of chaotic processes. The main element for the stage was constructed by Lorenz, the sensitivity. Transitivity and the existence of infinitely many periodic orbits were inherited from the analysis of Poincaré for homoclinic motions. Next, the sensitivity was combined with the idea of Poisson stability to make the basics of the Li–Yorke definition of chaos: frequent separation and proximality. The suggestions, and first of all sensitivity and the density of periodic solutions, provide the instruments for the observation of chaos such as positive Lyapunov exponents and bifurcation diagrams. These are the main steps that make scientists and engineers do research for chaotic dynamics without a knowledge of the dynamical systems theory, and accordingly there is the strong opinion that the chaos theory exists, which does not depend so much on the theorics of dynamical systems and differential equations. The state of the deals is easily explainable if one takes into account the complexity of the subject. H. Poincaré himself comprehended that there should be such a property like sensitivity (instability), but did not have time to formalize the intuition to rigorous construction similarly to his several other predictions. The suggestions of Lorenz as well as Li and Yorke's study entitled "Period three implies chaos" essentially simplified the discussion when one concerns the difficulties of analysis for celestial mechanics, but this simultaneously poses an obstacle for deepening the theory. Therefore, as usual for the development of science, if the most universal property of the chaos cannot be found, then more general ones or new counterparts of the same generality level have to be considered. Another way is to proceed exactly from that point which was left by the genius who designed the theory. Our approach is based on the latter case since, loosely speaking, we modernize the definition of the Poisson stable point, which was created by H. Poincaré. In this book, it is revealed that a special kind of Poisson stable point, which we call an *unpredictable point*, gives rise to the existence of chaos in the quasi-minimal set. The existing definitions of chaos are formulated in sets of motions. This is the first time that the description of chaos is initiated from a single motion, an unpredictable one. This is a new oscillation in the line created by Poincaré. One can consider it as a critical moment, when chaos appears as an irregular object next to the regular ones. However, we prefer to say that right now the chaos is placed on the line of oscillations, and therefore, it is a subject of study in the framework of the theories of dynamical systems and differential equations, as in this book.

In spite of intensive researches concerning chaos, there is still a long distance to massive applications of methods of differential equations for analysis of chaotic dynamics. Input–output theorems are simple in formulation and widespread in the literature for differential equations, when the input and the output are oscillations of the same type such as periodic, almost periodic, recurrent, etc. In Chaps. 8–10, we consider inputs as well as outputs as sets of chaotically behaving solutions. This method is called *replication of chaos* and has been introduced and developed through our papers. Since the input–output mechanism makes the distribution of the phenomenon from one system to another possible, we are busy with the extension of chaos. In this book, the latest results of the method are presented.

The comprehension of the world through fractals promises to be one of the effective ways for creative scientific, biological, social, medical, and industrial activities. We are confident that self-similarity extended to arbitrary small scales of real-world subjects is in the core of effectiveness of the universe research and helps to optimize many processes as well as structures of objects. Thus, it is of significant interest to consider *fractals in dynamics.* Surprisingly, even the task of deformation of the sets, which keeps the fractal structure invariant, have not been investigated in the literature despite the fact that the problem is of interest for geometry, measure theory, and topology. The first steps in that direction are performed in this book. This makes it possible to develop continuous and discrete dynamics which admit fractals as points of trajectories as well as orbits themselves. These achievements allow us to analyze several practically useful problems.

To provide strong arguments for the generosity of chaos in the real and abstract universe, we suggest the concept of *abstract similarity*. The self-similar space is equipped with chaos if a special property is assumed. This provides the way to prove chaos for fractals. The Sierpinski sets and Koch snowflake are among them. We believe that the concept and its developments can become a universal instrument for chaos and fractal investigation. It may unite different definitions of chaos as well as ways of chaos detection. This is true also for new methods to determine and analyze fractal structures.

The authors would like to express their sincere gratitude to Madina Tleubergenova and Akylbek Zhamanshin for the joint results, and the series editor Albert C. J. Luo for his interest in the monograph and patience during the publication of the book.

Ankara, Turkey — Marat Akhmet
Ankara, Turkey — Mehmet Onur Fen
Houn, Libya — Ejaily Milad Alejaily

Contents

Chapter 1
Introduction

The concept of chaos has been one of the attractive topics among scientists starting with the studies of Poincaré [93], Cartwright and Littlewood [37], Levinson [73], Lorenz [77], and Ueda [106]. The first mathematical definition of chaos was introduced by Li and Yorke [74] for discrete dynamical systems in a compact interval of the real line. The presence of an uncountable scrambled set is one of the main features of the Li–Yorke chaos. The original definition of Li and Yorke was extended to dimensions greater than one by Marotto [81]. It was mentioned in [81] that a multidimensional continuously differentiable map possesses generalized Li–Yorke chaos if it has a snap-back repeller. On the other hand, according to Devaney [42], a discrete map from a compact interval to itself possesses chaos if it has sensitive dependence on initial conditions (in short, sensitivity), it is topologically transitive, and its periodic points are dense in the interval. Another chaos type is homoclinic chaos [99], which is generated from the famous manuscript of Poincaré [93]. One of the most significant discoveries of Poincaré in the theory of dynamical systems is the presence of homoclinic orbits in the three body problem of celestial mechanics [28, 42, 91–93]. In any neighborhood of a structurally stable Poincaré homoclinic orbit there exist nontrivial hyperbolic sets containing a countable number of saddle periodic orbits and continuum of non-periodic Poisson stable orbits [56, 98, 100]. Therefore, the existence of a structurally stable Poincaré homoclinic orbit can be considered as a criterion for the presence of chaos [56]. Additionally, heteroclinic orbits are also important for the investigation of chaotic dynamics [34, 38].

When one says about homoclinic chaos it is assumed that there is the homoclinic structure, and moreover, there is instability. Lorenz [77] was the first who mentioned the divergence of neighbor motions as sensitivity, a specific sort of instability, and this was followed by Li and Yorke [74], when they use frequent separation and proximality for the same purpose. It was shown in paper [70] that if a map on a compact interval has a two point scrambled set, then it possesses an uncountable scrambled set. The existence of Li–Yorke chaos in a spatiotemporal chaotic system

M. Akhmet et al., *Dynamics with Chaos and Fractals*, Nonlinear Systems and Complexity 29, https://doi.org/10.1007/978-3-030-35854-9_1

was proved in [75] by means of Marotto theorem, and generalizations of Li–Yorke chaos to mappings in Banach spaces and complete metric spaces were provided in [69, 96, 97]. Blanchard et al. [36], on the other hand, proved that the presence of positive topological entropy implies chaos in the sense of Li–Yorke. Moreover, Li–Yorke chaos on several spaces in connection with the cardinality of its scrambled sets was studied within the scope of the paper [60]. Furthermore, Li–Yorke sensitivity, which links the Li–Yorke chaos with the notion of sensitivity, was studied in the paper [27].

The absence of the quantitative description of instability makes homoclinic chaos different from the other chaos definitions, and this causes some sort of inconvenience. This is because sensitivity is assumed to be one of the main ingredients of chaos in its modern comprehension [42, 57, 59, 94, 112]. Poincaré himself was aware about the divergence of initially nearby trajectories, but had not given exact prescriptions how it should be proceeded. Therefore, one can say that the puzzle construction, which was initiated and designed by the French genius is not still completed. In this book, we are trying to make a contribution to this puzzle working. For that purpose we have utilized open Poisson stable motions, which accompany homoclinic chaos [99].

From the applications point of view, the theory of differential and discrete equations focuses on equilibria, periodic, and almost periodic oscillations. They meet the needs of any real world problem related to mechanics, electronics, economics, biology, etc., if one searches for regular and stable dynamics of an isolated motion. However, they are not sufficient for many modern and prospective demands of robotics, computer techniques, and the internet, and chaotic dynamics as well as fractals comprise constructive properties for applications. This is the reason why it is important to join the power of deterministic chaos and fractals with the immensely rich source of methods for differential and discrete equations.

The content of this book consists of three main parts: *Unpredictability, replication of chaos*, and *fractals*.

Unpredictability In this book, we develop the concept of Poisson stable point to unpredictable point by utilizing unpredictability as *individual sensitivity* for a motion. Thus, issuing from the single point of a trajectory we use it as the Ariadne's thread to come to the phenomenon, which we call *Poincaré chaos*. This phenomenon makes all the types of chaos closer to each other since it is a new description of motions in dynamics with homoclinic structure, and it admits ingredients similar to the other chaos types. The presence of infinitely many periodic motions in the other definitions can be substituted by a continuum of Poisson stable orbits. Our main hope is that these suggestions may bring research of chaos back to the theory of classical dynamical systems and provide instruments to make chaos a routine object of analysis in the theories of differential and discrete equations as periodic, quasi-periodic, and almost periodic solutions are. The introduction of a new type of motion is the strong argument for this, that is, the already existing list of oscillations in dynamical systems from equilibrium to Poisson stable orbits is now prolonged with unpredictable motions. This enlargement will give a push for the further extension of dynamical systems theory as well as the theories of differential, discrete, and hybrid equations. In applications, some properties and/or laws of dynamical systems can be lost or ignored, for instance, when non-autonomous or non-smooth systems are considered. Then we can apply unpredictable functions, a new type of oscillations which immediately follow Poisson stable motions of differential equations in the row of bounded solutions. They can be investigated for any type of equations since by our results they can be treated by methods of qualitative theory of differential equations. To this end we provide to the readers Chaps. 2–7.

The results for unpredictable solutions of systems of differential and discrete equations have been obtained very recently, i.e., the basics of the theory are being developed. Consequently, we consider various definitions and ways of discussion depending on the rich theories which exist for differential equations. It is worth noting that due to the brand newness of the results we use different definitions of unpredictable functions based on both the metric of Bebutov dynamics and the topology of uniform convergence on compact subsets of the real axis in the functional space. The latter seems to be more effective in theoretical and practical senses. Moreover, there are two different approaches for the proof of unpredictability of solutions in Chaps. 3 and 7. An attentive reader will see this very easily. For this reason we suggest to choose the approach which is more convenient to the reader's preferences.

The starting point of Chap. 2 is the unpredictable point, a new object for the dynamical systems theory founded by Poincaré [92] and Birkhoff [35]. In that chapter, we develop the Poisson stability of a motion to unpredictability such that a new type of chaos, the Poincaré chaos, has been obtained.

In Chap. 3, we investigate unpredictable solutions of quasilinear systems of differential equations by means of the Bebutov dynamical system [95]. An unpredictable function is defined as an unpredictable point of the Bebutov dynamics, and the first theorems on the existence of unpredictable solutions are proved.

As a continuation of Chaps. 2 and 3, we focus on the construction of an unpredictable function which is continuous on the real axis in Chap. 4. As auxiliary results, unpredictable orbits for the symbolic dynamics and the logistic map are obtained. It has become clear according to the results of Chaps. 3 and 4 that the concept of unpredictable points can be easily extended to the object of analysis in the theory of differential equations by considering unpredictable functions as points moving by shifts of the time argument.

In Chap. 5, on the other hand, we apply the topology of uniform convergence on compact sets to define unpredictable functions. The topology is metrizable and easy for applications with integral operators. Thus, one can accept that we lay a corner stone to the foundation of differential equations theory related to unpredictable solutions, and consequently, chaos. Therefore, in accordance with the results of Chap. 5, a new field to analyze in the theory of differential equations has been discovered. Since many results of differential equations have their counterparts in discrete equations [71], one can suppose that theorems on the existence of unpredictable solutions can be proved for discrete equations. Chapter 5 is one to realize the both paradigms. The existence and uniqueness theorems for quasilinear delay differential equations and difference equations have been proved, when the perturbation is an unpredictable function or sequence. This is visualized as Poincaré chaos in simulations.

The novelty of Chap. 6 is the investigation of unpredictable solutions non-homogeneous hyperbolic linear systems of differential equations such that the matrix of coefficients can possess eigenvalues both with negative and positive real parts. The counterparts of these results for discrete equations are also mentioned in that chapter.

In Chap. 7, we take into account quasilinear systems of differential equations with perturbations depending on both time and space variables, but unpredictable only in the time variable. It is also revealed in Chap. 7 that all coordinates of the obtained solutions are unpredictable. The solutions are called *strongly unpredictable*. This will make the meaning of unpredictability stronger for applications.

One of the important tools to recognize chaos is the symbolic dynamics, whose phase space consists of one-sided or two-sided infinite sequences of symbols chosen from a finite alphabet [29, 42, 58, 62, 67, 112]. The set of allowed sequences is invariant under the shift map [112]. The earliest examples of symbolic dynamics can be found in the studies [61, 85]. The horseshoe map and the logistic map are examples in which symbolic dynamics arise [29, 42, 112]. In the present book, it is demonstrated that symbolic dynamics on two symbols possesses an unpredictable solution, and hence, Poincaré chaos, and this is used to show the existence of unpredictable solutions in the dynamics of the Hénon map and the logistic map with certain coefficients.

The visualization of unpredictable solutions of systems of differential equations is important for the identification of chaotic behavior. One can make use of asymptotic or exponential stability of an unpredictable solution to approximate it with an arbitrary small error by means of a solution of the system under consideration with an appropriate initial data. This approximation eventually displays the unpredictable solution and useful for the recognition of irregularity and chaos.

Since we rigorously prove that unpredictable inputs produce outputs of the same type in systems of differential and discrete equations and owing to the fact that unpredictable solutions are necessarily accompanied with chaotic dynamics, our results return the investigation of chaos into the main stream of classical dynamical systems and, consequently, a huge number of rigorous mathematical methods, numerical instruments, and applications will now be involved in investigations of chaotic processes.

Simulation Remarks One can be confused with the numerical visualization of unpredictable functions and solutions of differential/discrete equations since it is a difficult task or not possible to determine the initial data for the solution. For this task we first utilize in the present manuscript the rich experience of simulations for bounded solutions. We issue from the fact that mainly asymptotically stable unpredictable solutions are supposed to be visualized. For this reason one can simulate any solution from the domain of attraction of the considered dynamics being confident that it eventually approximates the needed motion. Other arguments for the effectiveness of our numerical illustrations rely on the shadowing principle and density of the unpredictable dynamics in the quasi-minimal set.

Replication of Chaos Another large part of the book is concerned with increaseness of the role of differential equations research for the investigation of chaotic behavior.

It is known that if a function $I(t)$ with a certain property such as boundedness, periodicity, or almost periodicity is considered as an input for an evolution equation $z'(t) = T[z] + I(t)$, where $T[z]$ is a linear operator with spectra placed in the left half of the complex plane, then the equation produces a solution, an output, with a similar property of boundedness, periodicity, or almost periodicity [40, 54]. On the basis of this input–output mechanism, it is natural to ask whether chaotic inputs produce chaotic outputs in systems of differential and discrete equations. Since the concept of chaos is much more complex than just single bounded, periodic or almost periodic solutions, to answer this question we initiated in papers [1–9] a new method. It was called *replication of chaos* and is widely applied and developed in the studies [13–22, 25, 50–53]. Replication of chaos consists of the verification of ingredients of chaos such as sensitivity, transitivity, proximality, and the existence of infinitely many unstable regular motions [42, 74, 112] for solutions of an equation with chaotic perturbation. We consider as an input first of all a single function, a member of a chaotic set, to obtain a solution which is a member of another chaotic set. Additionally, we consider chaotic sets of functions as the input and the output. We have been forced to consider sets of functions as inputs and outputs since Devaney or Li–Yorke chaos are indicated through relation of motions (sensitivity, transitiveness, proximality). Thus, we consider the input and the output not only as single functions, but also as collections of functions. The way of our investigation is arranged in the well accepted traditional mathematical fashion, but with a new and a more complex way of arrangement of the connections between the input and the output. The method of replication of chaos gives a very effective instrument for application of the accumulated knowledge in chaos research. This method is applied in Chaps. 8–10.

Chapter 8 deals with the presence of Li–Yorke chaos in dynamic equations on time scales. The mathematical description of the chaos is provided for dynamic equations on time scales, and the proximality and frequent separation features are theoretically proved.

The generation of homoclinic and heteroclinic motions in the continuous-time dynamics of economic models is provided in Chap. 9. Exogenous shocks are utilized in the model, and an example for the Kaldor model of the aggregate economy is presented.

Results on replication of chaos and unpredictable solutions of differential equations have been applied in Chap. 10 to study the chaotic behavior of the hydrosphere and its influence on global weather and climate. Coupled systems based on the Lorenz [77] and Vallis [107, 108] systems are utilized in Chap. 10 to model the global ocean–atmosphere dynamics.

The third part of the book is devoted to constructing dynamics and chaos for fractals. This consists of two tasks. First, a connection of the dynamics of differential equations with fractals is proposed by developing a fractal mapping iteration such that the motion associated with a differential equation is considered as a map with a fractal as an initial set. Secondly, a new concept of abstract self-similarity is developed to construct chaotic dynamics for fractals.

Fractals Geometry and Chaos Chaos and fractals are interesting fields of scientific research in mathematics, physics, engineering, and many other branches of sciences. They provide us with powerful mathematical tools to analyze and understand the irregularity and complexity of many natural and artificial phenomena. Fractal geometry is older than chaos theory, however, the mathematical terms *chaos* and *fractal* are Irish twins. Tien-Yien Li and James Yorke [74] use the word *chaos* in 1975 to describe an irregular behavior of certain type of dynamical systems, whereas the word *fractal* was coined by Benoit Mandelbrot [78] in the same year, refers to certain geometrical structures.

Fractal geometry is a mathematical tool used to describe many natural structures that adopt various degrees of *self-similarity*, as well as to design some artificial structures. A set that displays self-similarity and repeats the same patterns at every scale is usually called fractal [55]. Mandelbrot defined a fractal as a set for which the Hausdorff dimension strictly exceeds the topological dimension [79]. Dealing with fractals goes back to the seventeenth century when Gottfried Leibniz introduced notions of recursive self-similarity [114]. Since then, history has not recorded anything about self-similarity until the late nineteenth century when Karl Weierstrass introduced in 1872 a function that being everywhere continuous but nowhere differentiable. The graph of the Weierstrass function became an example of a self-similar fractal. The Cantor set, constructed by Georg Cantor in 1883, is considered as the most essential and influential fractal, since it is a simple and perfect example for theory and applications of fractals. The idea of self-similarity received more attention in the work of Helge von Koch. He devised in 1904 a continuous but non-differentiable curve that never intersects itself. The curve is considered as one of the simplest regular fractals. Waclaw Sierpinski was one of the mathematicians who made significant contributions in the field of fractals. He introduced the famous triangular fractal in 1916, known as the Sierpinski gasket. The fractal is generated by a recursive process of removing symmetrical parts from an initial triangle. In an analogous way to the gasket, Sierpinski developed a square fractal known as the Sierpinski carpet. Julia sets gained significance in being generated using the dynamics of iterative function. They are discovered by Gaston Julia and Pierre Fatou in 1917–19, where they studied independently the iteration of rational functions in the complex plane [80]. During the same period, the mathematician Felix Hausdorff formulated the notion of fractional dimension which became a very important tool for studying and characterizing the geometrical complexity of fractals. Paul Lévy studied the self-similar curves and surfaces, and in 1938 he described a new fractal curve, the Lévy C curve. Benoit Mandelbrot was one of the first to use computer to study and generate fractal shapes. He defines a fractal as a set whose Hausdorff dimension strictly larger than its topological dimension [79].

Several researches pointed out that a close relationship between chaos and fractal geometry can be observed. This can be seen, for instant, in the dynamics of Fatou-Julia iteration used to construct Julia and Mandelbrot sets where two neighbor points in the domain which are close to the boundary may have completely different behavior. That is, we can say about sensitivity in fractal structures. Chaos tells us about the state of irregularity and divergence of trajectories which depend in the nature of the dynamics, whereas the fractal concept can be used to study complex geometric structures. Therefore, the interlink between chaos and fractals is more clear when fractal dimension is used to measure the extent to which a trajectory fills its phase space. In other words, fractal dimension of the orbit in phase space implies the existence of a strange attractor [83]. The fundamental work on the chaotic nature of fractals has been done only for specific types categorized under the totally disconnected fractals [32, 39, 42]. In that work, the topological conjugacy concept was utilized to prove that these fractal sets are invariant for certain chaotic maps. Except for that, relatively few studies have been carried out on chaotic dynamical systems for fractals, and perhaps the most relevant one is what have been done on the Sierpinski carpet in [46]. In that research, the author shows that the dynamical system associated with a shift transformation defined on the Sierpinski carpet set is chaotic in the sense of topological mixing.

Dynamics Through Fractal Mappings The famous Laplace's Demon is not only of strict physical determinism, but also related to the power of differential equations. When deterministically extended structures are taken into consideration, it is admissible that fractals are dense both in the nature and in the dynamics. In particular, this is true because fractal structures are closely related to chaos. This implies that dynamics have to be an instrument of the extension. Oppositely, one can animate the arguments for the Demon if dynamics will be investigated with fractals. To make advances in the direction, first of all, one should consider fractals as states of dynamics. In other words, instead of single points and finite/infinite dimensional vectors, fractals should be points of trajectories as well as trajectories themselves. If one realizes this approach, fractals will be proved to be dense in the universe, since modeling the real world is based on differential equations and their developments. Our main goal is to initiate the involvement of fractals as states of dynamical systems, and in the first step we answer the simple question "How can fractals be mapped?".

Fractals are in the forefront of researches in many areas of science as well as for interdisciplinary investigations [32, 82]. One cannot say that motion is a strange concept for fractals. Dynamics are beside the fractals immediately as they are constructed by iterations. It is mentioned in the book [87] that it is inadequate to talk about fractals while ignoring the dynamical processes which created them. That is, iterations are in the basis of any fractal, but we still cannot say that differential equations are widely interrelated to fractals, for instance, as much as manifolds [103]. Our book is intended to open a gate for an inflow of methods of differential equations, discrete equations and any other methods of dynamics research to the realm of fractals. This will help not only to investigate fractals but also to make their diversity richer and ready for intensive applications. Formally speaking, in our investigation we join dynamics of iterations, which can be called inner dynamics, with outer dynamics of differential and discrete equations. The concept of fractals has already many applications, however, the range would be significantly enlarged if all the power of differential equations will be utilized for the structures. This is why our suggestions are crucial for fractals considered in biology, physics, city planning, economy, image recognition, artificial neural networks, brain activity study, chemistry, and all engineering disciplines [33, 66, 90, 105, 109, 113], i.e., in every place, where the geometrical objects in physical and/or abstract sense may appear [49]. It deserves to say that differential equations related to fractals were already discussed in [68, 104], where they are considered as domains of partial differential equations, but not as building blocks of trajectories. In this sense one can take advantage of the papers [68, 104] in the next development of our proposals. The same is true for the studies concerned with deep analyses of fractals growth performed in [31, 110].

In Chap. 11, we make a possible study of mapping fractals, which is simple from one side since it relates to classical functions, but from another side it is a developed one since we apply the mapping function in a new manner, which nevertheless still relates to the original ideas of Julia. Based on the constructed mapping iteration, dynamics for fractals are introduced. We proceed upon the fact that motion of fractals comprehend as deformation was not considered in literature at all. From the mathematical point of view, the deformation is equivalent to mapping. Why mapping for fractals was not investigated in the literature before? The reason lies in the complexity of fractals building. Theoretically, the procedure is complex and sophisticated since it contains infinitely many steps. Consequently, to map fractals, one has to combine a map's algorithm with the complex algorithm of a fractal construction. We have found the combination, and it is one of the main achievements of the Chap. 11. That is, in fact, we introduce a new type of functions, with fractal domains, and determine the properties for the functions to have fractal-images and approve this. Moreover, using the algorithm of mapping, we find conditions for discrete equations to admit trajectories consisting of fractals as well as conditions for continuous trajectories of autonomous differential equations (dynamical systems) to be of fractals. Furthermore, a discussion about admitting fractal-points for non-autonomous differential equations is provided. For the theoretical discussions, we apply results of the fractal geometry (dimension) and theorems and definitions from the theory of dynamical systems, differential and discrete equations theory. The results presented in Chap. 11 originally appeared in the paper [23, 24, 26].

Abstract Self-Similarity and Chaotic Dynamics for Fractals Similarity is an expression of sharing common properties. It is used to indicate that two or more objects are similar to each other. In mathematics, similarity is a relation between point sets. That is, two geometric figures are similar if there exists a similarity transformation that maps one onto the other [76, 111]. The notion of self-similarity is used to say about the similarity within a single object. Self-similarity became one of the most important concepts in modern science. From the geometrical point of view, it is defined as the property of objects whose parts, at all scales, are similar to the whole [31, 65]. Roots of the idea of self-similarity date back to the seventeenth century when Gottfried Leibniz introduced the notions of recursive self-similarity [114]. The last third of nineteenth century has witnessed a lot of work into the subject which has been embodied in several discoveries such as Weierstrass function by Karl Weierstrass, Cantor set by Georg Cantor, and space-filling curves described by Giuseppe Peano and David Hilbert. Other examples of self-similar sets are Koch curve discovered by Helge von Koch in 1904, Sierpinski gasket and carpet which are introduced by Waclaw Sierpinski in 1916, and Lévy C curve described by Paul Lévy in 1938. Benoit Mandelbrot invented the term *fractal* in 1975 [78] to describe certain geometrical structures that exhibit self-similarity. Since then, this word has been employed to denote all the above mentioned sets, and the field became known as *fractal geometry*. Consequently, the fractal concept is axiomatically linked with the notion of self-similarity which is considered to be one of the acceptable definitions of fractals. That is, a fractal can be defined as a set that display self-similarity at all scales. However, Mandelbrot defines a fractal as a set whose Hausdorff dimension strictly larger than its topological dimension [79]. To sum up, self-similarity and fractional dimension are the most two important features of fractals. The connection between them is that self-similarity is the easiest way to construct a set that has fractional dimension [41].

The idea of a self-similar set was first considered by Moran in 1946 [84] who gave a mathematical definition of a geometric construction as a collection of sets satisfying specific conditions. The Moran's construction has been studied extensively and developed by many researcher proposing various methods. In [88, 89, 102], the approach was generalized to build different classes of Moran-like geometric construction with the help of a symbolic space. The topological conjugacy to symbolic dynamical system was employed to describe the symbolic representation of the constructions. This type of self-similar sets is referred to as Cantor-like set. Interesting definitions of self-similarity and related problems of dimension and measure are discussed in the papers [30, 45, 47, 48, 63, 64, 72, 86, 101]. The manuscripts consider sets in Euclidean space $\mathbb{R}^n$ and define self-similar set as a union of its images under similarity transfunctions [47].

In Chap. 12, A new mathematical concept of abstract similarity is introduced. We develop self-similarity based on point-set structure in metric spaces and this is why we call it *abstract self-similarity*. The development does not rely on any special functions, and the *similarity map* used in our book is applied for chaos formation. The map is a natural consequence of the structure of the domain and it can be considered as a generalization of the Bernoulli shift on symbolic spaces. Our notion of similarity map, therefore, is different from the classical ones essentially considered in literature. In the present chapter, we are not concerned in analysis of dimension and measure but mostly rely on the distance, since our main goal is to discuss chaos problem for fractals. Nevertheless, we suppose that our suggestion may be useful for the next extension of the results obtained in

[30, 45, 47, 48, 63, 64, 72, 86, 101] for the abstract self-similarity case. First of all those which consider dimension and measure and corresponding problems of fractals.

The achieved results in Chap. 12 have a special importance that we do not utilize topological conjugacy to prove the presence of chaos in fractal sets. Moreover, we consider different types of chaos, namely Devaney, Li–Yorke, and Poincaré, and the results cover several kinds of self-similar fractals such as Cantor sets, Sierpinski fractals, and Koch curve. Chaos in fractals, particularly Sierpinski carpet and Koch curve, has not been considered in the literature before. Furthermore, our approach is applicable for sets in multidimensional spaces. A good example is the logistic map. It was shown that a chaos equivalent to Li–Yorke type can be extended to higher-dimensional discrete systems [43, 44, 81]. This requires employing special theorems like Marotto theorem [81]. Applying chaotic abstract similarity developed in our research, we have shown that Devaney, Li–Yorke, and Poincaré chaos can take place in the dynamics of n connected perturbed logistic maps. The results of Chap. 12 have been published in [10]. Applications of the method for neural networks and multidimensional chaos are realized in papers [11, 12].

References

1. M.U. Akhmet, Hyperbolic sets of impact systems, Dyn. Contin. Discrete Impuls. Syst. Ser. A Math. Anal. 15 (Suppl. S1), 1–2, in *Proceedings of the 5th International Conference on Impulsive and Hybrid Dynamical Systems and Applications* (Watan Press, Beijing, 2008)
2. M.U. Akhmet, Dynamical synthesis of quasi-minimal sets. Int. J. Bifurcat. Chaos **19**, 2423–2427 (2009)
3. M.U. Akhmet, Shadowing and dynamical synthesis. Int. J. Bifurcat. Chaos **19**, 3339–3346 (2009)
4. M.U. Akhmet, Devaney's chaos of a relay system. Commun. Nonlinear Sci. Numer. Simulat. **14**, 1486–1493 (2009)
5. M.U. Akhmet, Li-Yorke chaos in the system with impacts. J. Math. Anal. Appl. **351**, 804–810 (2009)
6. M.U. Akhmet, Creating a chaos in a system with relay. Int. J. Qualit. Th. Diff. Eqs. Appl. **3**, 3–7 (2009)
7. M.U. Akhmet, The complex dynamics of the cardiovascular system. Nonlinear Analysis **71**, e1922–e1931 (2009)
8. M.U. Akhmet, Homoclinical structure of the chaotic attractor. Commun. Nonlinear Sci. Numer. Simulat. **15**, 819–822 (2010)
9. M.U. Akhmet, *Principles of Discontinuous Dynamical Systems* (Springer, New York, 2010)
10. M. Akhmet, E.M. Alejaily, Abstract Similarity, Fractals and Chaos. ArXiv e-prints, arXiv:1905.02198, 2019 (submitted)
11. M. Akhmet, E.M. Alejaily, Domain-structured chaos in a Hopfield neural network. Int. J. Bifurc. Chaos, 2019 (in press)
12. M. Akhmet, E.M. Alejaily, Chaos on the Multi-Dimensional Cube. ArXiv e-prints, arXiv:1908.11194, 2019 (submitted)
13. M.U. Akhmet, M.O. Fen, Chaotic period-doubling and OGY control for the forced Duffing equation. Commun. Nonlinear Sci. Numer. Simul. **17**, 1929–1946 (2012)
14. M.U. Akhmet, M.O. Fen, Replication of chaos. Commun. Nonlinear Sci. Numer. Simul. **18**, 2626–2666 (2013)

15. M.U. Akhmet, M.O. Fen, Chaos generation in hyperbolic systems. Discontinuity Nonlinearity Complexity **1**, 353–365 (2012)
16. M.U. Akhmet, M.O. Fen, Shunting inhibitory cellular neural networks with chaotic external inputs. Chaos **23**, 023112 (2013)
17. M. Akhmet, M.O. Fen, Chaotification of impulsive systems by perturbations. Int. J. Bifurcat. Chaos **24**, 1450078 (2014)
18. M.U. Akhmet, M.O. Fen, Replication of discrete chaos. Chaotic Model. Simul. (CMSIM) **2**, 129–140 (2014)
19. M. Akhmet, M.O. Fen, Attraction of Li-Yorke chaos by retarded SICNNs. Neurocomputing **147**, 330–342 (2015)
20. M. Akhmet, M.O. Fen, A. Kıvılcım, Li-Yorke chaos generation by SICNNs with chaotic/almost periodic postsynaptic currents. Neurocomputing **173**, 580–594 (2016)
21. M. Akhmet, M.O. Fen, *Replication of Chaos in Neural Networks, Economics and Physics* (Higher Education Press, Beijing; Springer, Heidelberg, 2016)
22. M. Akhmet, M.O. Fen, Input-output mechanism of the discrete chaos extension, in *Complex Motions and Chaos in Nonlinear Systems*, ed. by V. Afraimovich, J.A.T. Machado, J. Zhang (Springer, Switzerland, 2016), pp. 203–233
23. M. Akhmet, M.O. Fen, E.M. Alejaily, Dynamics with fractals. Discontinuity Nonlinearity Complexity (in press)
24. M. Akhmet, M.O. Fen, E.M. Alejaily, Mapping Fatou-Julia Iterations. Proc. ICIME **2018**, 64–67 (2018)
25. M. Akhmet, M.O. Fen, E.M. Alejaily, Extension of sea surface temperature unpredictability. Ocean Dynamics **69**, 145–156 (2019)
26. M. Akhmet, M.O. Fen, E.M. Alejaily, Generation of fractals as Duffing equation orbits. Chaos **29**, 053113 (2019)
27. E. Akin, S. Kolyada, Li-Yorke sensitivity. Nonlinearity **16**, 1421–1433 (2003)
28. K.G. Andersson, Poincaré's discovery of homoclinic points. Arch. Hist. Exact Sci. **48**, 133–147 (1994)
29. B.-L. Hao, W.-M. Zheng, *Applied Symbolic Dynamics and Chaos* (World Scientific Publishing Company, 1998)
30. C. Bandt, S. Graf, Self-similar sets 7. A characterization of self-similar fractals with positive Hausdorff measure. Proc. Am. Math. Soc. **114**, 995–1001 (1992)
31. L. Barabasi, H.E. Stanley, *Fractal Concepts in Surface Growth* (Cambridge University Press, Cambridge, 1995)
32. M.F. Barnsley, *Fractals Everywhere* (Academic Press, London, 1988)
33. M. Batty, P.A. Longley, *Fractal Cities: A Geometry of Form and Function* (Academic Press, London, 1994)
34. A.L. Bertozzi, Heteroclinic orbits and chaotic dynamics in planar fluid flows. SIAM J. Math. Anal. **19**, 1271–1294 (1988)
35. G.D. Birkhoff, *Dynamical Systems*, vol. 9 (Amer. Math. Soc., Colloquium Publications, Providence, 1927)
36. F. Blanchard, E. Glasner, S. Kolyada, A. Maass, On Li-Yorke pairs. J. Reine Angew. Math. **2002**, 51–68 (2002)
37. M. Cartwright, J. Littlewood, On nonlinear differential equations of the second order I: The equation $\ddot{y} - k(1 - y^2)'y + y = bkcos(\lambda t + a)$, k large. J. Lond. Math. Soc. **20**, 180–189 (1945)
38. R. Chacon, J.D. Bejarano, Homoclinic and heteroclinic chaos in a triple-well oscillator. J. Sound Vib. **186**, 269–278 (1995)
39. G. Chen, Y. Huang, *Chaotic Maps: Dynamics, Fractals and Rapid Fluctuations, Synthesis Lectures on Mathematics and Statistics* (Morgan and Claypool Publishers, Texas, 2011)
40. C. Corduneanu, *Almost Periodic Functions* (Interscience Publishers, New York, London, Sydney, 1968)
41. R.M. Crownover, *Introduction to Fractals and Chaos* (Jones and Bartlett, Boston, MA, 1995)
42. R.L. Devaney, *An Introduction to Chaotic Dynamical Systems* (Addison-Wesley, USA, 1989)

43. P. Diamond, Chaotic behavior of systems of difference equations. Int. J. Syst. Sci. **7**, 953–956 (1976)
44. A. Dohtani, Occurrence of chaos in higher dimensional discrete time systems. SIAM J. Appl. Math. **52**, 1707–1721 (1992)
45. G.A. Edgar, *Measure, Topology, and Fractal Geometry* (Springer, New York, 1990)
46. C. Ercai, Chaos for the Sierpinski carpet. J. Stat. Phys. **88**, 979–984 (1997)
47. K.J. Falconer, Sub-self-similar sets. Trans. Amer. Math. Soc. **347**, 3121–3129 (1995)
48. K. J. Falconer, *The Geometry of Fractal Sets* (Cambridge Univ. Press, Cambridge, 1985)
49. P. Fatou, Sur les équations fonctionnelles, I, II, III. Bull. Soc. Math. France **47**, 161–271 (1919); **48**, 33–94 (1920); **48**, 208–314 (1920)
50. M.O. Fen, Persistence of chaos in coupled Lorenz systems. Chaos Solitons Fractals **95**, 200–205 (2017)
51. M.O. Fen, M. Akhmet, Impulsive SICNNs with chaotic postsynaptic currents. Discret. Contin. Dyn. Syst. Ser. B **21**, 1119–1148 (2016)
52. M.O. Fen, F. Tokmak Fen, SICNNs with Li-Yorke chaotic outputs on a time scale. Neurocomputing **237**, 158–165 (2017)
53. M.O. Fen, F. Tokmak Fen, Replication of period-doubling route to chaos in impulsive systems. Electron. J. Qual. Theory Differ. Equ. **2019**(58), 1–20 (2019)
54. A.M. Fink, *Almost Periodic Differential Equations* (Springer, New York, 1974)
55. G. Franceschetti, D. Riccio, *Scattering, Natural Surfaces and Fractals* (Academic Press, Burlington, 2007)
56. S.V. Gonchenko, L.P. Shil'nikov, D.V. Turaev, Dynamical phenomena in systems with structurally unstable Poincaré homoclinic orbits. Chaos **6**, 15–31 (1996)
57. J.M. Gonzáles-Miranda, *Synchronization and Control of Chaos* (Imperial College Press, London, 2004)
58. C. Grebogi, J.A. Yorke, *The Impact of Chaos on Science and Society* (United Nations University Press, Tokyo, 1997)
59. J. Guckenheimer, P.J. Holmes, *Nonlinear Oscillations, Dynamical Systems, and Bifurcations of Vector Fields* (Springer, New York, Heidelberg, Berlin, 1983)
60. J.L.G. Guirao, M. Lampart, Li and Yorke chaos with respect to the cardinality of the scrambled sets. Chaos Solitons Fractals **24**, 1203–1206 (2005)
61. J. Hadamard, Les surfaces à courbures opposées et leurs lignes géodésiques. J. Math. Pures et Appl. **4**, 27–74 (1898)
62. J. Hale, H. Koçak, *Dynamics and Bifurcations* (Springer, New York, 1991)
63. M. Hata, On the structure of self-similar sets. Jpn. J. Appl. Math. **2**, 381–414 (1985)
64. J. Hutchinson, Fractals and self-similarity. Indiana Univ. J. Math. **30**, 713–747 (1981)
65. A.K. Janahmadov, M. Javadov, *Fractal Approach to Tribology of Elastomers* (Springer, Switzerland, 2018)
66. J.A. Kaandorp, *Fractal Modelling: Growth and Form in Biology* (Springer, New York, 2012)
67. J. Kennedy, J.A. Yorke, Topological horseshoes. Trans. Am. Math. Soc. **353**, 2513–2530 (2001)
68. J. Kigami, *Analysis on Fractals* (Cambridge Univ. Press, Cambridge, 2001)
69. P. Kloeden, Z. Li, Li-Yorke chaos in higher dimensions: a review. J. Differ. Equ. Appl. **12**, 247–269 (2006)
70. M. Kuchta, J. Smítal, *Two Point Scrambled Set Implies Chaos. European Conference on Iteration Theory (ECIT 87)* (World Sci. Publishing, Singapore, 1989), pp. 427–430
71. V. Lakshmikantham, D. Trigiante, *Theory of Difference Equations: Numerical Methods and Applications* (Marcel Dekker, USA, 2002)
72. K.S. Lau, S.M. Ngai, H. Rao, Iterated function systems with overlaps and the self-similar measures. J. Lond. Math. Soc. **63**, 99–115 (2001)
73. N. Levinson, A second order differential equation with singular solutions. Ann. Math. **50**, 127–153 (1949)
74. T.Y. Li, J.A. Yorke, Period three implies chaos. Am. Math. Monthly **82**, 985–992 (1975)

75. P. Li, Z. Li, W.A. Halang, G. Chen, Li-Yorke chaos in a spatiotemporal chaotic system. Chaos Solitons Fractals **33**, 335–341 (2007)
76. S. Libeskind, *Euclidean and Transformational Geometry: A Deductive Inquiry* (Jones and Bartlett Publishers, Sudbury, MA, 2008)
77. E.N. Lorenz, Deterministic nonperiodic flow. J. Atmos. Sci. **20**, 130–141 (1963)
78. B.B. Mandelbrot, *Les Objets Fractals: Forme, Hasard, et Dimension* (Flammarion, Paris, 1975)
79. B.B. Mandelbrot, *The Fractal Geometry of Nature* (Freeman, New York, 1983)
80. B.B. Mandelbrot, *Fractals and Chaos: The Mandelbrot Set and Beyond* (Springer, New York, 2004)
81. F.R. Marotto, Snap-back repellers imply chaos in $\mathbb{R}^n$. J. Math. Anal. Appl. **63**, 199–223 (1978)
82. E.W. Mitchell, S.R. Murray (eds.), *Classification and Application of Fractals: New Research* (Nova Science Publishers, New York, 2012)
83. F.C. Moon, *Chaotic and Fractal Dynamics: An Introduction for Applied Scientists and Engineers* (Wiley, New York, 1992)
84. P.A.P. Moran, Additive functions of intervals and Hausdorff measure. Proc. Cambridge Philos. Soc. **42**, 15–23 (1946)
85. M. Morse, G.A. Hedlund, Symbolic dynamics. Am. J. Math. **60**, 815–866 (1938)
86. S.M. Ngai, Y. Wang, Hausdorff dimension of overlapping self-similar sets. J. Lond. Math. Soc. **63**, 655–672 (2001)
87. H-O. Peitgen, H. Jürgens, D. Saupe, *Chaos and Fractals* (Springer, New York, 2004)
88. Y. Pesin, *Dimension Theory in Dynamical Systems: Contemporary Views and Applications* (University of Chicago Press, Chicago, 1997)
89. Y. Pesin, H. Weiss, On the dimension of deterministic and random cantor-like sets, symbolic dynamics, and the Eckmann-Ruelle conjecture. Comm. Math. Phys. **182**, 105–153 (1996)
90. L. Pietronero, E. Tosatti, *Fractals in Physics* (North-Holland, Amsterdam, 2012)
91. H. Poincaré, Sur le probléme des trois corps et les équations de la dynamique. Acta Math. **13**, 1–270 (1880)
92. H. Poincaré, *Les méthodes nouvelles de la mécanique céleste*, Vol. 1, 2 (Gauthier-Villars, Paris, 1892)
93. H. Poincaré, *Les methodes nouvelles de la mecanique celeste*, Vol. III, Paris, 1899; reprint (Dover, New York, 1957)
94. C. Robinson, *Dynamical Systems: Stability, Symbolic Dynamics, and Chaos* (CRC Press, Boca Raton, 1995)
95. G.R. Sell, *Topological Dynamics and Ordinary Differential Equations* (Van Nostrand Reinhold Company, London, 1971)
96. Y. Shi, G. Chen, Chaos of discrete dynamical systems in complete metric spaces. Chaos Solitons Fractals **22**, 555–571 (2004)
97. Y. Shi, G. Chen, Discrete chaos in Banach spaces. Sci. China Ser. A Math. **48**, 222–238 (2005)
98. L.P. Shil'nikov, On a Poincaré-Birkhoff problem. Math. USSR-Sbornik **3**, 353–371 (1967)
99. L. Shilnikov, Homoclinic chaos, in *Nonlinear Dynamics, Chaotic and Complex Systems*, ed. by E. Infeld, R. Zelazny, A. Galkowski (Cambridge University Press, Cambridge, 1997), pp. 39–63
100. S. Smale, Diffeomorphisms with many periodic points, in *Differential and Combinatorial Topology*, ed. by S.S. Cairns (Princeton University Press, Princeton, 1965), pp. 63–80
101. D.W. Spear, Measure and self-similarity. Adv. Math. **91**, 143–157 (1992)
102. S. Stella, On Hausdorff dimension of recurrent net fractals. Proc. Am. Math. Soc. **116**, 389–400 (1992)
103. B.Yu. Sternin, V.E. Shatalov, *Differential Equations on Complex Manifolds* (Kluwer Academic Publishers, Dordrecht, 1994)
104. R.S. Strichartz, *Differential Equations on Fractals: A Tutorial* (Princeton University Press, Princeton, 2006)

105. M. Takayasu, H. Takayasu, Fractals and economics, in *Complex Systems in Finance and Econometrics*, ed. by R.A. Meyers (Springer, New York, 2011), pp. 444–463
106. Y. Ueda, Random phenomena resulting from non-linearity in the system described by Duffing's equation. Trans. Inst. Electr. Eng. Jpn. **98A**, 167–173 (1978)
107. G.K. Vallis, El Niño: A chaotic dynamical system? Science **232**, 243–245 (1986)
108. G.K. Vallis, Conceptual models of El Niño and the southern oscillation. J. Geophys. **93**, 13979–13991 (1988)
109. J.L. Véhel, E. Lutton, C. Tricot (Eds.), *Fractals in Engineering: From Theory to Industrial Applications* (Springer, New York, 1997)
110. T. Vicsek, *Fractal Growth Phenomena*, 2nd edn. (World Scientific, Singapore, 1992)
111. H.J. Vollrath, The understanding of similarity and shape in classifying tasks. Educ. Stud. Math. **8**, 211–224 (1977)
112. S. Wiggins, *Global Bifurcation and Chaos: Analytical Methods* (Springer, New York, Berlin, 1988)
113. L. Zhao, W. Li, L. Geng, Y. Ma, Artificial neural networks based on fractal growth, in *Advances in Automation and Robotics 2, Lecture Notes in Electrical Engineering*, vol. 123, ed. by G. Lee (Springer, Berlin, 2011), pp. 323–330
114. O. Zmeskal, P. Dzik, M. Vesely, Entropy of fractal systems. Comput. Math. Appl. **66**, 135–146 (2013)

Chapter 2
The Unpredictable Point and Poincaré Chaos

This chapter is an introductory one for the book. It is revealed that a special kind of Poisson stable point, which we call an unpredictable point, gives rise to the existence of chaos in the quasi-minimal set. The existing definitions of chaos are formulated in sets of motions. This is the first time that description of chaos is initiated from a single motion. The theoretical results are exemplified by means of the symbolic dynamics.

2.1 Preliminaries

The mathematical dynamics theory, which was founded by Poincaré [15] and significantly developed by the French genius and Birkhoff [4], was a source as well as the basis for the later discoveries and thorough investigations of complex dynamics [7, 9–11, 18]. The homoclinic chaos was discussed by Poincaré [14], and Lorenz [10] observed that a strange attractor contains a Poisson stable trajectory. Possibly, it was Hilmy [8, 13] who gave the first definition of a quasi-minimal set (as the closure of the hull of a Poisson stable motion). In [13, p. 361] one can find a theorem by Hilmy, which states the existence of an uncountable set of Poisson stable trajectories in a quasi-minimal set. We modify the Poisson stable points to unpredictable points such that the quasi-minimal set is chaotic.

Let (X, d) be a metric space and $\mathbb{T}$ refer to either the set of real numbers or the set of integers. A mapping $f : \mathbb{T} \times X \to X$ is a flow on X [17] if:

(i) $f(0, p) = p$ for all $p \in X$;
(ii) $f(t, p)$ is continuous in the pair of variables t and p;
(iii) $f(t_1, f(t_2, p)) = f(t_1 + t_2, p)$ for all $t_1, t_2 \in \mathbb{T}$ and $p \in X$.

M. Akhmet et al., *Dynamics with Chaos and Fractals*, Nonlinear Systems and Complexity 29, https://doi.org/10.1007/978-3-030-35854-9_2

If a mapping $f : \mathbb{T}_+ \times X \to X$, where $\mathbb{T}_+$ is either the set of non-negative real numbers or the set of non-negative integers, satisfies (i), (ii), and (iii), then it is called a semi-flow on X [17].

Suppose that f is a flow on X. A point $p \in X$ is stable P^+ (positively Poisson stable) if for any neighborhood $\mathscr{U}$ of p and for any $H_1 > 0$ there exists $t \geq H_1$ such that $f(t, p) \in \mathscr{U}$. Similarly, $p \in X$ is stable P^- (negatively Poisson stable) if for any neighborhood $\mathscr{U}$ of p and for any $H_2 < 0$ there exists $t \leq H_2$ such that $f(t, p) \in \mathscr{U}$. A point $p \in X$ is called stable P (Poisson stable) if it is both stable P^+ and stable P^- [13].

For a fixed $p \in X$, let us denote by Ω_p the closure of the trajectory $\mathscr{T}(p) = \{f(t, p) : t \in \mathbb{T}\}$, i.e., $\Omega_p = \overline{\mathscr{T}(p)}$. The set Ω_p is a quasi-minimal set if the point p is stable P and $\mathscr{T}(p)$ is contained in a compact subset of X [13]. We will also denote $\Omega_p^+ = \overline{\mathscr{T}^+(p)}$, where $\mathscr{T}^+(p) = \{f(t, p) : t \in \mathbb{T}_+\}$ is the positive semi-trajectory through p.

An essential result concerning quasi-minimal sets was given by Hilmy [8]. It was demonstrated that if the trajectory corresponding to a Poisson stable point p is contained in a compact subset of X and Ω_p is neither a rest point nor a cycle, then Ω_p contains an uncountable set of motions everywhere dense and Poisson stable. The following theorem can be proved by adapting the technique given in [8, 13].

Theorem 2.1 *Suppose that $p \in X$ is stable P^+ and $\mathscr{T}^+(p)$ is contained in a compact subset of X. If Ω_p^+ is neither a rest point nor a cycle, then it contains an uncountable set of motions everywhere dense and stable P^+.*

2.2 Dynamics with Unpredictable Points

In this section, we introduce unpredictable points and mention some properties of the corresponding motions. The results will be provided for semi-flows on X, but they are valid for flows as well. We will denote by $\mathbb{N}$ the set of natural numbers.

Definition 2.1 ([2]) A point $p \in X$ and the trajectory through it are *unpredictable* if there exist a positive number ϵ_0 (the unpredictability constant) and sequences $\{t_n\}$ and $\{\tau_n\}$, both of which diverge to infinity, such that $\lim_{n\to\infty} f(t_n, p) = p$ and $d[f(t_n + \tau_n, p), f(\tau_n, p)] \geq \epsilon_0$ for each $n \in \mathbb{N}$.

An important point to discuss is the sensitivity or unpredictability. In the famous research studies [6, 9, 10, 14, 15, 18], sensitivity has been considered as a property of a system on a certain set of initial data since it compares the behavior of at least couples of solutions. The above definition allows to formulate unpredictability for a single trajectory. Indicating an unpredictable point p, one can make an error by taking a point $f(t_n, p)$. Then $d[f(\tau_n, f(t_n, p)), f(\tau_n, p)] \geq \epsilon_0$, and this is unpredictability for the motion. Thus, we say about the unpredictability of a single trajectory, whereas the former definitions considered the property in a set of motions. In Sect. 2.3, it will be shown how to extend the unpredictability to a chaos.

The following assertion is valid.

Lemma 2.2.1 ([2]) *If $p \in X$ is an unpredictable point, then $\mathscr{T}^+(p)$ is neither a rest point nor a cycle.*

Proof Let the number ϵ_0 and the sequences $\{t_n\}$, $\{\tau_n\}$ be as in Definition 2.1. Assume that there exists a positive number ω such that $f(t+\omega, p) = f(t, p)$ for all $t \in \mathbb{T}_+$. According to the continuity of $f(t, p)$, there exists a positive number δ such that if $d[p, q] < \delta$ and $0 \le t \le \omega$, then $d[f(t, p), f(t, q)] < \epsilon_0$. Fix a natural number n such that $d[p_n, p] < \delta$, where $p_n = f(t_n, p)$. One can find an integer m and a number ω_0 satisfying $0 \le \omega_0 < \omega$ such that $\tau_n = m\omega + \omega_0$. In this case, we have that

$$d[f(\tau_n, p_n), f(\tau_n, p)] = d[f(\omega_0, p_n), f(\omega_0, p)] < \epsilon_0.$$

But, this is a contradiction since

$$d[f(\tau_n, p_n), f(\tau_n, p)] = d[f(t_n + \tau_n, p), f(\tau_n, p)] \ge \epsilon_0.$$

Consequently, $\mathscr{T}^+(p)$ is neither a rest point nor a cycle. □

It is seen from the next lemma that the unpredictability can be transmitted by the flow.

Lemma 2.2.2 ([2]) *If a point $p \in X$ is unpredictable, then every point of the trajectory $\mathscr{T}^+(p)$ is also unpredictable.*

Proof Suppose that the number ϵ_0 and the sequences $\{t_n\}$, $\{\tau_n\}$ are as in Definition 2.1. Fix an arbitrary point $q \in \mathscr{T}^+(p)$ such that $q = f(\bar{t}, p)$ for some $\bar{t} \in \mathbb{T}_+$. One can verify that

$$\lim_{n\to\infty} f(t_n, q) = \lim_{n\to\infty} f(t_n + \bar{t}, p) = \lim_{n\to\infty} f(\bar{t}, f(t_n, p)) = f(\bar{t}, p) = q.$$

Now, take a natural number n_0 such that $\tau_n > \bar{t}$ for each $n \ge n_0$. If we denote $\zeta_n = \tau_n - \bar{t}$, then we have for $n \ge n_0$ that

$$\begin{aligned} d[f(t_n + \zeta_n, q), f(\zeta_n, q)] &= d[f(t_n + \zeta_n, f(\bar{t}, p)), f(\zeta_n, f(\bar{t}, p))] \\ &= d[f(t_n + \tau_n, p), f(\tau_n, p)] \\ &\ge \epsilon_0. \end{aligned}$$

Clearly, $\zeta_n \to \infty$ as $n \to \infty$. Consequently, the point q is unpredictable. □

Remark 2.1 It is worth noting that the unpredictability constant ϵ_0 is common for each point on an unpredictable trajectory.

2.3 Chaos on the Quasi-Minimal Set

This section is devoted to the demonstration of chaotic dynamics on a quasi-minimal set. According to [6, 10], the dynamics on a set $S \subseteq X$ is sensitive if there exists a positive number ϵ_0 such that for each $u \in S$ and each positive number δ there exist a point $u_\delta \in S$ and a positive number τ_δ such that $d[u_\delta, u] < \delta$ and $d[f(\tau_\delta, u_\delta), f(\tau_\delta, u)] \geq \epsilon_0$.

The main result of this chapter is mentioned in the next theorem, and it is valid for both flows and semi-flows on X.

Theorem 2.2 ([2]) *The dynamics on Ω_p^+ is sensitive if $p \in X$ is an unpredictable point.*

Proof Let $\epsilon_0 > 0$ be the unpredictability constant corresponding to the point p. Fix an arbitrary positive number δ, and take a point $r \in \Omega_p^+$. First of all, consider the case $r \in \mathscr{T}^+(p)$. By Lemma 2.2.2, there exist sequences $\{t_n\}$ and $\{\tau_n\}$, both of which diverge to infinity, such that $\lim_{n\to\infty} f(t_n, r) = r$ and $d[f(t_n + \tau_n, r), f(\tau_n, r)] \geq \epsilon_0$ for each n. Fix a natural number n_0 such that $d[\overline{r}, r] < \delta$, where $\overline{r} = f(t_{n_0}, r)$. In this case, the inequality $d[f(\tau_{n_0}, \overline{r}), f(\tau_{n_0}, r)] \geq \epsilon_0$ is valid.

On the other hand, suppose that $r \in \Omega_p^+ \setminus \mathscr{T}^+(p)$. One can find a sequence $\{\eta_m\}$, $\eta_m \to \infty$ as $m \to \infty$, such that $\lim_{m\to\infty} r_m = r$, where $r_m = f(\eta_m, p)$. According to Lemma 2.2.2, for each $m \in \mathbb{N}$, there exist sequences $\{s_n^m\}$ and $\{\xi_n^m\}$, both of which diverge to infinity, such that $\lim_{n\to\infty} r_n^m = r_m$ and $d[f(\xi_n^m, r_n^m), f(\xi_n^m, r_m)] \geq \epsilon_0$, $n \in \mathbb{N}$, where $r_n^m = f(s_n^m, r_m)$.

Now, let m_0 be a natural number such that $d[r_{m_0}, r] < \delta/2$. Suppose that there exists a natural number n_1 satisfying

$$d[f(\xi_{n_1}^{m_0}, r_{m_0}), f(\xi_{n_1}^{m_0}, r)] \geq \epsilon_0/2.$$

If this is the case, then sensitivity is proved. Otherwise, fix $n_2 \in \mathbb{N}$ such that $d[r_{n_2}^{m_0}, r_{m_0}] < \delta/2$ so that $d[r_{n_2}^{m_0}, r] \leq d[r_{n_2}^{m_0}, r_{m_0}] + d[r_{m_0}, r] < \delta$. One can confirm that

$$\begin{aligned}
&d[f(\xi_{n_2}^{m_0}, r_{n_2}^{m_0}), f(\xi_{n_2}^{m_0}, r)] \\
&\geq d[f(\xi_{n_2}^{m_0}, r_{n_2}^{m_0}), f(\xi_{n_2}^{m_0}, r_{m_0})] - d[f(\xi_{n_2}^{m_0}, r_{m_0}), f(\xi_{n_2}^{m_0}, r)] \\
&> \epsilon_0/2.
\end{aligned}$$

The theorem is proved. □

In Theorem 2.2, we have proved the presence of sensitivity in the set Ω_p^+ if p is an unpredictable point in X. In the case that f is a flow on X, one can use the same proof for the verification of sensitivity in Ω_p. According to Theorem 2.1 and Lemma 2.2.1, if the positive semi-trajectory of an unpredictable point $p \in X$

is contained in a compact subset of X, then Ω_p^+ contains an uncountable set of everywhere dense stable P^+ motions. Additionally, since $\mathscr{T}^+(p)$ is dense in Ω_p^+, the transitivity is also valid in the dynamics.

On the basis of the last discussions, we propose the following definition.

Definition 2.2 ([2]) The dynamics on the quasi-minimal set Ω_p is called Poincaré chaotic if p is an unpredictable point.

In the framework of chaos there may be infinitely many periodic motions. For instance, chaos in the sense of Devaney [6] and Li–Yorke [9] admit a basis consisting of periodic motions. However, in our case, Poisson stable motions take place in the dynamics instead of periodic motions. Our definition does not contradict to this, and this possibility is exemplified in the next section.

2.4 Applications

In this section, we will mainly investigate symbolic dynamics [6, 19] and show the presence of unpredictable points as well as chaos on a quasi-minimal set in the sense mentioned in Sect. 2.3. Moreover, we will reveal by means of the topological conjugacy that the same is true for the logistic, Hénon and horseshoe maps.

Let us take into account the following space of bi-infinite sequences [19]:

$$\Sigma^2 = \{s = (\ldots s_{-2}s_{-1}.s_0s_1s_2\ldots) : s_j = 0 \text{ or } 1 \text{ for each } j\}$$

with the metric

$$d[s, \bar{s}] = \sum_{k=-\infty}^{\infty} \frac{|s_k - \bar{s}_k|}{2^{|k|}},$$

where $s = (\ldots s_{-2}s_{-1}.s_0s_1s_2\ldots)$, $\bar{s} = (\ldots \bar{s}_{-2}\bar{s}_{-1}.\bar{s}_0\bar{s}_1\bar{s}_2\ldots) \in \Sigma^2$. The shift map $\sigma : \Sigma^2 \to \Sigma^2$ is defined as

$$\sigma(\ldots s_{-2}s_{-1}.s_0s_1s_2\ldots) = (\ldots s_{-2}s_{-1}s_0.s_1s_2\ldots).$$

The map σ is continuous and the metric space Σ^2 is compact [19].

In order to show that the map σ possesses an unpredictable point in Σ^2, we need a collection of finite sequences s_i^m, $m \in \mathbb{N}$, $i = 1, 2, \ldots, 2^m$, consisting of 0's and 1's. Let us denote $s_1^1 = (0)$ and $s_2^1 = (1)$. For each $m \in \mathbb{N}$, we recursively define $s_{2i-1}^{m+1} = (s_i^m 0)$, $s_{2i}^{m+1} = (s_i^m 1)$, $i = 1, 2, \ldots, 2^m$. Here, for each m and i, the finite sequences s_{2i-1}^{m+1} and s_{2i}^{m+1} are obtained by, respectively, inserting 0 and 1 to the end of the sequence s_i^m of length m. For instance, the sequences with length 2 can be written as $s_1^2 = (s_1^1 0) = (00)$, $s_2^2 = (s_1^1 1) = (01)$, $s_3^2 = (s_2^1 0) = (10)$, $s_4^2 = (s_2^1 1) = (11)$.

Now, consider the following sequence:

$$s^* = (\ldots s_8^3 s_6^3 s_4^3 s_2^3 s_4^2 s_2^2 . s_1^1 s_1^2 s_3^2 s_1^3 s_3^3 s_5^3 s_7^3 \ldots).$$

It was demonstrated in [19] that the trajectory of s^* is dense in Σ^2. We will show that s^* is an unpredictable point of the dynamics (Σ^2, σ). For each $n \in \mathbb{N}$, one can find $j \in \mathbb{N}$ such that

$$s_{2j-1}^{2n+2} = (s_{-n}^* \ldots s_0^* \ldots s_n^* 0)$$

and

$$s_{2j}^{2n+2} = (s_{-n}^* \ldots s_0^* \ldots s_n^* 1).$$

Therefore, there exists a sequence $\{t_n\}$ with $t_n \geq n + \sum_{k=1}^{2n+1} k 2^{k-1}$, $n \in \mathbb{N}$, such that $\sigma^{t_n}(s^*) = s_i^*$ for $|i| \leq n$. Accordingly, the inequality $d[\sigma^{t_n}(s^*), s^*] \leq 1/2^{n-1}$ is valid so that $\sigma^{t_n}(s^*) \to s^*$ as $n \to \infty$. Hence, s^* is stable P^+. In a similar way, one can confirm that s^* is stable P^-. Note that Σ^2 is a quasi-minimal set since s^* is Poisson stable. On the other hand, suppose that there exists a natural number n such that

$$\sigma^{t_n+n+1}(s^*)_i = \sigma^{n+1}(s^*)_i$$

for each $i \geq 0$. Under this assumption we have that $\sigma^{t_n}(s^*)_i = s_i^*$ for $i \geq -n$. This is a contradiction since the sequence s^* is not eventually periodic. For this reason, for each $n \in \mathbb{N}$, there exists an integer $\tau_n \geq n+1$ such that $\sigma^{t_n+\tau_n}(s^*)_0 \neq \sigma^{\tau_n}(s^*)_0$. Hence, $d[\sigma^{t_n+\tau_n}(s^*), \sigma^{\tau_n}(s^*)] \geq 1$ for each $n \in \mathbb{N}$, and s^* is an unpredictable point in Σ^2.

One of the concepts that has a great importance in the theory of dynamical systems is the topological conjugacy, which allows us to make interpretation about complicated dynamics by using simpler ones. Let X and Y be metric spaces. A flow (semi-flow) f on X is topologically conjugate to a flow (semi-flow) g on Y if there exists a homeomorphism $h : X \to Y$ such that $h \circ f = g \circ h$ [6, 19]. The following theorem can be verified by using the arguments presented in [3].

Theorem 2.3 *Suppose that X and Y are metric spaces and a flow (semi-flow) f on X is topologically conjugate to a flow (semi-flow) g on Y. If there exists an unpredictable point whose trajectory is contained in a compact subset of X, then there also exists an unpredictable point whose trajectory is contained in a compact subset of Y.*

Since the shift map σ on Σ^2 is topologically conjugate to the Smale horseshoe [6, 19], one can conclude by using Theorem 2.3 that the horseshoe map possesses an unpredictable point and a trajectory. On the other hand, let us consider the Hénon map

$$x_{n+1} = \alpha - \beta y_n - x_n^2$$
$$y_{n+1} = x_n, \tag{2.4.1}$$

where $\beta \neq 0$ and $\alpha \geq (5+2\sqrt{5})(1+|\beta|)^2/4$. It was proved by Devaney and Nitecki [5] that the map (2.4.1) possesses a Cantor set in which the map is topologically conjugate to the shift map σ on Σ^2. Therefore, Theorem 2.3 also implies the presence of an unpredictable point and a trajectory in the dynamics of (2.4.1).

Next, as an example of a semi-flow, consider the following space of infinite sequences [6]:

$$\Sigma_2 = \{s = (s_0 s_1 s_2 \ldots) : s_j = 0 \text{ or } 1 \text{ for each } j\}$$

with the metric

$$d[s, \overline{s}] = \sum_{k=0}^{\infty} \frac{|s_k - \overline{s}_k|}{2^k},$$

where $s = (s_0 s_1 s_2 \ldots)$, $\overline{s} = (\overline{s}_0 \overline{s}_1 \overline{s}_2 \ldots) \in \Sigma_2$. The shift map $\sigma : \Sigma_2 \to \Sigma_2$ is defined as $\sigma(s_0 s_1 s_2 \ldots) = (s_1 s_2 s_3 \ldots)$. As in the case of the space of bi-infinite sequences, the metric space Σ_2 is compact and the map σ is continuous [6, 19].

Let us take into account the sequence

$$s^* = (\underbrace{\ 0\ 1\ }_{1\,blocks} \ |\underbrace{00\ 01\ 10\ 11}_{2\,blocks}|\underbrace{000\ 001\ 010\ 011\ \ldots}_{3\,blocks}|\ldots),$$

which is constructed by successively listing all blocks of 0's and 1's of length n, then length $n+1$, etc. This sequence is non-periodic and its trajectory $\mathscr{T}(s^*) = \{\sigma^n(s^*) : n = 0, 1, 2, \ldots\}$ is dense in Σ_2 [6]. Note that the number of all blocks of length n in s^* is 2^n. Based upon the construction of s^*, there exists a sequence $\{t_n\}$ satisfying $t_n \geq \sum_{j=1}^{n} j 2^j$, $n \in \mathbb{N}$, such that $s_i^* = \sigma^{t_n}(s^*)_i$ for each $i = 0, 1, 2, \ldots, n$. Clearly, $t_n \to \infty$ as $n \to \infty$ and $d[\sigma^{t_n}(s^*), s^*] \leq 1/2^n$ so that $\sigma^{t_n}(s^*) \to s^*$ as $n \to \infty$. Hence, s^* is stable P^+. In a very similar way to the bi-infinite sequences, one can show the existence of a sequence $\{\tau_n\}$, $\tau_n \to \infty$ as $n \to \infty$, such that $d[\sigma^{t_n+\tau_n}(s^*), \sigma^{\tau_n}(s^*)] \geq 1$ for each $n \in \mathbb{N}$. Thus, s^* is an unpredictable point in Σ_2.

It was shown in [16] that the logistic map $x_{n+1} = \mu x_n(1 - x_n)$ possesses an invariant Cantor set $\Lambda \subset [0, 1]$, and the map on Λ is topologically conjugate to σ on Σ_2 for $\mu > 4$. Therefore, the map with $\mu > 4$ possesses an unpredictable point and a trajectory in accordance with Theorem 2.3.

2.5 Notes

Emphasizing the ingredients of Devaney [6] and Li–Yorke [9] chaos, one can see that the definitions of chaos have been considered by means of sets of motions, but not through a single motion description. This is the first time in the literature that chaos is initiated from a single function. Thus, the line, equilibrium, periodic function, quasi-periodic function, almost periodic function, recurrent function, Poisson stable motion are prolonged with the new element-unpredictable motion. This supplement to the line creates the possibility of other functions behind the known ones.

An essential point to discuss is the sensitivity or unpredictability. In the literature, sensitivity has been considered through initially nearby different motions. However, we say about unpredictability as an interior property of a single trajectory. Then chaos appears in a neighborhood of the trajectory. The symbolic dynamics illustrates all the results and it is an important tool for the investigation of the complicated dynamics of continuous-time systems such as the Lorenz and Rössler equations [12, 20]. It is worth noting that unpredictable points can be replicated by the techniques summarized in [1].

One can see the proximity of chaos and quasi-minimal sets by comparing their definitions [6, 9, 13]. Transitivity is a common feature of them, and the closure of a Poisson stable trajectory contains infinitely many Poisson stable orbits. In its own turn we know that a periodic trajectory is also Poisson stable. Possibly, continuum of Poisson stable trajectories is an ultimate form of infinitely many cycles known for a chaos. One may also ask whether sensitivity is proper for any quasi-minimal set. These are the questions to discover more relations between quasi-minimal sets and chaos. The results of this chapter are published in paper [2].

References

1. M. Akhmet, M.O. Fen, *Replication of Chaos in Neural Networks, Economics and Physics* (Higher Education Press, Beijing; Springer, Heidelberg, 2016)
2. M. Akhmet, M.O. Fen, Unpredictable points and chaos. Commun. Nonlinear Sci. Numer. Simul. **40**, 1–5 (2016)
3. J. Banks, J. Brooks, G. Cairns, G. Davis, P. Stacey, On Devaney's definition of chaos. Am. Math. Monthly **99**, 332–334 (1992)
4. G.D. Birkhoff, *Dynamical Systems*, vol. 9 (Amer. Math. Soc., Colloquium Publications, Providence, 1927)
5. R. Devaney, Z. Nitecki, Shift automorphisms in the Hénon mapping. Commun. Math. Phys. **67**, 137–146 (1979)
6. R.L. Devaney, *An Introduction to Chaotic Dynamical Systems* (Addison-Wesley, USA, 1989)
7. M. Hénon, A two-dimensional mapping with a strange attractor. Comm. Math. Phys. **50**, 69–77 (1976)
8. H. Hilmy, Sur les ensembles quasi-minimaux dans les systèmes dynamiques. Ann. Math. **37**, 899–907 (1936)
9. T.Y. Li, J.A. Yorke, Period three implies chaos. Am. Math. Monthly **82**, 985–992 (1975)
10. E.N. Lorenz, Deterministic nonperiodic flow. J. Atmos. Sci. **20**, 130–141 (1963)
11. R. May, Simple mathematical models with very complicated dynamics. Nature **261**, 459–467 (1976)
12. K. Mischaikow, M. Mrozek, Chaos in the Lorenz equations: a computer assisted proof. Bull AMS **32**, 66–72 (1995)

13. V.V. Nemytskii, V.V. Stepanov, *Qualitative Theory of Differential Equations* (Princeton University Press, Princeton, New Jersey, 1960)
14. H. Poincaré, Sur le probléme des trois corps et les équations de la dynamique. Acta Math. **13**, 1–270 (1880)
15. H. Poincaré, *Les méthodes nouvelles de la mécanique céleste*, Vol. 1, 2 (Gauthier-Villars, Paris, 1892)
16. C. Robinson, *Dynamical Systems: Stability, Symbolic Dynamics, and Chaos* (CRC Press, Boca Raton, 1995)
17. G.R. Sell, *Topological Dynamics and Ordinary Differential Equations* (Van Nostrand Reinhold Company, London, 1971)
18. S. Smale, Diffeomorphisms with many periodic points, in *Differential and Combinatorial Topology*, ed. by S.S. Cairns (Princeton University Press, Princeton, 1965), pp. 63–80
19. S. Wiggins, *Global Bifurcation and Chaos: Analytical Methods* (Springer, New York, Berlin, 1988)
20. P. Zgliczyński, Computer assisted proof of chaos in the Rössler equations and in the Hénon map. Nonlinearity **10**, 243–252 (1997)

Chapter 3
Unpredictability in Bebutov Dynamics

In this chapter, an unpredictable solution is considered as a generator of the chaos in a quasilinear system by means of Bebutov dynamical system. The results can be easily extended to different types of differential equations. An example of an unpredictable function is provided. A proper irregular behavior in coupled Duffing equations is observed through simulations.

3.1 Introduction

The row of periodic, quasi-periodic, almost periodic, recurrent, Poisson stable motions had been successively developed in the theory of dynamical systems. Then, chaotic dynamics started to be considered, which is not a single motion phenomenon, since a prescribed set of motions is required for a definition [17, 23, 32]. This chapter serves for proceeding the row and involving chaos as a purely functional object in nonlinear dynamics. In our paper [13], we introduced unpredictable motions based on Poisson stability. This time, we introduce the concept of an unpredictable function as an unpredictable point in the Bebutov dynamics [30].

It was proved in [13] that an unpredictable point gives rise to the existence of chaos in the quasi-minimal set. Thus, if one shows the existence of an unpredictable solution of an equation, then the chaos exists. The present chapter as well as our previous results concerning replication of chaos [1] supports the opinion of Holmes [21] that the theory of chaos has to be a part of the theory of differential equations. Since the main body of the results on chaotic motions has been formulated in terms of differential and difference equations, we may suggest that all these achievements have to be embedded and developed in the theory of dynamical systems or more specifically, in the theory of differential equations or hybrid systems.

M. Akhmet et al., *Dynamics with Chaos and Fractals*, Nonlinear Systems and Complexity 29, https://doi.org/10.1007/978-3-030-35854-9_3

The rest of the chapter is organized as follows: In the next section, we give the auxiliary results from the paper [13]. Section 3.3 is concerned with the Bebutov dynamics and the description of unpredictable functions. The existence of unpredictable solutions in a quasilinear system is considered in Sect. 3.4. Section 3.5 is devoted to examples. Finally, some concluding remarks are given in Sect. 3.6.

3.2 Preliminaries

Throughout the chapter, we will denote by $\mathbb{R}$, $\mathbb{R}_+$, $\mathbb{N}$, and $\mathbb{Z}$ the sets of real numbers, non-negative real numbers, natural numbers, and integers, respectively. Moreover, we will make use of the usual Euclidean norm for vectors and the norm induced by the Euclidean norm for square matrices [22].

Let (X, d) be a metric space. A mapping $\pi : \mathbb{R}_+ \times X \to X$ is a semi-flow on X [30] if:

(i) $\pi(0, p) = p$ for all $p \in X$;
(ii) $\pi(t, p)$ is continuous in the pair of variables t and p;
(iii) $\pi(t_1, \pi(t_2, p)) = \pi(t_1 + t_2, p)$ for all $t_1, t_2 \in \mathbb{R}_+$ and $p \in X$.

Suppose that π is a semi-flow on X. A point $p \in X$ is stable P^+ (positively Poisson stable) if there exists a sequence $\{t_n\}$, $t_n \to \infty$ as $n \to \infty$, such that $\pi(t_n, p) \to p$ as $n \to \infty$ [25]. For a fixed $p \in X$, let us denote by Θ_p the closure of the trajectory $\mathscr{T}(p) = \{\pi(t, p) : t \in \mathbb{R}_+\}$, i.e., $\Theta_p = \overline{\mathscr{T}(p)}$.

It was demonstrated by Hilmy [20] that if the trajectory corresponding to a Poisson stable point p is contained in a compact subset of X and it is neither a rest point nor a cycle, then the quasi-minimal set contains an uncountable set of motions everywhere dense and Poisson stable. The following theorem can be proved by adapting the technique given in [20, 25].

Theorem 3.1 ([20, 25]) *Suppose that $p \in X$ is stable P^+ and $\mathscr{T}(p)$ is contained in a compact subset of X. If Θ_p is neither a rest point nor a cycle, then it contains an uncountable set of motions everywhere dense and stable P^+.*

The results of the present chapter are correct if one considers stable P^- (negatively Poisson stable) points for a semi-flow with negative time or both stable P^+ and stable P^- (Poisson stable) points for a flow. The definition of a quasi-minimal set is given for a Poisson stable point in [25].

The description of an unpredictable point and trajectory are as follows.

Definition 3.1 ([13]) A point $p \in X$ and the trajectory through it are *unpredictable* if there exist a positive number ϵ_0 (the unpredictability constant) and sequences $\{t_n\}$ and $\{\tau_n\}$, both of which diverge to infinity, such that $\lim_{n\to\infty} \pi(t_n, p) = p$ and $d(\pi(t_n + \tau_n, p), \pi(\tau_n, p)) \geq \epsilon_0$ for each $n \in \mathbb{N}$.

An important point to discuss is the sensitivity or unpredictability. In the famous research studies [17, 23, 24, 27, 31], sensitivity was considered as a property of a system on a certain set of initial data since it compares the behavior of at least couples of solutions. Definition 3.1 allows to formulate unpredictability for a single trajectory. Indicating an unpredictable point p, one can make an error by taking a point $\pi(t_n, p)$. Then $d(\pi(\tau_n, \pi(t_n, p)), \pi(\tau_n, p)) \geq \epsilon_0$, and this is unpredictability for the motion. Thus, we say about the unpredictability of a single trajectory, whereas the former definitions considered the property in a set of motions.

It was proved in [13] that if $p \in X$ is an unpredictable point, then $\mathscr{T}(p)$ is neither a rest point nor a cycle, and that if a point $p \in X$ is unpredictable, then every point of the trajectory $\mathscr{T}(p)$ is also unpredictable. It is worth noting that the unpredictability constant ϵ_0 is common for each point on an unpredictable trajectory.

The dynamics on a set $S \subseteq X$ is sensitive [17, 24] if there exists a positive number ϵ_0 such that for each $u \in S$ and each positive number δ there exist a point $u_\delta \in S$ and a positive number τ_δ such that $d(u_\delta, u) < \delta$ and $d(\pi(\tau_\delta, u_\delta), \pi(\tau_\delta, u)) \geq \epsilon_0$.

A result concerning sensitivity in a quasi-minimal set is given in the next theorem.

Theorem 3.2 ([13]) *The dynamics on Θ_p is sensitive if $p \in X$ is an unpredictable point.*

Theorem 3.2 mentions the presence of sensitivity in the set Θ_p if p is an unpredictable point in X. According to Theorem 3.1, if the trajectory $\mathscr{T}(p)$ of an unpredictable point $p \in X$ is contained in a compact subset of X, then Θ_p contains an uncountable set of everywhere dense stable P^+ motions. Additionally, since $\mathscr{T}(p)$ is dense in Θ_p, the transitivity is also valid in the dynamics. We have named this type of chaos as Poincaré chaos in the paper [13]. (See also Chap. 2.)

3.3 Unpredictable Functions

This section is devoted to the description of unpredictable functions and their connection with chaos. For that purpose the results provided in [30] will be utilized.

Let us denote by $C(\mathbb{R})$ the set of continuous functions defined on $\mathbb{R}$ with values in $\mathbb{R}^m$, and assume that $C(\mathbb{R})$ has the topology of uniform convergence on compact sets, i.e., a sequence $\{h_k\}$ in $C(\mathbb{R})$ is said to converge to a limit h if for every compact set $\mathscr{U} \subset \mathbb{R}$ the sequence of restrictions $\{h_k|_{\mathscr{U}}\}$ converges to $\{h|_{\mathscr{U}}\}$ uniformly.

One can define a metric d on $C(\mathbb{R})$ as [30]

$$d(h_1, h_2) = \sum_{k=1}^{\infty} 2^{-k} \rho_k(h_1, h_2), \tag{3.3.1}$$

where h_1, h_2 belong to $C(\mathbb{R})$ and

$$\rho_k(h_1, h_2) = \min\left\{1, \sup_{s\in[-k,k]} \|h_1(s) - h_2(s)\|\right\}, \quad k \in \mathbb{N}.$$

Let us define the mapping $\pi : \mathbb{R}_+ \times C(\mathbb{R}) \to C(\mathbb{R})$ by $\pi(t, h) = h_t$, where $h_t(s) = h(t + s)$. The mapping π defines a semi-flow on $C(\mathbb{R})$ and it is called the Bebutov dynamical system [30].

We describe an unpredictable function as follows.

Definition 3.2 ([14]) An unpredictable function is an unpredictable point of the Bebutov dynamical system.

According to Theorem $III.3$ [30], a motion $\pi(t, h)$ lies in a compact set if h is a bounded and uniformly continuous function. Assuming this, by means of Theorem 3.2, we obtain that an unpredictable function h determines chaos if it is bounded and uniformly continuous. On the basis of this result, one can say that if a differential equation admits an unpredictable solution which is uniformly continuous and bounded, then chaos is present in the set of solutions. In the next section, we will prove the existence of an unpredictable solution whose quasi-minimal set is a chaotic attractor.

3.4 Unpredictable Solutions of Quasilinear Systems

Consider the following quasilinear system:

$$x' = Ax + f(x) + g(t), \tag{3.4.2}$$

where the $m \times m$ constant matrix A has eigenvalues all with negative real parts, the function $f : \mathbb{R}^m \to \mathbb{R}^m$ is continuous and $g : \mathbb{R} \to \mathbb{R}^m$ is a uniformly continuous and bounded function.

Since the eigenvalues of the matrix A have negative real parts, there exist positive numbers K and ω such that $\|e^{At}\| \le Ke^{-\omega t}$, $t \ge 0$ [19].

The following conditions are required:

(C1) There exists a positive number M_f such that $\sup_{x\in\mathbb{R}^m} \|f(x)\| \le M_f$;

(C2) There exists a positive number L_f such that $\|f(x_1) - f(x_2)\| \le L_f \|x_1 - x_2\|$ for all $x_1, x_2 \in \mathbb{R}^m$;

(C3) $KL_f - \omega < 0$.

The main result of the present chapter is mentioned in the next theorem.

Theorem 3.3 ([14]) *Suppose that the conditions $(C1)$–$(C3)$ are valid. If the function $g(t)$ is unpredictable, then system (3.4.2) possesses a unique uniformly exponentially stable unpredictable solution, which is uniformly continuous and bounded on $\mathbb{R}$.*

Proof Using the technique for quasilinear equations [19], one can confirm under the conditions $(C1)$–$(C3)$ that system (3.4.2) possesses a unique bounded on $\mathbb{R}$ solution $\phi(t)$ which satisfies the relation

$$\phi(t) = \int_{-\infty}^{t} e^{A(t-u)}[f(\phi(u)) + g(u)]du. \tag{3.4.3}$$

Moreover, $\sup_{t\in\mathbb{R}} \|\phi(t)\| \le M_\phi$, where $M_\phi = \dfrac{K(M_f + M_g)}{\omega}$ and $M_g = \sup_{t\in\mathbb{R}} \|g(t)\|$. The solution $\phi(t)$ is uniformly continuous on $\mathbb{R}$ since $\sup_{t\in\mathbb{R}} \left\|\phi'(t)\right\| \le \|A\|\, M_\phi + M_f + M_g$.

Suppose that $x(t)$ is a solution of (3.4.2) such that $x(t_0) = x_0$ for some $t_0 \in \mathbb{R}$ and $x_0 \in \mathbb{R}^m$. It can be verified that

$$\|x(t) - \phi(t)\| \le K\, \|x_0 - \phi(t_0)\|\, e^{(KL_f-\omega)(t-t_0)},\ t \ge t_0,$$

and therefore, $\phi(t)$ is uniformly exponentially stable.

Since the function $g(t)$ is unpredictable, there exist a positive number $\epsilon_0 \le 1$ and sequences $\{t_n\}$, $\{\tau_n\}$, both of which diverge to infinity, such that $d(g_{t_n}, g) \to 0$ as $n \to \infty$ and $d(g_{t_n+\tau_n}, g_{\tau_n}) \ge \epsilon_0$ for all $n \in \mathbb{N}$, where the distance function d is given by (3.3.1).

First of all, we shall show that $d(\phi_{t_n}, \phi) \to 0$ as $n \to \infty$. Fix an arbitrary small positive number $\epsilon < 1$ and suppose that α is a positive number satisfying $\alpha \le \dfrac{\omega - KL_f}{2\omega + K - 2KL_f}$. Let k_0 be a sufficiently large natural number such that

$$k_0 \ge \max\left\{\frac{\ln(1/\alpha\epsilon)}{\ln 2}, \frac{1}{\omega - KL_f}\ln\left(\frac{2K(M_f + M_g)}{\omega\alpha\epsilon}\right)\right\}. \tag{3.4.4}$$

There exists a natural number n_0 such that if $n \ge n_0$ then $d(g_{t_n}, g) < 2^{-2k_0}\alpha\epsilon$. Therefore, for $n \ge n_0$, the inequality $\rho_{2k_0}(g_{t_n}, g) < \alpha\epsilon$ is valid. Since $\alpha\epsilon < 1$, we have that $\|g(t_n + s) - g(s)\| < \alpha\epsilon$ for $s \in [-2k_0, 2k_0]$.

Making use of the relation (3.4.3), one can obtain that

$$\phi(t_n + s) - \phi(s) = \int_{-\infty}^{s} e^{A(s-u)}[f(\phi(t_n + u)) - f(\phi(u)) + g(t_n + u) - g(u)]du.$$

Thus, if s belongs to the interval $[-2k_0, 2k_0]$, then it can be verified that

$$\|\phi(t_n+s)-\phi(s)\| \le \frac{2K(M_f+M_g)}{\omega}e^{-\omega(s+2k_0)} + \frac{K\alpha\epsilon}{\omega}\left(1-e^{-\omega(s+2k_0)}\right)$$
$$+KL_f\int_{-2k_0}^{s} e^{-\omega(s-u)}\,\|\phi(t_n+u)-\phi(u)\|\,du. \qquad (3.4.5)$$

Now, let us define the functions

$$\psi_n(s)=e^{\omega s}\,\|\phi(t_n+s)-\phi(s)\|\,,\ n\ge n_0.$$

Inequality (3.4.5) implies that

$$\psi_n(s)\le \frac{K\alpha\epsilon}{\omega}e^{\omega s}+\left(\frac{2K(M_f+M_g)-K\alpha\epsilon}{\omega}\right)e^{-2\omega k_0}+KL_f\int_{-2k_0}^{s}\psi_n(u)du.$$

Applying the Gronwall's Lemma [15], one can confirm that

$$\psi_n(s)\le \frac{K\alpha\epsilon}{\omega-KL_f}e^{\omega s}\left(1-e^{(KL_f-\omega)(s+2k_0)}\right)+\frac{2K(M_f+M_g)}{\omega}e^{KL_f s}e^{2(KL_f-\omega)k_0}.$$

Hence, the inequality

$$\|\phi(t_n+s)-\phi(s)\| < \frac{K\alpha\epsilon}{\omega-KL_f}+\frac{2K(M_f+M_g)}{\omega}e^{(KL_f-\omega)(s+2k_0)}$$

is valid. Since the number k_0 satisfies (3.4.4), we have $e^{(KL_f-\omega)k_0} \le \dfrac{\omega\alpha\epsilon}{2K(M_f+M_g)}$ so that

$$\|\phi(t_n+s)-\phi(s)\| < \left(1+\frac{K}{\omega-KL_f}\right)\alpha\epsilon$$

for $s\in[-k_0,k_0]$. Therefore, the inequality

$$\sup_{s\in[-k,k]}\|\phi(t_n+s)-\phi(s)\| < \left(1+\frac{K}{\omega-KL_f}\right)\alpha\epsilon$$

holds for each integer k with $1\le k\le k_0$. It is clear that $\left(1+\dfrac{K}{\omega-KL_f}\right)\alpha\epsilon<1$. Thus,

$$\rho_k(\phi_{t_n},\phi) < \left(1+\frac{K}{\omega-KL_f}\right)\alpha\epsilon,\ 1\le k\le k_0.$$

For $n\ge n_0$, it can be obtained by using (3.4.4) one more time that

$$\begin{aligned} d(\phi_{t_n}, \phi) &= \sum_{k=1}^{\infty} 2^{-k} \rho_k(\phi_{t_n}, \phi) \\ &< \left(1 + \frac{K}{\omega - KL_f}\right) \alpha\epsilon \sum_{k=1}^{k_0} 2^{-k} + \sum_{k=k_0+1}^{\infty} 2^{-k} \\ &< \left(2 + \frac{K}{\omega - KL_f}\right) \alpha\epsilon \\ &\leq \epsilon. \end{aligned}$$

Hence, $d(\phi_{t_n}, \phi) \to 0$ as $n \to \infty$.

Next, we will verify the presence of a positive number $\overline{\epsilon}_0$ and a sequence $\{\overline{\tau}_n\}$, $\overline{\tau}_n \to \infty$ as $n \to \infty$, such that $d(\phi_{t_n+\overline{\tau}_n}, \phi_{\overline{\tau}_n}) \geq \overline{\epsilon}_0$ for all $n \in \mathbb{N}$.

Let N be a natural number such that $\sum_{k=N+1}^{\infty} 2^{-k} \leq \frac{\epsilon_0}{2}$. One can confirm that

$$\sum_{k=1}^{N} 2^{-k} \rho_k(g_{t_n+\tau_n}, g_{\tau_n}) \geq \frac{\epsilon_0}{2}.$$

In this case, for each $n \in \mathbb{N}$, there exist integers k_0^n between 1 and N such that

$$\rho_{k_0^n}(g_{t_n+\tau_n}, g_{\tau_n}) \geq \frac{2^{k_0^n}\epsilon_0}{2N} \geq \frac{\epsilon_0}{N}.$$

Therefore, it can be verified that

$$\sup_{s\in[-k_0^n, k_0^n]} \|g(t_n + \tau_n + s) - g(\tau_n + s)\| \geq \frac{\epsilon_0}{N}, \ n \in \mathbb{N}.$$

The last inequality implies the existence of numbers $\eta_n \in [-k_0^n, k_0^n]$ satisfying

$$\|g(t_n + \tau_n + \eta_n) - g(\tau_n + \eta_n)\| \geq \frac{\epsilon_0}{N}, \ n \in \mathbb{N}. \tag{3.4.6}$$

Suppose that $g(s) = (g_1(s), g_2(s), \ldots, g_m(s))$, where each g_i, $1 \leq i \leq m$, is a real-valued function. In accordance with (3.4.6), for each $n \in \mathbb{N}$, there is an integer j_n, $1 \leq j_n \leq m$, with

$$\left|g_{j_n}(t_n + \overline{\tau}_n) - g_{j_n}(\overline{\tau}_n)\right| \geq \frac{\epsilon_0}{Nm},$$

where $\overline{\tau}_n = \tau_n + \eta_n$, $n \in \mathbb{N}$. Since the function g is uniformly continuous, there exists a positive number $\Delta \leq 1$, which does not depend on the sequences $\{t_n\}$ and

$\{\overline{\tau}_n\}$, such that both of the inequalities

$$\|g(t_n + \overline{\tau}_n) - g(t_n + \overline{\tau}_n + s)\| \leq \frac{\epsilon_0}{4Nm}$$

and

$$\|g(\overline{\tau}_n) - g(\overline{\tau}_n + s)\| \leq \frac{\epsilon_0}{4Nm}$$

are valid for $s \in [-\Delta, \Delta]$. Thus, we have for $s \in [-\Delta, \Delta]$ that

$$\begin{aligned}\left|g_{j_n}(t_n + \overline{\tau}_n + s) - g_{j_n}(\overline{\tau}_n + s)\right| &\geq \left|g_{j_n}(t_n + \overline{\tau}_n) - g_{j_n}(\overline{\tau}_n)\right| \\ &\quad - \left|g_{j_n}(t_n + \overline{\tau}_n) - g_{j_n}(t_n + \overline{\tau}_n + s)\right| \\ &\quad - \left|g_{j_n}(\overline{\tau}_n) - g_{j_n}(\overline{\tau}_n + s)\right| \\ &\geq \frac{\epsilon_0}{2Nm}. \end{aligned} \tag{3.4.7}$$

For each $n \in \mathbb{N}$, one can find numbers $s_1^n, s_2^n, \ldots, s_m^n \in [-\Delta, \Delta]$ such that

$$\begin{aligned}&\left\| \int_{-\Delta}^{\Delta} [g(t_n + \overline{\tau}_n + u) - g(\overline{\tau}_n + u)]\, du \right\| \\ &= 2\Delta \Big(\sum_{i=1}^{m} [g_i(t_n + \overline{\tau}_n + s_i^n) - g_i(\overline{\tau}_n + s_i^n)]^2 \Big)^{1/2}. \end{aligned} \tag{3.4.8}$$

Hence, it can be deduced by means of (3.4.7) and (3.4.8) that

$$\begin{aligned}\left\| \int_{-\Delta}^{\Delta} [g(t_n + \overline{\tau}_n + u) - g(\overline{\tau}_n + u)]\, du \right\| &\geq 2\Delta \left| g_{j_n}(t_n + \overline{\tau}_n + s_{j_n}^n) - g_{j_n}(\overline{\tau}_n + s_{j_n}^n) \right| \\ &\geq \frac{\Delta \epsilon_0}{Nm}.\end{aligned}$$

Now, using the equation

$$\begin{aligned}\phi(t_n + \overline{\tau}_n + s) - \phi(\overline{\tau}_n + s) &= \phi(t_n + \overline{\tau}_n - \Delta) - \phi(\overline{\tau}_n - \Delta) \\ &\quad + \int_{-\Delta}^{s} A[\phi(t_n + \overline{\tau}_n + u) - \phi(\overline{\tau}_n + u)]du \\ &\quad + \int_{-\Delta}^{s} [f(\phi(t_n + \overline{\tau}_n + u)) - f(\phi(\overline{\tau}_n + u))]du \\ &\quad + \int_{-\Delta}^{s} [g(t_n + \overline{\tau}_n + u) - g(\overline{\tau}_n + u)]du,\end{aligned}$$

we attain that

$$\|\phi(t_n+\overline{\tau}_n+\Delta)-\phi(\overline{\tau}_n+\Delta)\| \geq \left\| \int_{-\Delta}^{\Delta} [g(t_n+\overline{\tau}_n+u)-g(\overline{\tau}_n+u)]du \right\|$$
$$- \|\phi(t_n+\overline{\tau}_n-\Delta)-\phi(\overline{\tau}_n-\Delta)\|$$
$$- \int_{-\Delta}^{\Delta} \left(\|A\|+L_f\right) \|\phi(t_n+\overline{\tau}_n+u)-\phi(\overline{\tau}_n+u)\| \, du.$$

The last inequality implies that

$$\sup_{s\in[-\Delta,\Delta]} \|\phi(t_n+\overline{\tau}_n+s)-\phi(\overline{\tau}_n+s)\| \geq \overline{\epsilon}_0, \; n\in\mathbb{N},$$

where $\overline{\epsilon}_0 = \dfrac{\Delta\epsilon_0}{2Nm[1+\Delta(\|A\|+L_f)]}$. Therefore, we have $d(\phi_{t_n+\overline{\tau}_n}, \phi_{\overline{\tau}_n}) \geq \overline{\epsilon}_0$ for each $n \in \mathbb{N}$.

The theorem is proved. □

In the definition of Devaney chaos [17], periodic motions constitute a dense subset. However, in our case, instead of periodic motions, Poisson stable motions take place in the dynamics. More precisely, we say that the dynamics on the quasi-minimal set of functions on $\mathbb{R}$ is chaotic if the dynamics on it is sensitive, transitive and there exists a continuum of Poisson stable trajectories dense in the quasi-minimal set. Nevertheless, in the framework of chaos there may be infinitely many periodic motions. For instance, the symbolic dynamics of bi-infinite sequences possesses both an uncountable set of non-periodic Poisson stable motions as well as infinitely many cycles [13, 32].

3.5 Examples

Example 1 We will construct an unpredictable function in this example.

Consider the function $z(t)=(z_1(t), z_2(t))$ defined as $z_1(t)=p_i$, $z_2(t)=q_i$ for $t\in[i,i+1)$, $i\in\mathbb{Z}$, such that (p_i,q_i) is an unpredictable trajectory [13] of the Hénon map

$$\begin{aligned} p_{i+1} &= \alpha_0 - \beta_0 q_i - p_i^2 \\ q_{i+1} &= p_i, \end{aligned} \tag{3.5.1}$$

where $\beta_0 \neq 0$ and $\alpha_0 \geq (5+2\sqrt{5})(1+|\beta_0|)^2/4$. The unpredictable trajectory belongs to a Cantor set such that there exists a positive number R satisfying $\|(p_i,q_i)\| \leq R$ for each $i\in\mathbb{Z}$ [16, 28].

Define the continuous on $\mathbb{R}$ function $\psi(t)$ such that

$$\psi(t) = e^{-\gamma(t-i)}\psi(i) + \int_i^t e^{-\gamma(t-u)} z(u)du,\ t \in [i, i+1],\ i \in \mathbb{Z},$$

where γ is a positive number and $\psi(0) = \displaystyle\int_{-\infty}^{0} e^{\gamma u} z(u)du.$

Let us show that $\psi(t)$ is an unpredictable function. Fix an arbitrary small positive number $\epsilon < 1$, and let β be a positive number such that $\beta \leq \dfrac{\gamma}{2\gamma+1}$. Suppose that r_0 is a sufficiently large natural number such that $r_0 \geq \max\left\{\dfrac{\ln(1/\beta\epsilon)}{\ln 2}, \dfrac{1}{\gamma}\ln\left(\dfrac{2R}{\gamma\beta\epsilon}\right)\right\}$. Since (p_i, q_i) is an unpredictable trajectory of (3.5.1), there exist a positive number ϵ_0 and sequences $\{i_n\}$, $\{j_n\}$, both of which diverge to infinity, such that

$$\left\|(p_{i+i_n}, q_{i+i_n}) - (p_i, q_i)\right\| < \beta\epsilon,\ n \in \mathbb{N},\ i = -2r_0, -2r_0+1, \ldots, r_0 - 1,$$

and

$$\left\|(p_{i_n+j_n}, q_{i_n+j_n}) - (p_{j_n}, q_{j_n})\right\| \geq \epsilon_0,\ n \in \mathbb{N}.$$

If $s \in [-r_0, r_0]$, then one can confirm that $\|\psi(i_n + s) - \psi(s)\| < \left(1 + \dfrac{1}{\gamma}\right)\beta\epsilon$ for each $n \in \mathbb{N}$. Therefore, $\pi(i_n, \psi) \to \psi$ as $n \to \infty$ so that $\psi(t)$ is a positively Poisson stable point of the Bebutov dynamical system. On the other hand, we have $\sup\limits_{s\in[0,1]} \|\psi(i_n + j_n + s) - \psi(j_n + s)\| \geq \dfrac{(1 - e^{-\gamma})\epsilon_0}{\gamma(1 + e^{-\gamma})}$, and this proves that $\psi(t)$ is an unpredictable function.

Example 2 In this example, we will show how an unpredictable point may cause irregular dynamics. For that purpose, we will make use of coupled Duffing equations such that the first one is forced with a relay function and the second one is perturbed with the solutions of the former.

Let us consider the following forced Duffing equation:

$$x'' + 0.68x' + 1.6x + 0.008x^3 = \nu(t, \zeta, \lambda), \tag{3.5.2}$$

where the forcing term $\nu(t, \zeta, \lambda)$ is a relay function defined as

$$\nu(t, \zeta, \lambda) = \begin{cases} 1.2, & \text{if } \zeta_{2j}(\lambda) < t \leq \zeta_{2j+1}(\lambda),\ j \in \mathbb{Z}, \\ 0.4, & \text{if } \zeta_{2j-1}(\lambda) < t \leq \zeta_{2j}(\lambda),\ j \in \mathbb{Z}. \end{cases} \tag{3.5.3}$$

In (3.5.3), the sequence $\zeta = \left\{\zeta_j\right\}_{j\in\mathbb{Z}}$ of switching moments is defined through the equation $\zeta_j = j + \kappa_j$, $j \in \mathbb{Z}$, where the sequence $\left\{\kappa_j\right\}_{j\in\mathbb{Z}}$ is a solution of the logistic map

$$\kappa_{j+1} = \lambda \kappa_j (1 - \kappa_j). \tag{3.5.4}$$

By means of the variables $x_1 = x$ and $x_2 = x'$, Eq. (3.5.2) can be written as a system in the following form:

$$\begin{aligned} x_1' &= x_2 \\ x_2' &= -1.6x_1 - 0.68x_2 - 0.008x_1^3 + \nu(t, \zeta, \lambda). \end{aligned} \tag{3.5.5}$$

We suppose that the parameter λ in (3.5.4) is greater than 4 such that the map possesses an invariant Cantor set $\Lambda \subset [0, 1]$ [28], and it was demonstrated in [13] that for such values of the parameter the map possesses an unpredictable point in Λ. Let us consider system (3.5.5) with $\zeta_0 \in \Lambda$. For each natural number p, system (3.5.5) admits an unstable periodic solution with period $2p$ if p is odd and an unstable periodic solution with period p if p is even [4]. The reader is referred to [1–12] for more information about the dynamics of relay systems.

In order to illustrate the irregular dynamics of (3.5.5), we make use of the value $\lambda = 4.007$ in the system and depict the solution corresponding to the initial data $x_1(0.41) = 0.6$, $x_2(0.41) = 0.5$ and $\zeta_0 = 0.41$ in Fig. 3.1. The simulation results seen in Fig. 3.1 confirm the presence of irregular behavior in the dynamics of (3.5.5). It is worth noting that even if the logistic map (3.5.4) with $\lambda = 4.007$ has an invariant Cantor set, the chosen initial value of the sequence $\{\zeta_j\}$ allows to simulate the solution for $0.41 \leq t \leq 100$. Due to the instability, simulations of the system cannot be provided for large intervals of time.

Next, we take into account another Duffing equation,

$$y'' + 0.95y' + 1.8y + 0.005y^3 = 0. \tag{3.5.6}$$

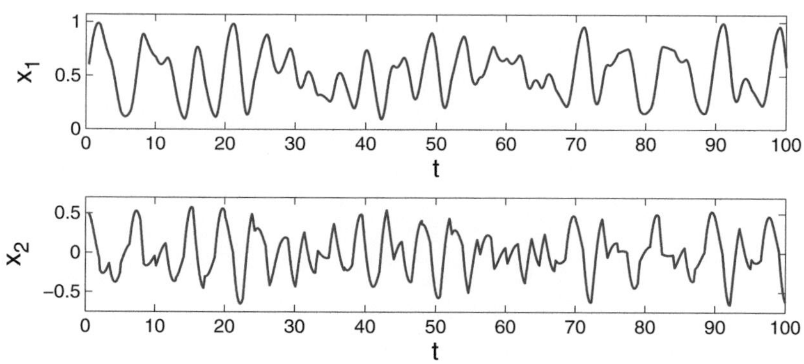

Fig. 3.1 The solution of (3.5.5) with $x_1(0.41) = 0.6$, $x_2(0.41) = 0.5$ and $\kappa_0 = 0.41$. The value $\lambda = 4.007$ is used in the simulation. The figure reveals the presence of chaos in the dynamics of (3.5.5)

Using the variables $y_1 = y$ and $y_2 = y'$, Eq. (3.5.6) can be reduced to the system

$$\begin{aligned} y_1' &= y_2 \\ y_2' &= -1.8y_1 - 0.95y_2 - 0.005y_1^3. \end{aligned} \tag{3.5.7}$$

We perturb (3.5.7) with the solutions of (3.5.5), and set up the system

$$\begin{aligned} z_1' &= z_2 + x_1(t) \\ z_2' &= -1.8z_1 - 0.95z_2 - 0.005z_1^3 + x_2(t). \end{aligned} \tag{3.5.8}$$

System (3.5.8) is in the form of (3.4.2), where $A = \begin{pmatrix} 0 & 1 \\ -1.8 & -0.95 \end{pmatrix}$, $f(z_1, z_2) = (0, -0.005z_1^3)$ and $g(t) = (x_1(t), x_2(t))$. Both eigenvalues of A have real parts -0.475, and the coefficient of the nonlinear term is chosen sufficiently small in absolute value so that the conditions $(C1)$–$(C3)$ are valid for (3.5.8).

Figure 3.2 shows the solution of (3.5.8) with $z_1(0.41) = 0.1$ and $z_2(0.41) = 0.2$. For the simulation, the solution $(x_1(t), x_2(t))$ which is represented in Fig. 3.1 is used. One can observe in Fig. 3.2 that the represented solution behaves irregularly.

Next, we will demonstrate the presence of periodic motions in system (3.5.8) by means of the Ott–Grebogi–Yorke (OGY) control technique [26]. Since the logistic map (3.5.4) is the main source of the chaotic behavior in the coupled system (3.5.5) + (3.5.8), we will apply the OGY method to the map. Let us explain briefly the method for the logistic map [29]. Suppose that the parameter λ in (3.5.4) is allowed to vary in the range $[4.007 - \varepsilon, 4.007 + \varepsilon]$, where ε is a given small positive number. Consider an arbitrary solution $\{\kappa_j\}$, $\kappa_0 \in \Lambda$, of the map and denote by $\kappa^{(i)}$, $i = 1, 2, \ldots, p$, the target p-periodic orbit to be stabilized. In the

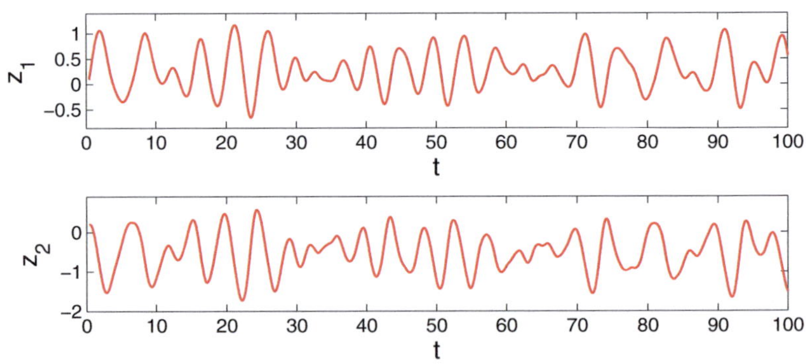

Fig. 3.2 Chaotic behavior in the dynamics of system (3.5.8). The solution $(x_1(t), x_2(t))$ represented in Fig. 3.1 is utilized as the perturbation in (3.5.8). The irregularity is observable in both z_1 and z_2 coordinates

OGY control method [29], at each iteration step j after the control mechanism is switched on, we consider the logistic map with the parameter value $\lambda = \bar{\lambda}_j$, where

$$\bar{\lambda}_j = 4.007 \left(1 + \frac{(2\kappa^{(i)} - 1)(\kappa_j - \kappa^{(i)})}{\kappa^{(i)}(1 - \kappa^{(i)})} \right), \tag{3.5.9}$$

provided that the number on the right-hand side of the formula (3.5.9) belongs to the interval $[4.007 - \varepsilon, 4.007 + \varepsilon]$. In other words, formula (3.5.9) is valid if the trajectory $\{\kappa_j\}$ is sufficiently close to the target periodic orbit. Otherwise, we take $\bar{\lambda}_j = 4.007$, so that the system evolves at its original parameter value, and wait until the trajectory $\{\kappa_j\}$ enters in a sufficiently small neighborhood of the periodic orbit $\kappa^{(i)}$, $i = 1, 2, \ldots, p$, such that the inequality $-\varepsilon \leq 4.007 \frac{(2\kappa^{(i)} - 1)(\kappa_j - \kappa^{(i)})}{\kappa^{(i)}(1 - \kappa^{(i)})} \leq \varepsilon$ holds. If this is the case, the control of chaos is not achieved immediately after switching on the control mechanism. Instead, there is a transition time before the desired periodic orbit is stabilized. The transition time increases if the number ε decreases [18].

Figure 3.3 shows the stabilization of an unstable 2-periodic solution of (3.5.8). Here, the OGY control method is used around the fixed point 3.007/4.007 of the logistic map (3.5.4), and the simulation is performed for the initial data $x_1(0.41) = 0.6$, $x_2(0.41) = 0.5$, $z_1(0.41) = 0.1$, $z_2(0.41) = 0.2$, $\zeta_0 = 0.41$. The control is switched on at $t = \zeta_{20}$ and the value $\varepsilon = 0.095$ is utilized. One can confirm that even if the control is switched on at $t = \zeta_{20}$ there is a transition time before the stabilization such that the control becomes dominant approximately at $t = 76$. Figure 3.3 reveals that the OGY control technique is appropriate for the stabilization of the unstable periodic motions of system (3.5.8).

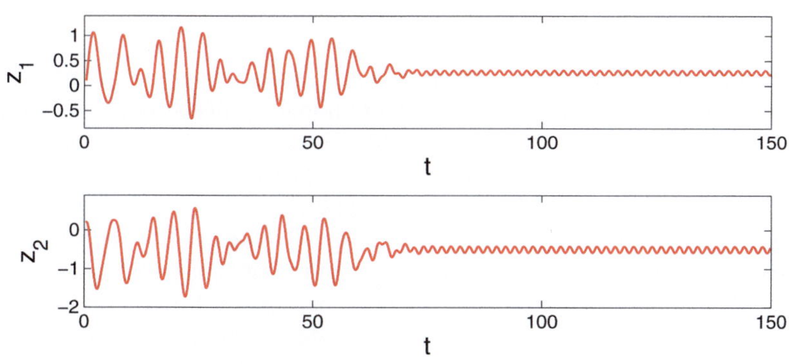

Fig. 3.3 The stabilization of the 2-periodic solution of (3.5.8) corresponding to the fixed point 3.007/4.007 of the logistic map (3.5.4). The value $\varepsilon = 0.095$ is used and the control is switched on at $t = \zeta_{20}$

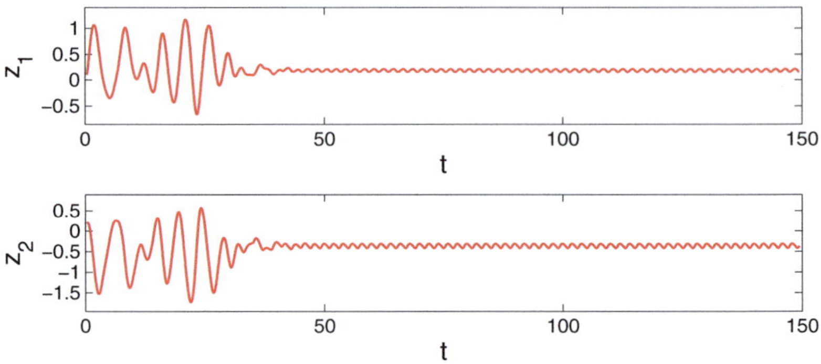

Fig. 3.4 The stabilization of the 2-periodic solution of (3.5.8) corresponding to the 2-periodic orbit $\kappa^{(1)} \approx 0.34459$, $\kappa^{(2)} \approx 0.90497$ of (3.5.4). The value $\varepsilon = 0.072$ is used and the control is switched on at $t = \zeta_{25}$

On the other hand, Fig. 3.4 shows the simulation result for (3.5.8) when the OGY method is applied around the 2-periodic orbit $\kappa^{(1)} \approx 0.34459$, $\kappa^{(2)} \approx 0.90497$ of (3.5.4). The represented solution corresponds again to the initial data $x_1(0.41) = 0.6$, $x_2(0.41) = 0.5$, $z_1(0.41) = 0.1$, $z_2(0.41) = 0.2$, $\zeta_0 = 0.41$. The value $\varepsilon = 0.072$ is used and the control is switched on at $t = \zeta_{25}$. The presence of a transition time before the stabilization is observable in Fig. 3.4 such that the control becomes dominant approximately at $t = 46$. One can observe that the stabilized 2-periodic solutions seen in Figs. 3.3 and 3.4 are different, and this reveals the presence of periodic motions.

3.6 Notes

The unpredictable function has been defined as an unpredictable point of the Bebutov dynamics, and chaos in the quasi-minimal set of the function is verified. This is the first time in the literature that the existence of an unpredictable solution for a quasilinear ordinary differential equation is proved.

The concept of unpredictable solutions can be useful for finding more delicate features of chaos in systems with complicated dynamics. Researches based on unpredictable functions may pave the way for the functional analysis of chaos to involve the operator theory results. Hopefully, our approach will give a basis for a deeper comprehension and possibility to unite different appearances of chaos. In this framework, the results can be developed for partial differential equations, integro-differential equations, functional differential equations, and evolution systems. The results of this chapter are published in paper [14].

References

1. M.U. Akhmet, Devaney's chaos of a relay system. Commun. Nonlinear Sci. Numer. Simulat. **14**, 1486–1493 (2009)
2. M.U. Akhmet, Li-Yorke chaos in the system with impacts. J. Math. Anal. Appl. **351**, 804–810 (2009)
3. M.U. Akhmet, Creating a chaos in a system with relay. Int. J. Qualit. Th. Diff. Eqs. Appl. **3**, 3–7 (2009)
4. M.U. Akhmet, *Principles of Discontinuous Dynamical Systems* (Springer, New York, 2010)
5. M.U. Akhmet, M.O. Fen, Chaotic period-doubling and OGY control for the forced Duffing equation. Commun. Nonlinear Sci. Numer. Simul. **17**, 1929–1946 (2012)
6. M.U. Akhmet, M.O. Fen, Replication of chaos. Commun. Nonlinear Sci. Numer. Simul. **18**, 2626–2666 (2013)
7. M.U. Akhmet, M.O. Fen, Shunting inhibitory cellular neural networks with chaotic external inputs. Chaos **23**, 023112 (2013)
8. M. Akhmet, M.O. Fen, Chaotification of impulsive systems by perturbations. Int. J. Bifurcat. Chaos **24**, 1450078 (2014)
9. M. Akhmet, M.O. Fen, Generation of cyclic/toroidal chaos by Hopfield neural networks. Neurocomputing **145**, 230–239 (2014)
10. M. Akhmet, M.O. Fen, Attraction of Li-Yorke chaos by retarded SICNNs. Neurocomputing **147**, 330–342 (2015)
11. M. Akhmet, M.O. Fen, A. Kıvılcım, Li-Yorke chaos generation by SICNNs with chaotic/almost periodic postsynaptic currents. Neurocomputing **173**, 580–594 (2016)
12. M. Akhmet, M.O. Fen, *Replication of Chaos in Neural Networks, Economics and Physics* (Higher Education Press, Beijing; Springer, Heidelberg, 2016)
13. M. Akhmet, M.O. Fen, Unpredictable points and chaos. Commun. Nonlinear Sci. Numer. Simul. **40**, 1–5 (2016)
14. M. Akhmet, M.O. Fen, Existence of unpredictable solutions and chaos. Turk. J. Math. **41**, 254–266 (2017)
15. C. Corduneanu, *Principles of Differential and Integral Equations* (Allyn and Bacon, Inc., Boston, 1971)
16. R. Devaney, Z. Nitecki, Shift automorphisms in the Hénon mapping. Commun. Math. Phys. **67**, 137–146 (1979)
17. R.L. Devaney, *An Introduction to Chaotic Dynamical Systems* (Addison-Wesley, USA, 1989)
18. J.M. Gonzáles-Miranda, *Synchronization and Control of Chaos* (Imperial College Press, London, 2004)
19. J.K. Hale, *Ordinary Differential Equations* (Krieger Publishing Company, Malabar, Florida, 1980)
20. H. Hilmy, Sur les ensembles quasi-minimaux dans les systèmes dynamiques. Ann. Math. **37**, 899–907 (1936)
21. P. Holmes, Poincaré, celestial mechanics, dynamical-systems theory and "chaos". Physics Reports, Review section of Physics Letters **193**, 137–163 (1990)
22. R.A. Horn, C.R. Johnson, *Matrix Analysis* (Cambridge University Press, USA, 1992)
23. T.Y. Li, J.A. Yorke, Period three implies chaos. Am. Math. Monthly **82**, 985–992 (1975)
24. E.N. Lorenz, Deterministic nonperiodic flow. J. Atmos. Sci. **20**, 130–141 (1963)
25. V.V. Nemytskii, V.V. Stepanov, *Qualitative Theory of Differential Equations* (Princeton University Press, Princeton, New Jersey, 1960)
26. E. Ott, C. Grebogi, J.A. Yorke, Controlling chaos. Phys. Rev. Lett. **64**, 1196–1199 (1990)
27. H. Poincaré, *Les méthodes nouvelles de la mécanique céleste*, Vol. 1, 2 (Gauthier-Villars, Paris, 1892)
28. C. Robinson, *Dynamical Systems: Stability, Symbolic Dynamics, and Chaos* (CRC Press, Boca Raton, 1995)

29. H.G. Schuster, *Handbook of Chaos Control* (Wiley-Vch, Weinheim, 1999)
30. G.R. Sell, *Topological Dynamics and Ordinary Differential Equations* (Van Nostrand Reinhold Company, London, 1971)
31. S. Smale, Diffeomorphisms with many periodic points, in *Differential and Combinatorial Topology*, ed. by S.S. Cairns (Princeton University Press, Princeton, 1965), pp. 63–80
32. S. Wiggins, *Global Bifurcation and Chaos: Analytical Methods* (Springer, New York, Berlin, 1988)

Chapter 4
Nonlinear Unpredictable Perturbations

The results of this chapter are continuation of the research of Poincaré chaos initiated in Chaps. 2 and 3. We focus on the construction of an unpredictable function, continuous on the real axis. This is the first time that perturbations depend nonlinearly on unpredictable functions. (See also Chap. 3, Definition 3.2.) As auxiliary results, unpredictable orbits for the symbolic dynamics and the logistic map are obtained. By shaping the unpredictable function as well as Poisson function we have performed the first step in the development of the theory of unpredictable solutions for differential and discrete equations. The results are preliminary ones for deep analysis of chaos existence in differential and hybrid systems. Illustrative examples concerning unpredictable solutions of differential equations are provided.

4.1 Preliminaries

It is useless to say that the theory of dynamical systems is a research of oscillations, and the latest motion of the theory is the Poisson stable trajectory [17]. In paper [8] inspired by chaos investigation we have introduced a new type of oscillation, next to the Poisson stable one, and called the initial point for it the *unpredictable point* and the trajectory itself the *unpredictable orbit*. These novelties make a connection of the homoclinic chaos and the latest types of chaos possible through the concept of *unpredictability*, that is sensitivity assigned to a single orbit. Thus, the *Poincaré chaos* concept has been eventually shaped in our paper [8]. We have also determined the unpredictable function on the real axis as an unpredictable point of the Bebutov dynamics in paper [9] to involve widely differential and discrete equations to the chaos investigation. Nonetheless, we need a more precise description of what one understands as unpredictable function. The present chapter is devoted to this constructive duty. One can say that the analysis became productive, since we have learnt that the unpredictable functions can be bounded, and this is also

M. Akhmet et al., *Dynamics with Chaos and Fractals*, Nonlinear Systems and Complexity 29, https://doi.org/10.1007/978-3-030-35854-9_4

true for Poisson functions, newly introduced in this chapter. Thus, the discussion is focused on chaotic attractors bounded in the space variables. This is important for applications, and it remains legal through our suggestions.

We utilize the topology of uniform convergence on any compact subset of the real axis to introduce the unpredictable functions. More precisely, the Bebutov dynamical system [18] has been applied. Additionally, using the same dynamics we have introduced Poisson functions. All these make our duty of incorporating chaos investigation to theory of differential equations initiated in papers [1–6] and in the book [7] seems to proceed in the correct way.

The main goal of this chapter is the construction of a concrete unpredictable function and its application to differential equations. To give the procedure we start with unpredictable sequences as motions of symbolic dynamics and the logistic map. Then an unpredictable function is determined as an improper convolution integral with a relay function. Finally, we demonstrated in examples unpredictable functions as solutions of differential equations.

Let (X, d) be a metric space and $\pi : \mathbb{T}_+ \times X \to X$, where $\mathbb{T}_+$ is either the set of non-negative real numbers or the set of non-negative integers, be a semi-flow on X, i.e., $\pi(0, x) = x$ for all $x \in X$, $\pi(t, x)$ is continuous in the pair of variables t and x, and $\pi(t_1, \pi(t_2, x)) = \pi(t_1 + t_2, x)$ for all $t_1,\ t_2 \in \mathbb{T}_+$, $x \in X$.

A point $x \in X$ is called positively Poisson stable (stable P^+) if there exists a sequence $\{t_n\}$ satisfying $t_n \to \infty$ as $n \to \infty$ such that $\lim_{n\to\infty} \pi(t_n, x) = x$ [16]. For a given point $x \in X$, let us denote by Θ_x the closure of the trajectory $T(x) = \{\pi(t, x) : t \in \mathbb{T}_+\}$, i.e., $\Theta_x = \overline{T(x)}$. The set Θ_x is a quasi-minimal set if the point x is stable P^+ and $T(x)$ is contained in a compact subset of X [16].

It was demonstrated by Hilmy [15] that if the trajectory corresponding to a Poisson stable point x is contained in a compact subset of X and it is neither a rest point nor a cycle, then the quasi-minimal set contains an uncountable set of motions everywhere dense and Poisson stable. The following theorem can be proved by adapting the technique given in [15, 16].

Theorem 4.1 ([8]) *Suppose that $x \in X$ is stable P^+ and $T(x)$ is contained in a compact subset of X. If Θ_x is neither a rest point nor a cycle, then it contains an uncountable set of motions everywhere dense and stable P^+.*

The definitions of an unpredictable point and unpredictable trajectory are as follows.

Definition 4.1 ([8]) A point $x \in X$ and the trajectory through it are unpredictable if there exist a positive number ϵ_0 (the unpredictability constant) and sequences $\{t_n\}$ and $\{\tau_n\}$, both of which diverge to infinity, such that $\lim_{n\to\infty} \pi(t_n, x) = x$ and $d(\pi(t_n + \tau_n, x), \pi(\tau_n, x)) \geq \epsilon_0$ for each $n \in \mathbb{N}$.

Markov [16] proved that a trajectory stable in both Poisson and Lyapunov (uniformly) senses must be an almost periodic one. Since Definition 4.1 implies instability, an unpredictable motion cannot be almost periodic. In particular, it is neither an equilibrium nor a cycle.

Based on unpredictable points, a new chaos definition was provided in the paper [8] (see also Chap. 2) as follows.

Definition 4.2 ([8]) The dynamics on the quasi-minimal set Θ_x is called Poincaré chaotic if x is an unpredictable point.

It is worth noting that Poincaré chaos admits properties similar to the ingredients of Devaney chaos [12]. In the paper [8], it was proved that if x is an unpredictable point, then the dynamics on Θ_x is sensitive. That is, there exists a positive number $\widetilde{\epsilon}_0$ such that for each $x_1 \in \Theta_x$ and for each positive number δ there exist a point $x_2 \in \Theta_x$ and a positive number $\bar{t}$ such that $d(x_1, x_2) < \delta$ and $d(f(\bar{t}, x_1), f(\bar{t}, x_2)) \geq \widetilde{\epsilon}_0$. Besides, since the trajectory $T(x)$ is dense in Θ_x, transitivity is also present in the dynamics. Moreover, according to Theorem 4.1, there exists a continuum of stable P^+ orbits in Θ_x. In our paper [8] we gave the definition of Poincaré chaos for flows, but in this chapter we give the definition for semi-flows, since the discussion in the paper [8] is valid also for the latter case.

Let us denote by $C(\mathbb{R})$ the set of continuous functions defined on $\mathbb{R}$ with values in $\mathbb{R}^m$, and assume that $C(\mathbb{R})$ has the topology of uniform convergence on compact sets, i.e., a sequence $\{h_k\}$ in $C(\mathbb{R})$ is said to converge to a limit h if for every compact set $\mathcal{U} \subset \mathbb{R}$ the sequence of restrictions $\{h_k|_{\mathcal{U}}\}$ converges to $\{h|_{\mathcal{U}}\}$ uniformly.

One can define a metric ρ on $C(\mathbb{R})$ as [18]

$$\rho(h_1, h_2) = \sum_{k=1}^{\infty} 2^{-k} \rho_k(h_1, h_2), \tag{4.1.1}$$

where h_1, h_2 belong to $C(\mathbb{R})$ and

$$\rho_k(h_1, h_2) = \min\left\{1, \sup_{s\in[-k,k]} \|h_1(s) - h_2(s)\|\right\}, \quad k \in \mathbb{N}.$$

Let us define the mapping $\pi : \mathbb{R}_+ \times C(\mathbb{R}) \to C(\mathbb{R})$ by $\pi(t, h) = h_t$, where $h_t(s) = h(t + s)$. The mapping π is a semi-flow on $C(\mathbb{R})$, and it is called the Bebutov dynamical system [18].

Using the Bebutov dynamical system, we give the descriptions of a Poisson function and an unpredictable function in the next definitions.

Definition 4.3 ([9]) A Poisson function is a Poisson stable point of the Bebutov dynamical system.

Definition 4.4 ([9]) An unpredictable function is an unpredictable point of the Bebutov dynamical system.

It is clear that an unpredictable function is a Poisson function. According to Theorem $III.3$ [18], a motion $\pi(t, h)$ lies in a compact set if h is a bounded and uniformly continuous function. Therefore, an unpredictable function h determines Poincaré chaos in the Bebutov dynamical system if it is bounded and uniformly continuous. Moreover, any system of differential equations which admits uniformly continuous and bounded unpredictable solution has a Poincaré chaos. For differential equations, we say that a solution is an unpredictable one if it is uniformly continuous and bounded on the real axis.

Let us consider the system

$$x'(t) = Ax(t) + f(x(t)) + g(t), \tag{4.1.2}$$

where all eigenvalues of the constant $p \times p$ matrix A have negative real parts, the function $f : \mathbb{R}^p \to \mathbb{R}^p$ is bounded, and the function $g : \mathbb{R} \to \mathbb{R}^p$ is a uniformly continuous and bounded. Since the eigenvalues of the matrix A have negative real parts, there exist positive numbers K_0 and ω such that $\left\|e^{At}\right\| \leq K_0 e^{\omega t}$ for $t \geq 0$ [13].

The presence of an unpredictable solution in the dynamics of (4.1.2) is mentioned in the next theorem.

Theorem 4.2 ([9]) *If $g(t)$ is an unpredictable function and the function $f(x)$ is Lipschitzian with a sufficiently small Lipschitz constant L_f such that $K_0 L_f - \omega < 0$, then system (4.1.2) possesses a unique uniformly exponentially stable unpredictable solution.*

In the next two sections we will consider unpredictable functions whose domain consists of all integers, that is, *unpredictable sequences*.

4.2 An Unpredictable Sequence of the Symbolic Dynamics

In this section, we will show the presence of an unpredictable point in the symbolic dynamics [12, 21] with a distinguishing feature.

Let us consider the space $\Sigma_2 = \left\{s = (s_0 s_1 s_2 \ldots) \mid s_j = 0 \text{ or } 1\right\}$ of infinite sequences of 0's and 1's with the metric

$$d(s,t) = \sum_{k=0}^{\infty} \frac{|s_k - t_k|}{2^k},$$

where $s = (s_0 s_1 s_2 \ldots)$, $t = (t_0 t_1 t_2 \ldots) \in \Sigma_2$. The Bernoulli shift $\sigma : \Sigma_2 \to \Sigma_2$ is defined as $\sigma(s_0 s_1 s_2 \ldots) = (s_1 s_2 s_3 \ldots)$. The map σ is continuous and Σ_2 is a compact metric space [12, 21].

In the book [12], the sequence

$$s^* = (\underbrace{0\ 1}_{1\,blocks}\ |\underbrace{00\ 01\ 10\ 11}_{2\,blocks}|\underbrace{000\ 001\ 010\ 011\ \ldots}_{3\,blocks}|\ldots), \tag{4.2.3}$$

was considered, which is constructed by successively listing all blocks of 0's and 1's of length n, then length $n+1$, etc. In the proof of the next lemma, an element

$s^{**} = (s_0^{**}s_1^{**}s_2^{**}\ldots)$ of Σ_2 will be constructed in a similar way to s^* with the only difference that the order of the blocks will be chosen in a special way. An extension of the constructed sequence to the left-hand side will also be provided.

Lemma 4.2.1 ([10]) *For each increasing sequence $\{m_n\}$ of positive integers, there exist a sequence $s^{**} \in \Sigma_2$ and sequences $\{\alpha_n\}$, $\{\beta_n\}$ of positive integers, both of which diverge to infinity, such that*

(i) $d(\sigma^{\alpha_n+r}(s^{**}), \sigma^r(s^{**})) \le 2^{-m_n}, \ r = -n, -n+1, \ldots, n,$
(ii) $d(\sigma^{\alpha_n+\beta_n}(s^{**}), \sigma^{\beta_n}(s^{**})) \ge 1$ *for each* $n \in \mathbb{N}$.

Proof Fix an arbitrary increasing sequence $\{m_n\}$ of positive integers. For each $n \in \mathbb{N}$, define $\alpha_n = \sum_{k=1}^{n+m_n} k2^k$ and $\beta_n = n + m_n + 1$. Clearly, both of the sequences $\{\alpha_n\}$ and $\{\beta_n\}$ diverge to infinity. We will construct a sequence $s^{**} \in \Sigma_2$ such that the inequalities (i) and (ii) are valid.

First of all, we choose the terms $s_0^{**}, s_1^{**}, \ldots, s_{\alpha_1-1}^{**}$ by successively placing the blocks of 0's and 1's in an increasing length, starting from the blocks of length 1 till the end of the ones with length $m_1 + 1$. The order of the blocks with the same length can be arbitrary without any repetitions. Let us take $s_{\alpha_1+k}^{**} = s_k^{**}$ for $k = 0, 1, \ldots, m_1 + 1$, i.e., the terms of the first block $(s_{\alpha_1}^{**}s_{\alpha_1+1}^{**}\cdots s_{\alpha_1+m_1+1}^{**})$ of length $m_1 + 2$ is chosen the same as the first $m_1 + 2$ terms of the sequence s^{**}. Moreover, we take the second block of length m_1+2 in a such a way that its first term $s_{\alpha_1+\beta_1}^{**}$ is different from $s_{\beta_1}^{**}$. After that we continue placing the remaining blocks of length $m_1 + 2$ and the ones with length greater than $m_1 + 2$ till the last block of length $m_2 + 2$, and again, the blocks of the same length can be in any order without repetitions.

For each $n \ge 2$, we set the last block of length $m_n + n$ such that $s_{\alpha_n-k}^{**} = s_{\alpha_k-k}^{**}$ for each $k = 1, 2, \ldots, n-1$. Then, the terms of the first block of length $m_n + n + 1$ are constituted by taking $s_{\alpha_n+k}^{**} = s_k^{**}$, $k = 0, 1, \ldots, m_n + n$. Moreover, the second block of length $m_n + n + 1$ is chosen such that $s_{\alpha_n+\beta_n}^{**} \ne s_{\beta_n}^{**}$. Lastly, the remaining blocks of 0's and 1's are successively placed similar to the case mentioned above so that the lengths of the blocks in s^{**} are in an increasing order and there are no repetitions within the blocks of the same length. By this way the construction of the sequence $s^{**} \in \Sigma_2$ is completed. We fix the extension of the sequence s^{**} to the left by choosing $\sigma^{-k}(s^{**})_0 = s_{\alpha_k-k}^{**}$, $k \in \mathbb{N}$. For each $n \in \mathbb{N}$, we have $\sigma^{\alpha_n-n}(s^{**})_k = \sigma^{-n}(s^{**})_k$, $k = 0, 1, \ldots, m_n + 2n$ and $\sigma^{\alpha_n+\beta_n}(s^{**})_0 \ne \sigma^{\beta_n}(s^{**})_0$ so that the inequalities (i) and (ii) are valid. □

The technique presented in [12] can be used to show that the trajectory $T(s^{**}) = \{\sigma^i(s^{**}) : i \in \mathbb{Z}\}$ is dense in Σ_2, i.e., $\Theta_{s^{**}} = \Sigma_2$. By Lemma 2.2 in [8] any sequence $\sigma^i(s^{**})$, $i \in \mathbb{Z}$, is an unpredictable point of the Bernoulli dynamics on Σ_2, and Σ_2 is a quasi-minimal set. It implies from the last theorem that $T(s^{**})$ is an unpredictable function on $\mathbb{Z}$, i.e., an unpredictable sequence. According to Theorem 3.1 presented in [8], the dynamics on Σ_2 is Poincaré chaotic. Moreover, there are infinitely many unpredictable sequences in the set.

4.3 An Unpredictable Solution of the Logistic Map

In this section, we will demonstrate the presence of an unpredictable solution of the equation

$$\eta_{n+1} = F_\mu(\eta_n), \tag{4.3.4}$$

where $F_\mu(s) = \mu s(1-s)$ is the logistic map.

The result is provided in the next theorem.

Theorem 4.3 ([10]) *For each $\mu \in [3+(2/3)^{1/2}, 4]$ and sequence of positive numbers $\delta_n \to 0$, there exists a solution $\{\eta_n\}$, $n \in \mathbb{Z}$, of Eq. (4.3.4) such that*

(i) $\left|\eta_{i_n+r} - \eta_r\right| < \delta_n,\ r = -h_0 n, -h_0 n + 1, \ldots, h_0 n,$
(ii) $\left|\eta_{i_n+j_n} - \eta_{j_n}\right| \geq \epsilon_0$ *for each* $n \in \mathbb{N}$,

where ϵ_0 is a positive number, $h_0 > 4$ is a natural number, and $\{i_n\}$, $\{j_n\}$ are integer valued sequences both of which diverge to infinity.

Proof Fix $\mu \in [3+(2/3)^{1/2}, 4]$ and a sequence $\{\delta_n\}$ of positive real numbers with $\delta_n \to 0$ as $n \to \infty$. Take a neighborhood $U \subset [0,1]$ of the point $1-1/\mu$. According to Theorem 6 of paper [19], there exist a natural number $h_0 > 4$ and a Cantor set $\Lambda \subset U$ such that the map $F_\mu^{h_0}$ on Λ is topologically conjugate to the Bernoulli shift σ on Σ_2. Therefore, there exists a homeomorphism $S : \Sigma_2 \to \Lambda$ such that $S \circ \sigma = F_\mu^{h_0} \circ S$. Since S is uniformly continuous on Σ_2, for each $n \in \mathbb{N}$, there exists a number $\overline{\delta}_n > 0$ such that for each $s^1, s^2 \in \Sigma_2$ with $d(s^1, s^2) < \overline{\delta}_n$, we have $\left|\eta^1 - \eta^2\right| < \delta_n/\mu^{h_0-1}$, where $\eta^1 = S(s^1)$, $\eta^2 = S(s^2)$.

Let $\{m_n\}$ be an increasing sequence of natural numbers such that $2^{-m_n} < \overline{\delta}_n$ for each $n \in \mathbb{N}$. According to Lemma 4.2.1, there exist a sequence $s^{**} \in \Sigma_2$ and sequences $\{\alpha_n\}$, $\{\beta_n\}$ both of which diverge to infinity such that

$$d(\sigma^{\alpha_n+r}(s^{**}), \sigma^r(s^{**})) \leq 2^{-m_n},\ r = -n, -n+1, \ldots, n,$$

and $d(\sigma^{\alpha_n+\beta_n}(s^{**}), \sigma^{\beta_n}(s^{**})) \geq 1$ for each $n \in \mathbb{N}$.

Now, let $\{\eta_n\}$, $n \in \mathbb{Z}$, be the solution of (4.3.4) with $\eta_{h_0 k} = S(\sigma^k(s^{**}))$, $k \in \mathbb{Z}$. Since the inequality $\left|F_\mu(u_1) - F_\mu(u_2)\right| \leq \mu\,|u_1 - u_2|$ is valid for every $u_1, u_2 \in [0,1]$, we have for each $n \in \mathbb{N}$ that $\left|\eta_{i_n+r} - \eta_r\right| < \delta_n$, $r = -h_0 n, -h_0 n + 1, \ldots, h_0 n$, where $i_n = h_0 \alpha_n$. Besides, using the arguments presented in [11], one can verify the existence of a positive number ϵ_0 such that $\left|\eta_{i_n+j_n} - \eta_{j_n}\right| \geq \epsilon_0$, $n \in \mathbb{N}$, where $j_n = h_0 \beta_n$. □

By the topological equivalence and results on Σ_2 of the last section, one can make several observations from the proved theorem. Any number η_n, $n \in \mathbb{Z}$, is an unpredictable point of the logistic map dynamics, and the Cantor set Λ mentioned in the proof of Theorem 4.3 is a quasi-minimal set. Moreover, the sequence $\{\eta_n\}$ is unpredictable. By Theorem 3.1 mentioned in [8], the dynamics on the quasi-minimal set is Poincaré chaotic, and there are infinitely many unpredictable sequences in the set. The last observation will be applied in the next section to construct an unpredictable function.

4.4 An Unpredictable Function

In this section, we provide an example of an unpredictable function benefiting from the dynamics of the logistic map (4.3.4).

Let us fix two different points d_1 and d_2 in $\mathbb{R}^p$, and suppose that γ is a positive number. Take a sequence $\{k_n\}$ of positive integers satisfying both of the inequalities $2^{-k_n} \leq \dfrac{1}{2n}$ and $e^{-\gamma k_n} \leq \dfrac{\gamma}{4\,\|d_1 - d_2\|\,n}$ for each $n \in \mathbb{N}$. Fix $\mu \in [3 + (2/3)^{1/2}, 4]$ and a sequence $\{\delta_n\}$ of positive numbers such that $\delta_n \leq \dfrac{1}{12\,\|d_1 - d_2\|\,nk_n}$, $n \in \mathbb{N}$. In a similar way to the items (i) and (ii) of Theorem 4.3, one can verify that there exist a positive number ϵ_0, a sequence $\{i_n\}$ of even positive integers, a sequence $\{j_n\}$ of positive integers, and a solution $\{\eta_n\}$, $n \in \mathbb{Z}$, of the logistic map (4.3.4) such that the inequalities

$$\left|\eta_{i_n+r} - \eta_r\right| \leq \delta_n, \; r = -2k_n, -2k_n + 1, \ldots, k_n - 1, \tag{4.4.5}$$

and

$$\left|\eta_{i_n+j_n} - \eta_{j_n}\right| \geq \epsilon_0 \tag{4.4.6}$$

hold for each $n \in \mathbb{N}$.

It is easy to observe that the constructed sequence $\{\eta_n\}$ is unpredictable and consequently, it generates a quasi-minimal set and Poincaré chaos similar to that of the last section.

Now, consider the function $\phi : \mathbb{R} \to \mathbb{R}^p$ defined as

$$\phi(t) = \int_{-\infty}^{t} e^{-\gamma(t-s)} \nu(s) ds, \tag{4.4.7}$$

where the function $\nu(t)$ is defined as

$$\nu(t) = \begin{cases} d_1, \text{ if } \zeta_{2j} < t \leq \zeta_{2j+1}, \; j \in \mathbb{Z}, \\ d_2, \text{ if } \zeta_{2j-1} < t \leq \zeta_{2j}, \; j \in \mathbb{Z}, \end{cases}$$

and the sequence $\{\zeta_j\}$, $j \in \mathbb{Z}$, of switching moments is defined through the equation $\zeta_j = j + \eta_j$ for each j, in which $\{\eta_j\}$ is the solution of (4.3.4) satisfying (4.4.5) and (4.4.6). The function $\phi(t)$ is bounded such that $\sup_{t\in\mathbb{R}} \|\phi(t)\| \le \frac{\max\{\|d_1\|, \|d_2\|\}}{\gamma}$. Moreover, $\phi(t)$ is uniformly continuous since its derivative is bounded.

In the proof of the following theorem, we will denote by $\widehat{(a,b]}$ the oriented interval such that $\widehat{(a,b]} = (a,b]$ if $a < b$ and $\widehat{(a,b]} = (b,a]$ if $a > b$.

Theorem 4.4 ([10]) *The function $\phi(t)$ is unpredictable.*

Proof First of all, we will show that $\rho(\phi_{i_n}, \phi) \to 0$ as $n \to \infty$, where ρ is the metric defined by Eq. (4.1.1). Let us fix an arbitrary natural number n. The functions $\phi(i_n + s)$ and $\phi(s)$ satisfy the equation

$$\phi(i_n + s) - \phi(s) = \int_{-\infty}^{s} e^{-\gamma(s-u)} \left(\nu(i_n + u) - \nu(u)\right) du.$$

It is worth noting that for each $r \in \mathbb{Z}$ both of the points ζ_r and $\zeta_{i_n+r} - i_n$ belong to the interval $(r, r+1)$. Moreover, $\|\nu(i_n + s) - \nu(s)\| = \|d_1 - d_2\|$ for $s \in \bigcup_{r=-\infty}^{\infty} \widehat{(\zeta_r, \zeta_{i_n+r} - i_n]}$, and $\|\nu(i_n + s) - \nu(s)\| = 0$, otherwise.

Since for each $r = -2k_n, -2k_n + 1, \ldots, k_n - 1$ the distance between the points ζ_r and $\zeta_{i_n+r} - i_n$ are at most δ_n, one can verify for each $s \in [-k_n, k_n]$ that

$$\begin{aligned}
\|\phi(i_n + s) - \phi(s)\| &\le \int_{-\infty}^{-2k_n} e^{-\gamma(s-u)} \|\nu(i_n + u) - \nu(u)\| \, du \\
&\quad + \sum_{r=-2k_n}^{k_n - 1} \left| \int_{\zeta_{i_n+r} - i_n}^{\zeta_r} \|\nu(i_n + u) - \nu(u)\| \, du \right| \\
&\le \frac{\|d_1 - d_2\|}{\gamma} e^{-\gamma k_n} + 3k_n \delta_n \|d_1 - d_2\| \\
&\le \frac{1}{2n}.
\end{aligned}$$

Hence, we have

$$\rho(\phi_{i_n}, \phi) = \sum_{k=1}^{\infty} 2^{-k} \rho_k(\phi_{i_n}, \phi) \le \sum_{k=1}^{k_n} 2^{-k} \rho_k(\phi_{i_n}, \phi) + 2^{-k_n} \le \frac{1}{n}.$$

The last inequality implies that $\rho(\phi_{i_n}, \phi) \to 0$ as $n \to \infty$, i.e., $\phi(t)$ is a Poisson function.

Now, let us show the existence of a positive number $\overline{\epsilon}_0$ satisfying $\overline{\epsilon}_0 \to 0$ as $\epsilon_0 \to 0$ such that $\rho(\phi_{i_n+j_n}, \phi_{j_n}) \geq \overline{\epsilon}_0$ for each $n \in \mathbb{N}$. For a fixed natural number n, using the equations

$$\phi(i_n + j_n + s) = e^{-\gamma s}\phi(i_n + j_n) + \int_0^s e^{-\gamma(s-u)}\nu(i_n + j_n + u)du$$

and

$$\phi(j_n + s) = e^{-\gamma s}\phi(j_n) + \int_0^s e^{-\gamma(s-u)}\nu(j_n + u)du,$$

we obtain that

$$\begin{aligned}\|\phi(i_n + j_n + 1) - \phi(j_n + 1)\| \geq & \left\| \int_{\zeta_{i_n+j_n}-i_n-j_n}^{\zeta_{j_n}-j_n} e^{-\gamma(1-u)}(d_1 - d_2)du \right\| \\ & -e^{-\gamma} \|\phi(i_n + j_n) - \phi(j_n)\| \\ \geq & \frac{e^{\gamma\epsilon_0} - 1}{\gamma e^{\gamma}} \|d_1 - d_2\| - e^{-\gamma} \|\phi(i_n + j_n) - \phi(j_n)\|.\end{aligned}$$

Therefore, it can be verified for each $k \in \mathbb{N}$ that

$$\sup_{s\in[-k,k]} \|\phi(i_n + j_n + s) - \phi(j_n + s)\| \geq \frac{(e^{\gamma\epsilon_0} - 1) \|d_1 - d_2\|}{\gamma (1 + e^{\gamma})}. \tag{4.4.8}$$

Let us denote

$$\overline{\epsilon}_0 = \min\left\{1, \frac{(e^{\gamma\epsilon_0} - 1) \|d_1 - d_2\|}{\gamma (1 + e^{\gamma})}\right\}.$$

It can be confirmed by means of inequality (4.4.8) that $\rho_k(\phi_{i_n+j_n}, \phi_{j_n}) \geq \overline{\epsilon}_0$, $k \in \mathbb{N}$. Thus, $\rho(\phi_{i_n+j_n}, \phi_{j_n}) \geq \overline{\epsilon}_0$ for each $n \in \mathbb{N}$. Consequently, the function $\phi(t)$ is unpredictable. □

One of the possible ways useful for applications to generate unpredictable functions from a given one is provided in the next theorem.

Theorem 4.5 ([10]) *Let $\phi : \mathbb{R} \to \mathscr{H}$ be an unpredictable function, where $\mathscr{H}$ is a bounded subset of $\mathbb{R}^p$. If $h : \mathscr{H} \to \mathbb{R}^q$ is a function such that there exist positive numbers L_1 and L_2 satisfying $L_1 \|u - \overline{u}\| \leq \|h(u) - h(\overline{u})\| \leq L_2 \|u - \overline{u}\|$ for all $u, \overline{u} \in \mathscr{H}$, then the function $\psi : \mathbb{R} \to \mathbb{R}^q$ defined as $\psi(t) = h(\phi(t))$ is also unpredictable.*

Proof Since $\phi(t)$ is an unpredictable function, there exist a positive number ϵ_0 and sequences $\{t_n\}$ and $\{\tau_n\}$, both of which diverge to infinity, such that $\lim_{n\to\infty} \rho(\phi_{t_n}, \phi) = 0$ and $\rho(\phi_{t_n+\tau_n}, \phi_{\tau_n}) \geq \epsilon_0$ for each $n \in \mathbb{N}$.

Firstly, we will show that $\lim_{n\to\infty} \rho(\psi_{t_n}, \psi) = 0$. Fix an arbitrary positive number ϵ, and let us denote $\alpha = \max\{1, L_2\}$. There exists a natural number n_0 such that for all $n \geq n_0$ the inequality $\rho(\phi_{t_n}, \phi) < \epsilon/\alpha$ is valid. For each $k \in \mathbb{N}$, one can confirm that

$$\begin{aligned}
\rho_k(\psi_{t_n}, \psi) &= \min\left\{1, \sup_{s\in[-k,k]} \|h(\phi(t_n+s)) - h(\phi(s))\|\right\} \\
&\leq \min\left\{1, L_2 \sup_{s\in[-k,k]} \|\phi(t_n+s) - \phi(s)\|\right\} \\
&\leq \alpha\rho_k(\phi_{t_n}, \phi).
\end{aligned}$$

Therefore, it can be verified for each $n \geq n_0$ that

$$\rho(\psi_{t_n}, \psi) \leq \alpha\rho(\phi_{t_n}, \phi) < \epsilon.$$

Hence, $\lim_{n\to\infty} \rho(\psi_{t_n}, \psi) = 0$.

Next, we will show the existence of a positive number $\overline{\epsilon}_0$ such that $\rho(\psi_{t_n+\tau_n}, \psi_{\tau_n}) \geq \overline{\epsilon}_0$ for each $n \in \mathbb{N}$. Denote $\beta = \min\{1, L_1\}$. For each $k \in \mathbb{N}$, we have that

$$\begin{aligned}
\rho_k(\psi_{t_n+\tau_n}, \psi_{\tau_n}) &= \min\left\{1, \sup_{s\in[-k,k]} \|h(\phi(t_n+\tau_n+s)) - h(\phi(\tau_n+s))\|\right\} \\
&\geq \min\left\{1, L_1 \sup_{s\in[-k,k]} \|\phi(t_n+\tau_n+s) - \phi(\tau_n+s)\|\right\} \\
&\geq \beta\rho_k(\phi_{t_n+\tau_n}, \phi_{\tau_n}).
\end{aligned}$$

Thus, the inequality

$$\rho(\psi_{t_n+\tau_n}, \psi_{\tau_n}) \geq \beta\rho(\phi_{t_n+\tau_n}, \phi_{\tau_n}) \geq \overline{\epsilon}_0$$

holds for each $n \in \mathbb{N}$, where $\overline{\epsilon}_0 = \beta\epsilon_0$. Consequently, the function $\psi(t)$ is unpredictable. □

A corollary of Theorem 4.5 is as follows.

Corollary 4.1 ([10]) *If $\phi : \mathbb{R} \to \mathscr{H}$ is an unpredictable function, where $\mathscr{H}$ is a bounded subset of $\mathbb{R}^p$, then the function $\psi : \mathbb{R} \to \mathbb{R}^p$ defined as $\psi(t) = P\phi(t)$, where P is a constant, nonsingular, $p \times p$ matrix, is also an unpredictable function.*

Proof The function $h : \mathscr{H} \to \mathbb{R}^p$ defined as $h(u) = Pu$ satisfies the inequality

$$L_1 \|u_1 - u_2\| \le \|h(u_1) - h(u_2)\| \le L_2 \|u_1 - u_2\|,$$

for $u_1, u_2 \in \mathscr{H}$ with $L_1 = 1/\left\|P^{-1}\right\|$ and $L_2 = \|P\|$. Therefore, by Theorem 4.5, the function $\psi(t)$ is unpredictable. □

In the next section, the existence of Poincaré chaos in the dynamics of differential equations will be presented.

4.5 Unpredictable Solutions of Differential Equations

Consider the differential equation

$$x'(t) = -\frac{3}{2}x(t) + \nu(t), \tag{4.5.9}$$

where the function $\nu(t)$ is defined as

$$\nu(t) = \begin{cases} 0.7, & \text{if } \zeta_{2j} < t \le \zeta_{2j+1}, \ j \in \mathbb{Z}, \\ -0.4, & \text{if } \zeta_{2j-1} < t \le \zeta_{2j}, \ j \in \mathbb{Z}. \end{cases} \tag{4.5.10}$$

In (4.5.10), the sequence $\{\zeta_j\}$ is defined as $\zeta_j = j + \eta_j, \ j \in \mathbb{Z}$, and $\{\eta_j\}$ is the unpredictable sequence determined in Sect. 4.4 for the map (4.3.4) with $\mu = 3.91$.

According to Theorem 4.4,

$$\psi(t) = \int_{-\infty}^{t} e^{-3(t-s)/2}\nu(s)ds$$

is a globally asymptotically stable unpredictable solution of (4.5.9). We represent a solution of (4.5.9) corresponding to the initial data $x(\zeta_0) = 0.3, \ \zeta_0 = 0.4$ in Fig. 4.1.

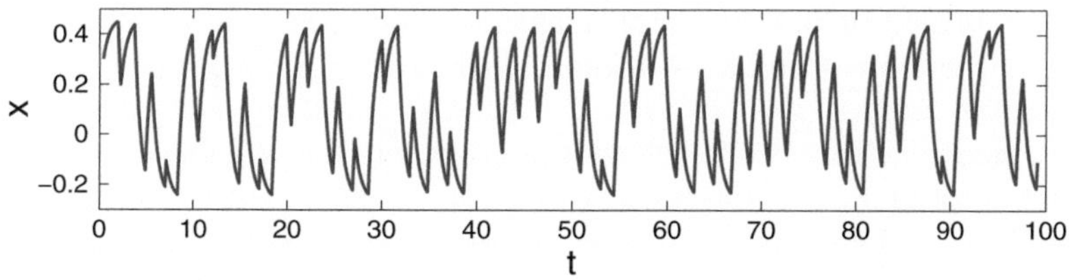

Fig. 4.1 Chaotic behavior in equation (4.5.9). The figure confirms that Poincaré chaos takes place in the dynamics of equation (4.5.9)

The choice of the coefficient $\mu = 3.91$ and the initial value $\zeta_0 = 0.4$ is approved by the shadowing analysis in paper [14]. The simulation seen in Fig. 4.1 supports the result of Theorem 4.4 such that the Eq. (4.5.9) behaves chaotically.

Next, we will demonstrate the chaotic behavior of a multidimensional system of differential equations.

Let us take into account the system

$$x'(t) = Ax(t) + \nu(t), \tag{4.5.11}$$

where $x(t) = (x_1(t), x_2(t), x_3(t)) \in \mathbb{R}^3$, $A = \begin{pmatrix} -3 & 0 & -1 \\ -2 & 1 & -2 \\ -2 & 4 & -4 \end{pmatrix}$, and the function $\nu : \mathbb{R} \to \mathbb{R}^3$ is defined as

$$\nu(t) = \begin{cases} (-1, 1, 2), & \text{if } \zeta_{2j} < t \le \zeta_{2j+1},\ j \in \mathbb{Z}, \\ (3, 1, -1), & \text{if } \zeta_{2j-1} < t \le \zeta_{2j},\ j \in \mathbb{Z}. \end{cases} \tag{4.5.12}$$

Similarly to equation (4.5.10), in (4.5.12), the sequence $\{\zeta_j\}$ of switching moments is defined through the equation $\zeta_j = j + \eta_j$, where $\{\eta_j\}$ is the unpredictable sequence determined in Sect. 4.4 for the map (4.3.4) with $\mu = 3.86$.

By means of the transformation $y = P^{-1}x$, where $P = \begin{pmatrix} 1 & 2 & 1 \\ 0 & 1 & -1 \\ -1 & 0 & -2 \end{pmatrix}$, system (4.5.11) can be written as

$$y'(t) = Dy(t) + \overline{\nu}(t), \tag{4.5.13}$$

where $D = \begin{pmatrix} -2 & 0 & 0 \\ 0 & -3 & 0 \\ 0 & 0 & -1 \end{pmatrix}$, and

$$\overline{\nu}(t) = \begin{cases} (0, 0, -1), & \text{if } \zeta_{2j} < t \le \zeta_{2j+1},\ j \in \mathbb{Z}, \\ (1, 1, 0), & \text{if } \zeta_{2j-1} < t \le \zeta_{2j},\ j \in \mathbb{Z}. \end{cases}$$

System (4.5.13) admits an unpredictable solution $\overline{\psi}(t)$ in accordance with Theorem 4.4. Therefore, Corollary 4.1 implies that $\psi(t) = P\overline{\psi}(t)$ is an unpredictable solution of (4.5.11).

To demonstrate the chaotic behavior, we depict in Fig. 4.2 the x_2-coordinate of the solution of (4.5.11) corresponding to the initial data $x_1(\zeta_0) = 0.17$, $x_2(\zeta_0) = 0.51$, $x_3(\zeta_0) = 0.48$, $\zeta_0 = 0.4$. The coefficient $\mu = 3.86$ and the initial value $\zeta_0 = 0.4$ are considered for shadowing in [14]. The chaotic behavior is also valid in the remaining coordinates, which are not just pictured here. Moreover, Fig. 4.3 shows the 3-dimensional chaotic trajectory of the same solution. It is worth noting that the chaotic solutions of (4.5.11) take place inside the compact region

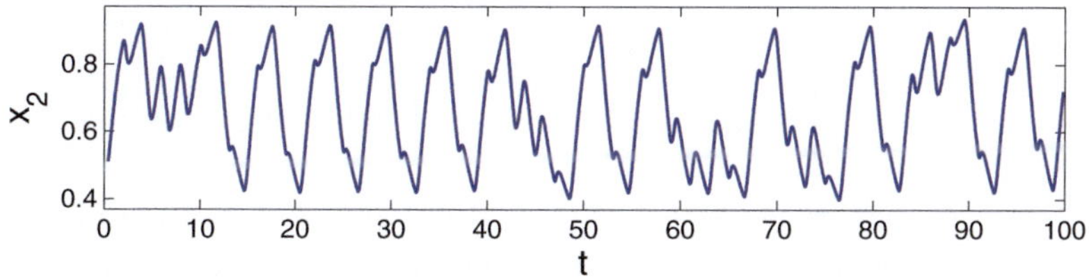

Fig. 4.2 The x_2-coordinate of the chaotic solution of system (4.5.11)

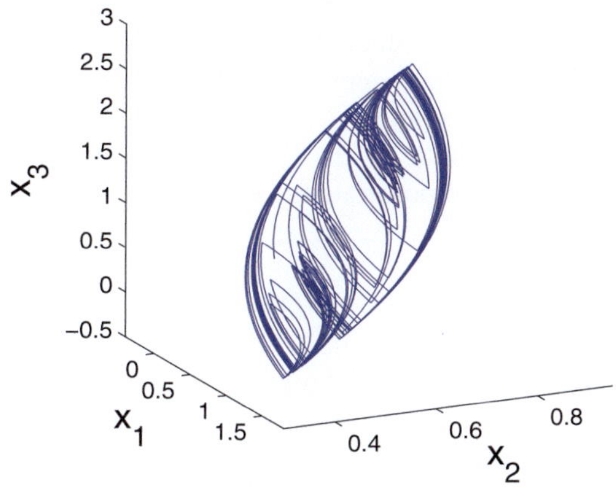

Fig. 4.3 Chaotic trajectory of system (4.5.11). The figure reveals the presence of Poincaré chaos in system (4.5.11)

$$\mathscr{H} = \big\{(x_1, x_2, x_3) \in \mathbb{R}^3 \mid -1 \le x_1 \le 1.1,$$
$$0.3 \le x_2 \le 1, \ -0.4 \le x_3 \le 1.9\big\}. \tag{4.5.14}$$

Figures 4.2 and 4.3 support the results of Theorem 4.4 and Corollary 4.1 such that the represented solution behaves chaotically.

Now, we will demonstrate the extension of unpredictable solutions and Poincaré chaos. For that purpose, we consider the system

$$z'(t) = Bz(t) + f(z(t)) + h(\psi(t)), \tag{4.5.15}$$

where $z(t) = (z_1(t), z_2(t), z_3(t)) \in \mathbb{R}^3$, $B = \begin{pmatrix} -3 & 2 & -1 \\ 0 & -5/2 & 0 \\ 2 & -11/2 & -1 \end{pmatrix}$, and $\psi(t)$ is the unpredictable solution of (4.5.11). In system (4.5.15), the function $f : \mathbb{R}^3 \to \mathbb{R}^3$, $f(u) = (f_1(u), f_2(u), f_3(u))$, is defined as $f_1(u) = 0.03 \sin u_1$, $f_2(u) = 0.04u_2^2$ for $|u_2| \le 1$, $f_2(u) = 0.04$ for $|u_2| > 1$, $f_3(u) = 0.06 \tanh u_3$, and the function $h : \mathbb{R}^3 \to \mathbb{R}^3$, $h(u) = (h_1(u), h_2(u), h_3(u))$, is defined as $h_1(u) = 2 \arctan(u_1)$, $h_2(u) = u_3 + 0.1u_3^3$, $h_3(u) = 0.5u_2^2$, where $u = (u_1, u_2, u_3)$.

The eigenvalues of the matrix B are $-5/2$, $-2 \pm i$, and $e^{Bt} = Pe^{Jt}P^{-1}$, where $J = \begin{pmatrix} -5/2 & 0 & 0 \\ 0 & -2 & -1 \\ 0 & 1 & -2 \end{pmatrix}$ and $P = \begin{pmatrix} 1 & 0 & -1 \\ 1/2 & 0 & 0 \\ 1/2 & 1 & 1 \end{pmatrix}$. One can verify that $\|e^{Bt}\| \leq K_0 e^{-\omega t}$, $t \geq 0$, with $K_0 = \|P\| \, \|P^{-1}\| \approx 7.103$ and $\omega = 2$. The function $f(u)$ is bounded and it satisfies the Lipschitz condition $\|f(u) - f(\overline{u})\| \leq L_f \|u - \overline{u}\|$, $u, \overline{u} \in \mathbb{R}^3$, with $L_f = 0.08$ such that the inequality $K_0 L_f - \omega < 0$ is valid.

On the other hand, the function $h(u)$ satisfies the conditions of Theorem 4.5 with $L_1 = 0.3$ and $L_2 = 2.083$ inside the region $\mathscr{H}$ defined by (4.5.14) so that $h(x(t))$ is an unpredictable function if $x(t)$ is an unpredictable function. Therefore, according to Theorem 4.2, system (4.5.15) possesses a unique uniformly exponentially stable unpredictable solution.

Let us denote by $\theta(t)$ the solution of (4.5.11) whose trajectory is shown in Fig. 4.3. To demonstrate the extension of Poincaré chaos, we take into account the system

$$z'(t) = Bz(t) + f(z(t)) + h(\theta(t)). \tag{4.5.16}$$

We depict in Fig. 4.4 the z_3-coordinate of the solution of (4.5.16) with $z_1(\zeta_0) = 0.53$, $z_2(\zeta_0) = 0.57$, $z_3(\zeta_0) = -1.51$, where $\zeta_0 = 0.4$. Moreover, the trajectory of the same solution is represented in Fig. 4.5. The simulations shown in Figs. 4.4 and 4.5 support Theorem 4.2 such that the Poincaré chaos of system (4.5.11) is extended by (4.5.16).

4.6 Notes

In Chap. 3, an unpredictable function was defined as an unpredictable point of the Bebutov dynamical system. In the present chapter, we have obtained samples of unpredictable functions and sequences, which are in the basis of Poincaré chaos. The unpredictable sequences are utilized in the construction of piecewise continuous functions, and such functions in their own turn are utilized for the construction

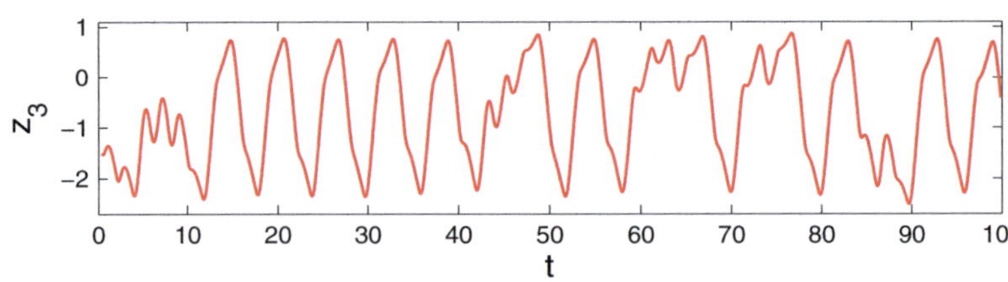

Fig. 4.4 The z_3-coordinate of the chaotic solution of system (4.5.16). The figure manifests the presence of Poincaré chaos in system (4.5.16)

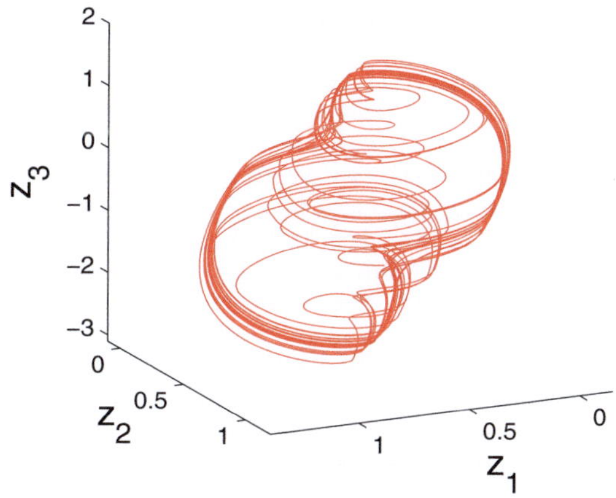

Fig. 4.5 Chaotic trajectory of system (4.5.16). It is seen in the figure that the Poincaré chaos of system (4.5.11) is extended by (4.5.16)

of continuous unpredictable functions. Thus, one can claim that the basics of the new theory of unpredictable functions have been lied in the present chapter. Automatically, the results concerning the analyses of functions and sequences make it possible to formulate new problems of the existence of unpredictable solutions for differential equations of different types as well as for discrete and hybrid systems of equations, similar to the results for periodic, almost periodic, and other types of solutions. These are all strong arguments for the insertion of chaos research to the theory of differential equations. In addition to the role of the present chapter for the theory of differential equations, the concept of unpredictable points and orbits introduced in our studies [8, 9] and the additional results of the present chapter will bring the chaos research to the scope of the classical theory of dynamical systems. Moreover, not less importantly, these concepts extend the boundaries of the theory of dynamical systems significantly, since we are dealing with a new type of motions, which are behind and next to Poisson stable trajectories. From another point of view, the results of this chapter request the development of techniques to determine whether a point is an unpredictable one in concrete dynamics. For that purpose one can apply the results which exist for the indication of Poisson stable points [20]. One more interesting study depending on the present results can be performed if one tries to find a numerical approach for the recognition of unpredictable points. The results of this chapter are published in paper [10].

References

1. M.U. Akhmet, Dynamical synthesis of quasi-minimal sets. Int. J. Bifurcat. Chaos **19**, 2423–2427 (2009)
2. M.U. Akhmet, Shadowing and dynamical synthesis. Int. J. Bifurcat. Chaos **19**, 3339–3346 (2009)
3. M.U. Akhmet, Li-Yorke chaos in the system with impacts. J. Math. Anal. Appl. **351**, 804–810 (2009)

4. M.U. Akhmet, M.O. Fen, Replication of chaos. Commun. Nonlinear Sci. Numer. Simul. **18**, 2626–2666 (2013)
5. M.U. Akhmet, M.O. Fen, Shunting inhibitory cellular neural networks with chaotic external inputs. Chaos **23**, 023112 (2013)
6. M.U. Akhmet, M.O. Fen, Entrainment by chaos. J. Nonlinear Sci. **24**, 411–439 (2014)
7. M. Akhmet, M.O. Fen, *Replication of Chaos in Neural Networks, Economics and Physics* (Higher Education Press, Beijing; Springer, Heidelberg, 2016)
8. M. Akhmet, M.O. Fen, Unpredictable points and chaos. Commun. Nonlinear Sci. Numer. Simul. **40**, 1–5 (2016)
9. M. Akhmet, M.O. Fen, Existence of unpredictable solutions and chaos. Turk. J. Math. **41**, 254–266 (2017)
10. M. Akhmet, M.O. Fen, Poincaré chaos and unpredictable functions. Commun. Nonlinear Sci. Numer. Simul. **48**, 85–94 (2017)
11. J. Banks, J. Brooks, G. Cairns, G. Davis, P. Stacey, On Devaney's definition of chaos. Am. Math. Monthly **99**, 332–334 (1992)
12. R.L. Devaney, *An Introduction to Chaotic Dynamical Systems* (Addison-Wesley, USA, 1989)
13. J.K. Hale, *Ordinary Differential Equations* (Krieger Publishing Company, Malabar, Florida, 1980)
14. S.M. Hammel, J.A. Yorke, C. Grebogi, Do numerical orbits of chaotic dynamical processes represent true orbits? J. Complexity **3**, 136–145 (1987)
15. H. Hilmy, Sur les ensembles quasi-minimaux dans les systèmes dynamiques. Ann. Math. **37**, 899–907 (1936)
16. V.V. Nemytskii, V.V. Stepanov, *Qualitative Theory of Differential Equations* (Princeton University Press, Princeton, New Jersey, 1960)
17. H. Poincaré, *Les methodes nouvelles de la mecanique celeste*, Vol. III, Paris, 1899; reprint (Dover, New York, 1957)
18. G.R. Sell, *Topological Dynamics and Ordinary Differential Equations* (Van Nostrand Reinhold Company, London, 1971)
19. Y. Shi, P. Yu, On chaos of the logistic maps. Dynam. Contin. Discrete Impuls. Syst. Ser. B **14**, 175–195 (2007)
20. L. Shilnikov, Homoclinic chaos, in *Nonlinear Dynamics, Chaotic and Complex Systems*, ed. by E. Infeld, R. Zelazny, A. Galkowski (Cambridge University Press, Cambridge, 1997), pp. 39–63
21. S. Wiggins, *Global Bifurcation and Chaos: Analytical Methods* (Springer, New York, Berlin, 1988)

Chapter 5
Unpredictability in Topological Dynamics

In this chapter, the topology of uniform convergence on compact sets is applied to define unpredictable functions. The unpredictable sequence is defined as a specific unpredictable function on the set of integers. The definitions are convenient to be verified as solutions of differential and discrete equations. The topology is metrizable and easy for applications with integral operators. To demonstrate the effectiveness of the approach, the existence and uniqueness of the unpredictable solution for a delay differential equation is proved as well as for quasilinear discrete systems. As a corollary of the theorem, a similar assertion for a quasilinear ordinary differential equation is formulated. The results are demonstrated numerically, and an application to Hopfield neural networks is provided. In particular, Poincaré chaos near periodic orbits is observed. The completed research contributes to the theory of chaos as well as to the theory of differential and discrete equations, considering unpredictable solutions.

5.1 Introduction

In this chapter, another step in the adaptation of unpredictable functions to the theory of differential equations has been made. We apply the uniform convergence on compact subsets of the real axis to determine unpredictable functions for two reasons. The first reason is that the topology is easily metrizable, in particular, to the metric for Bebutov dynamical system [28], and consequently, the unpredictable functions and solutions immediately imply the presence of Poincaré chaos according to our results in [5]. The second one is the easy verification of the convergence. Thus, the present chapter is useful for the theory of differential equations as well as chaos researches. For the construction of unpredictable functions we have applied the results on the equivalence of discrete dynamics obtained in papers [29, 30]. Moreover, an application to Hopfield neural networks [17, 23] is provided.

M. Akhmet et al., *Dynamics with Chaos and Fractals*, Nonlinear Systems and Complexity 29, https://doi.org/10.1007/978-3-030-35854-9_5

In the instrumental sense, discreteness has been the main object in chaos investigation. To check this, it is sufficient to recall definitions of chaos [10, 22, 32], which are based on sequences and maps, as well as Smale horseshoe and logistic maps, Bernoulli shift [32], which are in the core of the chaos theory. One can say that stroboscopic observation of a motion was the single way to indicate the irregularity in a continuous dynamics. The definitions of chaos for continuous dynamics, which are not related to discreteness [1–4], are requested for embedding the research to the theory of differential equations. The research as well as the origin of the chaos [25] gave us strong arguments for the development of motions in classical dynamical systems theory [9] by proceeding behind Poisson stable points to unpredictable points [5]. Then, the dynamics has been specified such that a function that is bounded on the real axis is an unpredictable point [6, 7]. In the papers [6, 7], we have demonstrated that unpredictable functions are easy to be analyzed as solutions of differential equations. This paradigm is not completed, if one does not consider discrete equations. Therefore, in this chapter we also deliver discrete analogues for unpredictable functions, calling them unpredictable sequences, and prove assertions on the existence and uniqueness of unpredictable solutions of discrete equations for the first time in the literature. The results can be useful for applications and theoretical analyses, in particular, for the modern development of computer technologies, software, and robotics [11, 31].

5.2 Quasilinear Delay Differential Equations

Let us introduce the following definition.

Definition 5.1 ([8]) A uniformly continuous and bounded function $\vartheta : \mathbb{R} \to \mathbb{R}^m$ is unpredictable if there exist positive numbers ϵ_0, δ and sequences $\{t_n\}$, $\{u_n\}$ both of which diverge to infinity such that $\|\vartheta(t + t_n) - \vartheta(t)\| \to 0$ as $n \to \infty$ uniformly on compact subsets of $\mathbb{R}$ and $\|\vartheta(t + t_n) - \vartheta(t)\| \geq \epsilon_0$ for each $t \in [u_n - \delta, u_n + \delta]$ and $n \in \mathbb{N}$.

To create Poincaré chaos [5], uniform continuity is not a necessary condition for an unpredictable function $\vartheta(t)$, and instead of the condition $\|\vartheta(t + t_n) - \vartheta(t)\| \geq \epsilon_0$ for each $t \in [u_n - \delta, u_n + \delta]$ and $n \in \mathbb{N}$, one can request that $\|\vartheta(t_n + u_n) - \vartheta(u_n)\| \geq \epsilon_0$ for each $n \in \mathbb{N}$. For the needs of verification of theorems on the existence of unpredictable solutions of differential equations we apply Definition 5.1, but for the future studies the following definitions may also be beneficial.

Definition 5.2 ([8]) A continuous and bounded function $\vartheta : \mathbb{R} \to \mathbb{R}^m$ is unpredictable if there exists a positive number ϵ_0 and sequences $\{t_n\}$, $\{u_n\}$ both of which diverge to infinity such that $\|\vartheta(t + t_n) - \vartheta(t)\| \to 0$ as $n \to \infty$ uniformly on compact subsets of $\mathbb{R}$ and $\|\vartheta(t_n + u_n) - \vartheta(u_n)\| \geq \epsilon_0$ for each $n \in \mathbb{N}$.

Definition 5.3 ([8]) A continuous and bounded function $\vartheta : \mathbb{R} \to \mathbb{R}^m$ is unpredictable if there exist a positive number ϵ_0 and sequences $\{t_n\}$, $\{u_n\}$ both of which diverge to infinity such that $\|\vartheta(t_n) - \vartheta(0)\| \to 0$ as $n \to \infty$ and $\|\vartheta(t_n + u_n) - \vartheta(u_n)\| \geq \epsilon_0$ for each $n \in \mathbb{N}$.

The main object of the present section is the following system of delay differential equations:

$$x'(t) = Ax(t) + f(x(t-\tau)) + g(t), \tag{5.2.1}$$

where τ is a positive number, the eigenvalues of the matrix $A \in \mathbb{R}^{m\times m}$ have negative real parts, $f : \mathbb{R}^m \to \mathbb{R}^m$ is a continuous function, and $g : \mathbb{R} \to \mathbb{R}^m$ is a uniformly continuous and bounded function. Our purpose is to prove that system (5.2.1) possesses a unique unpredictable solution which is uniformly exponentially stable, provided that the function $g(t)$ is unpredictable in accordance with Definition 5.1.

In the remaining parts of the paper, we will make use of the usual Euclidean norm for vectors and the norm induced by the Euclidean norm for square matrices.

Since the eigenvalues of the matrix A in system (5.2.1) have negative real parts, there exist numbers $K \geq 1$ and $\omega > 0$ such that $\left\|e^{At}\right\| \leq Ke^{-\omega t}$ for $t \geq 0$.

The following conditions are required.

(C1) There exists a positive number M_f such that $\sup\limits_{x\in\mathbb{R}^m} \|f(x)\| \leq M_f$;

(C2) There exists a positive number L_f such that $\|f(x_1) - f(x_2)\| \leq L_f \|x_1 - x_2\|$ for all $x_1, x_2 \in \mathbb{R}^m$;

(C3) $\omega - 2KL_f e^{\omega\tau/2} > 0$.

The following theorem is concerned with the unpredictable solution of system (5.2.1).

Theorem 5.1 ([8]) *Suppose that conditions $(C1)$–$(C3)$ are valid. If the function $g(t)$ is unpredictable, then system (5.2.1) possesses a unique uniformly exponentially stable unpredictable solution.*

Proof Under the conditions $(C1)$–$(C3)$, one can verify using the techniques for delay differential equations [12] that there exists a unique solution $\phi(t)$ of (5.2.1) which is bounded on the whole real axis and satisfies the relation

$$\phi(t) = \int_{-\infty}^{t} e^{A(t-s)}[f(\phi(s-\tau)) + g(s)]ds.$$

It is clear that $\sup\limits_{t\in\mathbb{R}} \|\phi(t)\| \leq M_\phi$, where $M_\phi = \dfrac{K(M_f + M_g)}{\omega}$ and $M_g = \sup\limits_{t\in\mathbb{R}} \|g(t)\|$. According to the results of [12], the solution $\phi(t)$ is uniformly exponentially stable. We will show that the solution $\phi(t)$ is unpredictable.

Since $g(t)$ is an unpredictable function, there exist a positive number ϵ_0 and sequences $\{t_n\}$, $\{u_n\}$ both of which diverge to infinity such that

$\|g(t+t_n)-g(t)\| \to 0$ as $n \to \infty$ on compact subsets of $\mathbb{R}$, and $\|g(t_n+u_n)-g(u_n)\| \geq \epsilon_0$ for each $n \in \mathbb{N}$.

Fix an arbitrary $\epsilon > 0$, and denote $R_1 = \dfrac{2\omega M_\phi K}{\omega - 2KL_f e^{\omega\tau/2}}$, $R_2 = \dfrac{K}{\omega - KL_f}$. Condition $(C3)$ implies that both R_1 and R_2 are positive numbers. Take a positive number γ satisfying $\gamma < \dfrac{1}{R_1+R_2}$, and suppose that E is a positive number such that $E \geq \dfrac{2}{\omega}\ln\left(\dfrac{1}{\gamma\epsilon}\right)$.

Let α and β be fixed real numbers with $\beta > \alpha$. There exists a natural number n_0 such that for each $t \in [\alpha - E, \beta]$ and each $n \geq n_0$ the inequality $\|g(t+t_n)-g(t)\| < \gamma\epsilon$ holds.

Fix an arbitrary natural number $n \geq n_0$, and define the function $\xi(t) = \phi(t) - \phi(t+t_n)$. This function satisfies the delay equation

$$\begin{aligned}\xi'(t) = {} & A\xi(t) + f(\xi(t-\tau)+\phi(t+t_n-\tau)) \\ & -f(\phi(t+t_n-\tau)) + g(t) - g(t+t_n).\end{aligned} \tag{5.2.2}$$

By applying arguments of equivalent integral equations for the last system, one can find that for $t \geq \alpha - E$, $\xi(t)$ satisfies the relation

$$\begin{aligned}\xi(t) = {} & e^{A(t-\alpha+E)}[\phi(\alpha-E)-\phi(t_n+\alpha-E)] \\ & + \int_{\alpha-E}^{t} e^{A(t-s)}[f(\xi(s-\tau)+\phi(s+t_n-\tau)) - f(\phi(s+t_n-\tau))]ds \\ & + \int_{\alpha-E}^{t} e^{A(t-s)}[g(s)-g(s+t_n)]ds.\end{aligned}$$

Let us denote by $\mathscr{C}$ the set of continuous functions $\xi(t)$ defined on $\mathbb{R}$ such that

$$\|\xi(t)\| \leq R_1 e^{-\omega(t-\alpha+E)/2} + R_2\gamma\epsilon$$

for $\alpha - E - \tau \leq t \leq \beta$ and $\|\xi\|_\infty \leq 2K\left(M_\phi + \dfrac{M_f+M_g}{\omega}\right)$, where $\|\xi\|_\infty = \sup\limits_{t\in\mathbb{R}}\|\xi(t)\|$.

Define on $\mathscr{C}$ the operator Π by

$$\Pi\xi(t) = \begin{cases} \phi(t) - \phi(t+t_n), \ t < \alpha - E, \\ e^{A(t-\alpha+E)}\,[\phi(\alpha-E)-\phi(t_n+\alpha-E)] + \displaystyle\int_{\alpha-E}^{t} e^{A(t-s)}[g(s)-g(s+t_n)]ds \\ + \displaystyle\int_{\alpha-E}^{t} e^{A(t-s)}[f(\xi(s-\tau)+\phi(s+t_n-\tau)) - f(\phi(s+t_n-\tau))]ds, \ t \geq \alpha - E. \end{cases}$$

First of all, we will show that $\Pi(\mathscr{C}) \subseteq \mathscr{C}$. If $\xi(t)$ belongs to $\mathscr{C}$, then we have for $t \in [\alpha - E, \beta]$ that

$$\begin{aligned}
\|\Pi\xi(t)\| &\leq Ke^{-\omega(t-\alpha+E)}\,\|\phi(\alpha - E) - \phi(t_n + \alpha - E)\| + \int_{\alpha-E}^{t} K\gamma\epsilon e^{-\omega(t-s)}ds \\
&\quad + \int_{\alpha-E}^{t} KL_f e^{-\omega(t-s)}\,\|\xi(s-\tau)\|\,ds \\
&< \left(2M_\phi K + \frac{2KL_f R_1 e^{\omega\tau/2}}{\omega}\right) e^{-\omega(t-\alpha+E)/2} + \frac{K\gamma\epsilon(1 + L_f R_2)}{\omega} \\
&= R_1 e^{-\omega(t-\alpha+E)/2} + R_2\gamma\epsilon.
\end{aligned}$$

The inequality $\|\Pi\xi(t)\| < R_1 e^{-\omega(t-\alpha+E)/2} + R_2\gamma\epsilon$ is valid also for $t \in [\alpha - E - \tau, \alpha - E)$ since $R_1 > 2M_\phi$. On the other hand, if $\xi(t)$ belongs to $\mathscr{C}$, then one can confirm that $\|\Pi\xi\|_\infty \leq 2K\left(M_\phi + \dfrac{M_f + M_g}{\omega}\right)$. Hence, $\Pi(\mathscr{C}) \subseteq \mathscr{C}$.

Now, let us take two functions $\xi(t), \overline{\xi}(t) \in \mathscr{C}$. Clearly, $\Pi\xi(t) - \Pi\overline{\xi}(t) = 0$ for each $t < \alpha - E$. It can be verified for $t \geq \alpha - E$ that

$$\begin{aligned}
\left\|\Pi\xi(t) - \Pi\overline{\xi}(t)\right\| &\leq \int_{\alpha-E}^{t} KL_f e^{-\omega(t-s)}\left\|\xi(s-\tau) - \overline{\xi}(s-\tau)\right\| ds \\
&\leq \frac{KL_f}{\omega}\left(1 - e^{-\omega(t-\alpha+E)}\right)\sup_{t\geq\alpha-E-\tau}\left\|\xi(t) - \overline{\xi}(t)\right\|.
\end{aligned}$$

The last inequality yields $\left\|\Pi\xi - \Pi\overline{\xi}\right\|_\infty \leq \dfrac{KL_f}{\omega}\left\|\xi - \overline{\xi}\right\|_\infty$. Thus, the operator Π is contractive by means of condition $(C3)$.

According to the uniqueness of solutions for (5.2.2), $\xi(t) = \phi(t) - \phi(t + t_n)$ is the unique fixed point of the operator Π. Consequently, the sequence $\xi_k(t)$, $\xi_{k+1} = \Pi(\xi_k)$, $k = 0, 1, 2, \ldots$, where

$$\xi_0(t) = \begin{cases} \phi(t) - \phi(t + t_n), \ t < \alpha - E, \\ \phi(\alpha - E) - \phi(\alpha - E + t_n), \ t \geq \alpha - E, \end{cases}$$

which belongs to $\mathscr{C}$, is converging to $\phi(t) - \phi(t + t_n)$ on $\mathbb{R}$. Therefore,

$$\|\phi(t + t_n) - \phi(t)\| \leq R_1 e^{\omega(t-\alpha+E)/2} + R_2\gamma\epsilon$$

for $t \in [\alpha - E, \beta]$.

Since the number E is sufficiently large such that $E \geq \dfrac{2}{\omega}\ln\left(\dfrac{1}{\gamma\epsilon}\right)$, we have for each $t \in [\alpha, \beta]$ that

$$\|\phi(t+t_n)-\phi(t)\| \leq (R_1+R_2)\gamma\epsilon < \epsilon.$$

Hence, $\|\phi(t+t_n)-\phi(t)\| \to 0$ as $n \to \infty$ uniformly on the compact interval $[\alpha, \beta]$.

In the remaining part of the proof, we will show the existence of a sequence $\{\overline{u}_n\}$, $\overline{u}_n \to \infty$ as $n \to \infty$, and positive numbers $\overline{\epsilon}_0$, δ such that $\|\phi(t+t_n)-\phi(t)\| \geq \overline{\epsilon}_0$ for $t \in \left[\overline{u}_n - \delta, \overline{u}_n + \delta\right]$.

Since the function $g(t)$ is uniformly continuous, there exists a positive number $\widetilde{\delta}$, which is independent of t_n and u_n, $n \in \mathbb{N}$, such that both of the inequalities

$$\|g(t+t_n)-g(t_n+u_n)\| \leq \frac{\epsilon_0}{4\sqrt{m}}$$

and

$$\|g(t)-g(u_n)\| \leq \frac{\epsilon_0}{4\sqrt{m}}$$

hold for every $t \in [u_n - \widetilde{\delta}, u_n + \widetilde{\delta}]$ and $n \in \mathbb{N}$.

Fix an arbitrary natural number n, and suppose that $g(t) = (g_1(t), g_2(t), \ldots, g_m(t))$, where each $g_k(t)$, $1 \leq k \leq m$, is a real valued function. One can confirm that there exists an integer j_n, $1 \leq j_n \leq m$, such that

$$\left|g_{j_n}(t_n+u_n)-g_{j_n}(u_n)\right| \geq \frac{\epsilon_0}{\sqrt{m}}.$$

Therefore, using the last inequality, we obtain for $t \in \left[u_n - \widetilde{\delta}, u_n + \widetilde{\delta}\right]$ that

$$\begin{aligned}\left|g_{j_n}(t+t_n)-g_{j_n}(t)\right| &\geq \left|g_{j_n}(t_n+u_n)-g_{j_n}(u_n)\right| \\ &\quad - \left|g_{j_n}(t+t_n)-g_{j_n}(t_n+u_n)\right| \\ &\quad - \left|g_{j_n}(t)-g_{j_n}(u_n)\right| \\ &\geq \frac{\epsilon_0}{2\sqrt{m}}. \end{aligned} \tag{5.2.3}$$

There exist numbers $s_1^n, s_2^n, \ldots, s_m^n \in \left[u_n - \widetilde{\delta}, u_n + \widetilde{\delta}\right]$ such that

$$\left\| \int_{u_n-\widetilde{\delta}}^{u_n+\widetilde{\delta}} (g(s+t_n)-g(s))\, ds \right\| = 2\widetilde{\delta} \left[\sum_{i=1}^{m} \left(g_i(s_i^n+t_n)-g_i(s_i^n)\right)^2 \right]^{1/2}.$$

Accordingly, the inequality (5.2.3) implies that

$$\left\| \int_{u_n-\widetilde{\delta}}^{u_n+\widetilde{\delta}} (g(s+t_n) - g(s))\, ds \right\| \geq 2\widetilde{\delta} \left| g_{j_n}(s_{j_n}^n + t_n) - g_{j_n}(s_{j_n}^n) \right| \geq \frac{\widetilde{\delta}\epsilon_0}{\sqrt{m}}.$$

Now, using the relation

$$\begin{aligned}
\phi\left(t_n + u_n + \widetilde{\delta}\right) - \phi\left(u_n + \widetilde{\delta}\right) = \phi\left(t_n + u_n - \widetilde{\delta}\right) - \phi\left(u_n - \widetilde{\delta}\right) \\
+ \int_{u_n-\widetilde{\delta}}^{u_n+\widetilde{\delta}} A[\phi(s+t_n) - \phi(s)]ds \\
+ \int_{u_n-\widetilde{\delta}}^{u_n+\widetilde{\delta}} [f(\phi(s+t_n-\tau)) - f(\phi(s-\tau))]ds \\
+ \int_{u_n-\widetilde{\delta}}^{u_n+\widetilde{\delta}} [g(s+t_n) - g(s)]ds,
\end{aligned}$$

one can verify that

$$\begin{aligned}
\left\|\phi\left(t_n + u_n + \widetilde{\delta}\right) - \phi\left(u_n + \widetilde{\delta}\right)\right\| \geq \frac{\widetilde{\delta}\epsilon_0}{\sqrt{m}} - \left(1 + 2\widetilde{\delta}\,\|A\|\right) \sup_{t\in[u_n-\widetilde{\delta},u_n+\widetilde{\delta}]} \|\phi(t+t_n) - \phi(t)\| \\
-2\widetilde{\delta}L_f \sup_{t\in[u_n-\widetilde{\delta}-\tau,u_n+\widetilde{\delta}-\tau]} \|\phi(t+t_n) - \phi(t)\|.
\end{aligned}$$

Hence, we have

$$\sup_{t\in[u_n-\widetilde{\delta}-\tau,u_n+\widetilde{\delta}]} \|\phi(t+t_n) - \phi(t)\| \geq \frac{\widetilde{\delta}\epsilon_0}{2\sqrt{m}\left(1 + \widetilde{\delta}\,\|A\| + \widetilde{\delta}L_f\right)}.$$

Let $\overline{u}_n$ be a point that belongs to the interval $\left[u_n - \widetilde{\delta} - \tau,\, u_n + \widetilde{\delta}\right]$ satisfying

$$\sup_{t\in[u_n-\widetilde{\delta}-\tau,u_n+\widetilde{\delta}]} \|\phi(t+t_n) - \phi(t)\| = \|\phi(t_n + \overline{u}_n) - \phi(\overline{u}_n)\|.$$

Define the numbers

$$\overline{\epsilon}_0 = \frac{\widetilde{\delta}\epsilon_0}{4\sqrt{m}\left(1 + \widetilde{\delta}\,\|A\| + \widetilde{\delta}L_f\right)}$$

and

$$\delta = \frac{\widetilde{\delta}\epsilon_0}{8\sqrt{m}\left(1 + \widetilde{\delta}\,\|A\| + \widetilde{\delta}L_f\right)\left(M_\phi\,\|A\| + M_\phi L_f + M_g\right)}.$$

If t belongs to the interval $[\overline{u}_n - \delta, \overline{u}_n + \delta]$, then it can be obtained that

$$\begin{aligned}\|\phi(t+t_n) - \phi(t)\| &\geq \|\phi(t_n+\eta_n) - \phi(\eta_n)\| - \left| \int_{\eta_n}^{t} \|A\| \, \|\phi(s+t_n) - \phi(s)\| \, ds \right| \\ &\quad - \left| \int_{\eta_n}^{t} L_f \, \|\phi(s+t_n-\tau) - \phi(s-\tau)\| \, ds \right| \\ &\quad - \left| \int_{\eta_n}^{t} \|g(s+t_n) - g(s)\| \, ds \right| \\ &\geq \frac{\widetilde{\delta}\epsilon_0}{2\sqrt{m}\left(1+\widetilde{\delta}\,\|A\| + \widetilde{\delta} L_f\right)} - 2\delta\left(M_\phi \, \|A\| + M_\phi L_f + M_g\right) \\ &= \overline{\epsilon}_0.\end{aligned}$$

Hence, $\|\phi(t+t_n) - \phi(t)\| \geq \overline{\epsilon}_0$ for each t from the intervals $[\overline{u}_n - \delta, \overline{u}_n + \delta]$, $n \in \mathbb{N}$. Clearly, $\overline{u}_n \to \infty$ as $n \to \infty$. Consequently, the bounded solution $\phi(t)$ is unpredictable. □

Remark 5.1 The result of Theorem 5.1 is valid also for the case $\tau = 0$. More precisely, if $(C1)$, $(C2)$ are valid and $\omega - KL_f > 0$, then the system

$$x'(t) = Ax(t) + f(x(t)) + g(t)$$

possesses a unique uniformly exponentially stable unpredictable solution provided that $g(t)$ is an unpredictable function.

5.3 Quasilinear Discrete Equations

The definition of an unpredictable sequence is as follows.

Definition 5.4 ([8]) A bounded sequence $\{\kappa_i\}$, $i \in \mathbb{Z}$, in $\mathbb{R}^p$ is called unpredictable if there exist a positive number ϵ_0 and sequences $\{\zeta_n\}$, $\{\eta_n\}$, $n \in \mathbb{N}$, of positive integers both of which diverge to infinity such that $\left\|\kappa_{i+\zeta_n} - \kappa_i\right\| \to 0$ as $n \to \infty$ for each i in bounded intervals of integers and $\left\|\kappa_{\zeta_n+\eta_n} - \kappa_{\eta_n}\right\| \geq \epsilon_0$ for each $n \in \mathbb{N}$.

Definition 5.4 is of main use in the present section. It is requested by the method of the proof. Nevertheless, in future analyses, there may be needs for the following other definition, which can be considered as an analogue of Definition 5.3.

Definition 5.5 ([8]) A bounded sequence $\{\kappa_i\}$, $i \in \mathbb{Z}$, in $\mathbb{R}^p$ is called unpredictable if there exist a positive number ϵ_0 and sequences $\{\zeta_n\}$, $\{\eta_n\}$, $n \in \mathbb{N}$, of positive integers both of which diverge to infinity such that $\left\|\kappa_{\zeta_n} - \kappa_0\right\| \to 0$ as $n \to \infty$ and $\left\|\kappa_{\zeta_n+\eta_n} - \kappa_{\eta_n}\right\| \geq \epsilon_0$ for each $n \in \mathbb{N}$.

It is worth noting that the topologies in Definitions 5.1 and 5.4 are metrizable [28]. Consequently, the existence of an unpredictable sequence in the sense of Definition 5.4 indicates the presence of Poincaré chaos. Throughout the section, an unpredictable sequence and an unpredictable solution are understood as mentioned in Definition 5.4.

In this section, we will consider the following discrete equation:

$$z_{i+1} = Bz_i + h(z_i) + \psi_i, \tag{5.3.4}$$

where $i \in \mathbb{Z}$, $B \in \mathbb{R}^{p\times p}$ is a nonsingular matrix, $h : \mathbb{R}^p \to \mathbb{R}^p$ is a continuous function, and $\{\psi_i\}$, $i \in \mathbb{Z}$, is an unpredictable sequence.

The following assumptions on Eq. (5.3.4) are required.

(C4) There exists a positive number M_h such that $\sup_{x\in\mathbb{R}^p} \|h(x)\| \leq M_h$;

(C5) There exists a positive number L_h such that $\|h(x) - h(y)\| \leq L_h \|x - y\|$ for all $x, y \in \mathbb{R}^p$;

(C6) $\|B\| + L_h < 1$.

According to the results of [21], if conditions $(C4)$–$(C6)$ hold, then Eq. (5.3.4) possesses a unique bounded solution $\{\varphi_i\}$, $i \in \mathbb{Z}$, which satisfies the relation

$$\varphi_i = \sum_{j=-\infty}^{i} B^{i-j}\left(h(\varphi_{j-1}) + \psi_{j-1}\right). \tag{5.3.5}$$

One can show under the same conditions that the bounded solution attracts all other solutions of (5.3.4). More precisely, the inequality

$$\|z_i - \varphi_i\| \leq (\|B\| + L_h)^{(i-i_0)} \left\|z^0 - \varphi_{i_0}\right\|$$

is valid for all $i \geq i_0$, where $\{z_i\}$, $i \in \mathbb{Z}$, is a solution of (5.3.4) with $z_{i_0} = z^0$ for some integer i_0 and $z^0 \in \mathbb{R}^p$.

The following theorem is concerned with the existence of an unpredictable solution of the discrete equation (5.3.4).

Theorem 5.2 ([8]) *The bounded solution* $\{\varphi_i\}$, $i \in \mathbb{Z}$, *of Eq. (5.3.4) is unpredictable under the conditions* $(C4)$–$(C6)$.

Proof Fix an arbitrary positive number ϵ, and suppose that γ is a positive number satisfying

$$\gamma \leq \left[\frac{1}{1 - \|B\| - L_h} + \frac{2(M_h + M_\psi)}{1 - \|B\|}\right]^{-1}.$$

Let i_1 and i_2 be integers such that $i_2 > i_1$, and take a natural number E with

$$E \geq \frac{\ln(\gamma\epsilon)}{\ln(\|B\| + L_h)} - 1. \tag{5.3.6}$$

Since $\{\psi_i\}$, $i \in \mathbb{Z}$, is an unpredictable sequence, there exist a positive number ϵ_0 and sequences $\{\zeta_n\}$, $\{\eta_n\}$, $n \in \mathbb{N}$, of positive integers, both of which diverge to infinity, such that $\left\|\psi_{i+\zeta_n} - \psi_i\right\| \to 0$ as $n \to \infty$ for each i with $i_1 - E - 1 \leq i \leq i_2 - 1$ and $\left\|\psi_{\zeta_n+\eta_n} - \psi_{\eta_n}\right\| \geq \epsilon_0$ for each $n \in \mathbb{N}$.

First of all, we will show that $\left\|\varphi_{i+\zeta_n} - \varphi_i\right\| \to 0$ as $n \to \infty$ for each i with $i_1 \leq i \leq i_2$. There exists a natural number n_0, independent of i, such that for each $n \geq n_0$ the inequality $\left\|\psi_{i+\zeta_n} - \psi_i\right\| < \gamma\epsilon$ is valid whenever $i_1 - E - 1 \leq i \leq i_2 - 1$.

Fix an arbitrary integer $n \geq n_0$. One can obtain using the relation (5.3.5) that

$$\varphi_{i+\zeta_n} - \varphi_i = \sum_{j=-\infty}^{i} B^{i-j}\left(h(\varphi_{j+\zeta_n-1}) - h(\varphi_{j-1}) + \psi_{j+\zeta_n-1} - \psi_{j-1}\right).$$

Therefore, for $i_1 - E \leq i \leq i_2$, we have

$$\begin{aligned}\left\|\varphi_{i+\zeta_n} - \varphi_i\right\| < & \frac{2(M_h + M_\psi)}{1 - \|B\|}\|B\|^{i-i_1+E+1} \\ & + \frac{\gamma\epsilon}{1 - \|B\|}\left(1 - \|B\|^{i-i_1+E+1}\right) \\ & + L_h \sum_{j=i_1-E}^{i} \|B\|^{i-j}\left\|\varphi_{j+\zeta_n-1} - \varphi_{j-1}\right\|.\end{aligned} \tag{5.3.7}$$

Let us denote

$$r_i = \|B\|^{-i}\left\|\varphi_{i+\zeta_n} - \varphi_i\right\|$$

and

$$q_i = \frac{2(M_h + M_\psi)}{1 - \|B\|}\|B\|^{-i_1+E+1} + \frac{\gamma\epsilon}{1 - \|B\|}\left(\|B\|^{-i} - \|B\|^{-i_1+E+1}\right).$$

The inequality (5.3.7) yields

$$r_i < q_i + \frac{L_h}{\|B\|}\sum_{j=i_1-E}^{i} r_{j-1}.$$

It can be verified by applying the discrete analogue of Gronwall inequality that

$$r_i \leq q_i + \frac{L_h}{\|B\|}\sum_{j=i_1-E}^{i} q_{j-1}\left(1 + \frac{L_h}{\|B\|}\right)^{i-j}.$$

Thus, for $i_1 - E \le i \le i_2$, we have

$$r_i \le \frac{2(M_h + M_\psi)}{1 - \|B\|} \|B\|^{-i} (\|B\| + L_h)^{i-i_1+E+1} + \frac{\gamma\epsilon}{1 - \|B\| - L_h} \|B\|^{-i} \left[1 - (\|B\| + L_h)^{i-i_1+E+1}\right].$$

The last inequality implies that

$$\left\|\varphi_{i+\zeta_n} - \varphi_i\right\| < \frac{2(M_h + M_\psi)}{1 - \|B\|} (\|B\| + L_h)^{i-i_1+E+1} + \frac{\gamma\epsilon}{1 - \|B\| - L_h}.$$

One can confirm using (5.3.6) that $(\|B\| + L_h)^{i-i_1+E+1} \le \gamma\epsilon$ for $i_1 \le i \le i_2$. Therefore, for each $n \ge n_0$, the inequality

$$\left\|\varphi_{i+\zeta_n} - \varphi_i\right\| < \left[\frac{2(M_h + M_\psi)}{1 - \|B\|} + \frac{1}{1 - \|B\| - L_h}\right] \gamma\epsilon \le \epsilon$$

is valid for $i_1 \le i \le i_2$. Hence, $\left\|\varphi_{i+\zeta_n} - \varphi_i\right\| \to 0$ as $n \to \infty$ for each i with $i_1 \le i \le i_2$.

Next, we will show the existence of a positive number $\overline{\epsilon}_0$ and a sequence $\{\widetilde{\eta}_n\}$ with $\widetilde{\eta}_n \to \infty$ as $n \to \infty$ such that $\left\|\varphi_{\zeta_n+\widetilde{\eta}_n} - \varphi_{\widetilde{\eta}_n}\right\| \ge \overline{\epsilon}_0$ for each $n \in \mathbb{N}$.

Using the relations

$$\varphi_{\zeta_n+\eta_n+1} = B\varphi_{\zeta_n+\eta_n} + h(\varphi_{\zeta_n+\eta_n}) + \psi_{\zeta_n+\eta_n}$$

and

$$\varphi_{\eta_n+1} = B\varphi_{\eta_n} + h(\varphi_{\eta_n}) + \psi_{\eta_n}$$

we obtain for $n \in \mathbb{N}$ that

$$\left\|\varphi_{\zeta_n+\eta_n+1} - \varphi_{\eta_n+1}\right\| \ge \epsilon_0 - (\|B\| + L_h) \left\|\varphi_{\zeta_n+\eta_n} - \varphi_{\eta_n}\right\|.$$

Therefore,

$$\max\left\{\left\|\varphi_{\zeta_n+\eta_n+1} - \varphi_{\eta_n+1}\right\|, \left\|\varphi_{\zeta_n+\eta_n} - \varphi_{\eta_n}\right\|\right\} \ge \overline{\epsilon}_0, \tag{5.3.8}$$

where $\overline{\epsilon}_0 = \dfrac{\epsilon_0}{1 + \|B\| + L_h}$.

For each $n \in \mathbb{N}$, let us take $\widetilde{\eta}_n = \eta_n + 1$ if $\left\|\varphi_{\zeta_n+\eta_n+1} - \varphi_{\eta_n+1}\right\| \ge \left\|\varphi_{\zeta_n+\eta_n} - \varphi_{\eta_n}\right\|$, and we set $\widetilde{\eta}_n = \eta_n$ otherwise. Clearly, $\widetilde{\eta}_n \to \infty$ as $n \to \infty$. According to inequality (5.3.8), we have $\left\|\varphi_{\zeta_n+\widetilde{\eta}_n} - \varphi_{\widetilde{\eta}_n}\right\| \ge \overline{\epsilon}_0$ for each $n \in \mathbb{N}$. Consequently, the bounded solution $\{\varphi_i\}$, $i \in \mathbb{Z}$, of (5.3.4) is unpredictable. □

A possible way to obtain a different unpredictable sequence from a given one is mentioned in the following theorem.

Theorem 5.3 ([8]) *Suppose that* $\{\kappa_i\}$, $i \in \mathbb{Z}$, *is an unpredictable sequence such that* $\kappa_i \in \Lambda$ *for each* i, *where* Λ *is a bounded subset of* $\mathbb{R}^p$. *If* $\Phi : \Lambda \to \mathbb{R}^q$ *is a function such that there exist positive numbers* L_1 *and* L_2 *with* $L_1 \|s_1 - s_2\| \leq \|\Phi(s_1) - \Phi(s_2)\| \leq L_2 \|s_1 - s_2\|$ *for all* $s_1, s_2 \in \Lambda$, *then the sequence* $\{\overline{\kappa}_i\}$ *defined through the equation* $\overline{\kappa}_i = \Phi(\kappa_i)$, $i \in \mathbb{Z}$, *is also unpredictable.*

Proof Since $\{\kappa_i\}$, $i \in \mathbb{Z}$, is an unpredictable sequence, there exist a positive number ϵ_0 and sequences $\{\zeta_n\}$, $\{\eta_n\}$, $n \in \mathbb{N}$, of positive integers, both of which diverge to infinity, such that $\left\|\kappa_{i+\zeta_n} - \kappa_i\right\| \to 0$ as $n \to \infty$ for each i in bounded intervals of integers and $\left\|\kappa_{\zeta_n+\eta_n} - \kappa_{\eta_n}\right\| \geq \epsilon_0$ for each $n \in \mathbb{N}$.

Fix an arbitrary positive number ϵ, and let i_1 and i_2 be any two integers such that $i_2 > i_1$. One can find a natural number n_0, which does not depend on i, such that for each $n \geq n_0$ we have $\left\|\kappa_{i+\zeta_n} - \kappa_i\right\| < \epsilon/L_2$ whenever $i_1 \leq i \leq i_2$. Therefore, the inequality

$$\left\|\overline{\kappa}_{i+\zeta_n} - \overline{\kappa}_i\right\| \leq L_2 \left\|\kappa_{i+\zeta_n} - \kappa_i\right\| < \epsilon$$

is satisfied for each $n \geq n_0$ and each i with $i_1 \leq i \leq i_2$. This shows that $\left\|\overline{\kappa}_{i+\zeta_n} - \overline{\kappa}_i\right\| \to 0$ as $n \to \infty$ on bounded intervals of integers. On the other hand, for each $n \in \mathbb{N}$, we have that

$$\left\|\overline{\kappa}_{\zeta_n+\eta_n} - \overline{\kappa}_{\eta_n}\right\| \geq L_1 \left\|\kappa_{\zeta_n+\eta_n} - \kappa_{\eta_n}\right\| \geq L_1\epsilon_0.$$

Consequently, $\{\overline{\kappa}_i\}$, $i \in \mathbb{Z}$, is an unpredictable sequence. □

5.4 A Continuous Unpredictable Function via the Logistic Map

Consider the space $\Sigma_2 = \left\{s = (s_0 s_1 s_2 \ldots) \mid s_j = 0 \text{ or } 1\right\}$ of infinite sequences of 0's and 1's with the metric

$$d(s, \overline{s}) = \sum_{k=0}^{\infty} \frac{|s_k - \overline{s}_k|}{2^k},$$

where $s = (s_0 s_1 s_2 \ldots)$, $\overline{s} = (\overline{s}_0 \overline{s}_1 \overline{s}_2 \ldots) \in \Sigma_2$. The Bernoulli shift $\sigma : \Sigma_2 \to \Sigma_2$ is defined as $\sigma(s_0 s_1 s_2 \ldots) = (s_1 s_2 s_3 \ldots)$. The map σ is continuous and Σ_2 is a compact metric space [10, 32].

Through the proof of Lemma 3.1 [7], we constructed an element

$$s^{**} = (s_0^{**} s_1^{**} s_2^{**} \ldots)$$

of Σ_2 which is unpredictable by placing all blocks of 0's and 1's in a specific order without any repetitions and extending it to the left-hand side by appropriately choosing the terms s_i^{**} for negative values of i.

Let us take into account the logistic map

$$\lambda_{i+1} = F_\mu(\lambda_i), \tag{5.4.9}$$

where $i \in \mathbb{Z}$ and $F_\mu(s) = \mu s(1-s)$. The interval $[0, 1]$ is invariant under the iterations of (5.4.9) for $\mu \in (0, 4]$ [15].

It was proved by Shi and Yu [30] that for each $\mu \in [3 + (2/3)^{1/2}, 4]$, there exist a natural number $m_0 > 4$ and a Cantor set $\Lambda \subset [0, 1]$ such that the map $F_\mu^{m_0}$ on Λ is topologically conjugate to the Bernoulli shift σ on Σ_2. Using the results of paper [30] and Lemma 3.1 [7], a property of the logistic map (5.4.9) was given in Theorem 4.1 of paper [7]. According to Theorem 4.1 [7], for each $\mu \in [3 + (2/3)^{1/2}, 4]$, the logistic map (5.4.9) possesses an unpredictable solution.

Now, let us denote by $\{\psi_i\}$, $i \in \mathbb{Z}$, an unpredictable solution of the logistic map (5.4.9) with $\mu = 3.91$ inside the unit interval $[0, 1]$, and consider the function

$$\Theta(t) = \int_{-\infty}^{t} e^{-2(t-s)} \Omega(s) ds, \tag{5.4.10}$$

where the function $\Omega(t)$ is defined by $\Omega(t) = \psi_i$ for $t \in [i, i+1)$, $i \in \mathbb{Z}$.

It can be verified that the function $\Theta(t)$ is the unique globally exponentially stable solution of the differential equation

$$v'(t) = -2v(t) + \Omega(t). \tag{5.4.11}$$

Moreover, $\Theta(t)$ is bounded on the whole real axis such that $\sup_{t \in \mathbb{R}} |\Theta(t)| \leq \frac{1}{2}$, and it is uniformly continuous since its derivative is bounded.

Because the sequence $\{\psi_i\}$, $i \in \mathbb{Z}$, is unpredictable, there exist a positive number ϵ_0 and sequences $\{\zeta_n\}$, $\{\eta_n\}$ both of which diverge to infinity such that $\left|\psi_{i+\zeta_n} - \psi_i\right| \to 0$ as $n \to \infty$ for each i in bounded intervals of integers and

$$\left|\psi_{\zeta_n+\eta_n} - \psi_{\eta_n}\right| \geq \epsilon_0$$

for each n.

Let us fix an arbitrary positive number ϵ and take an arbitrary compact subset $[\alpha, \beta] \subset \mathbb{R}$. Suppose that N is a sufficiently large positive integer satisfying $N \geq \frac{1}{2} \ln \left(\frac{3}{2\epsilon} \right)$. There exists a natural number n_0 such that for each $n \geq n_0$ the inequality

$$\left|\psi_{i+\zeta_n} - \psi_i\right| < \frac{2\epsilon}{3}$$

is valid for $i = \lfloor\alpha\rfloor - N, \lfloor\alpha\rfloor - N + 1, \ldots, \lfloor\beta\rfloor$, where $\lfloor\alpha\rfloor$ and $\lfloor\beta\rfloor$ denote the largest integers which are not greater than α and β, respectively.

Fix a natural number $n \geq n_0$. Using the relation

$$\Theta(t+\zeta_n) - \Theta(t) = e^{-2(t-\lfloor\alpha\rfloor+N)}(\Theta(\lfloor\alpha\rfloor - N + \zeta_n) - \Theta(\lfloor\alpha\rfloor - N)) + \int_{\lfloor\alpha\rfloor-N}^{t} e^{-2(t-s)}[\Omega(s+\zeta_n) - \Omega(s)]ds,$$

one can verify for $t \in [\lfloor\alpha\rfloor, \lfloor\beta\rfloor + 1)$ that $|\Theta(t+\zeta_n) - \Theta(t)| < \epsilon$. Hence, $\Theta(t+\zeta_n) \to \Theta(t)$ as $n \to \infty$ uniformly on $[\alpha, \beta]$.

On the other hand, one can show that $\sup\limits_{t\in[\eta_n,\eta_n+1]} |\Theta(t+\zeta_n) - \Theta(t)| \geq \frac{\epsilon_0}{4}$ for each $n \in \mathbb{N}$. The last inequality implies that the function $\Theta(t)$ is unpredictable.

It is still difficult to simulate this unpredictable solution, but there is chaos because of unpredictability, and this is why in what follows we visualize the chaotic behavior.

Let us take into account the differential equation

$$v'(t) = -2v(t) + \widetilde{\Omega}(t), \tag{5.4.12}$$

where the function $\widetilde{\Omega}(t)$ is defined by $\widetilde{\Omega}(t) = \lambda_i$ for $t \in [i, i+1)$, $i \in \mathbb{Z}$, in which $\{\lambda_i\}$ is the solution of (5.4.9) with $\lambda_0 = 0.4$.

We depict in Fig. 5.1 the solution of (5.4.12) corresponding to the initial data $v(0) = 0.37$. The choice of the parameter $\mu = 3.91$ of the logistic map (5.4.9) and the value $\lambda_0 = 0.4$ were considered for shadowing in the paper [16]. It is seen in Fig. 5.1 that the dynamics of (5.4.12) is chaotic, and this supports that the function $\Theta(t)$ is unpredictable.

Illustrative examples that support the theoretical results are provided in the next section.

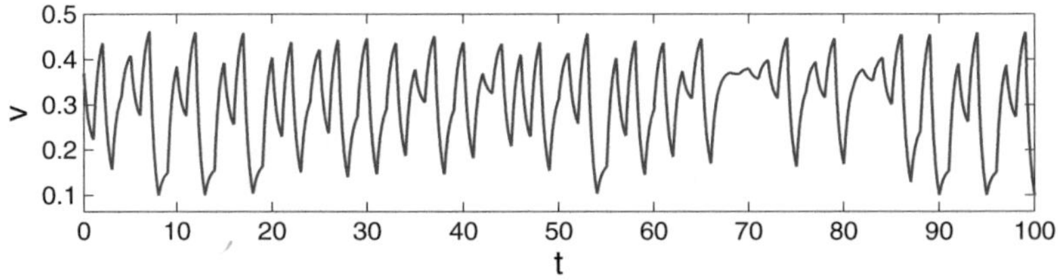

Fig. 5.1 Chaotic behavior of Eq. (5.4.12). The initial data $v(0) = 0.37$ is utilized

5.5 Examples

Example 1 In this example, we take into account the retarded non-autonomous differential equation

$$x''(t) + 4x'(t) + 1.5x(t) + 0.02x^2(t - 0.1) = \Theta(t), \tag{5.5.13}$$

where $\Theta(t)$ is the unpredictable function defined by (5.4.10).

Using the variables $x_1(t) = x(t)$ and $x_2(t) = x'(t)$, Eq. (5.5.13) can be written as

$$\begin{aligned} x_1'(t) &= x_2(t), \\ x_2'(t) &= -1.5x_1(t) - 4x_2(t) - 0.02x_1^2(t - 0.1) + \Theta(t). \end{aligned} \tag{5.5.14}$$

System (5.5.14) is in the form of (5.2.1) with $\tau = 0.1$, $f(x_1, x_2) = \left(0, -0.02x_1^2\right)$, and $A = \begin{pmatrix} 0 & 1 \\ -1.5 & -4 \end{pmatrix}$. The eigenvalues of the matrix A are $-2 + \sqrt{10}/2$ and $-2 - \sqrt{10}/2$. One can show that

$$e^{At} = P \begin{pmatrix} e^{(-2+\sqrt{10}/2)t} & 0 \\ 0 & e^{(-2-\sqrt{10}/2)t} \end{pmatrix} P^{-1},$$

where $P = \begin{pmatrix} 1 & (-4+\sqrt{10})/3 \\ (-4+\sqrt{10})/2 & 1 \end{pmatrix}$. Thus, the inequality $\left\|e^{At}\right\| \leq Ke^{-\omega t}$ is valid for $t \geq 0$ with $K = \|P\| \left\|P^{-1}\right\| \approx 2.0685$ and $\omega = 2 - \sqrt{10}/2$.

One can verify numerically that the solutions of (5.5.14) eventually enter the compact region

$$\mathscr{D} = \left\{(x_1, x_2) \in \mathbb{R}^2 : 0.14 \leq x_1 \leq 0.26, \ -0.06 \leq x_2 \leq 0.05\right\}$$

as t increases. Therefore, it is reasonable to consider the conditions $(C1)$ and $(C2)$ inside the region $\mathscr{D}$.

Conditions $(C1)$–$(C3)$ are valid for system (5.5.14) with $M_f = 0.001352$ and $L_f = 0.0104$. According to Theorem 5.1, system (5.5.14) possesses a unique uniformly exponentially stable unpredictable solution.

To demonstrate the chaos appearance due to the unpredictable function $\Theta(t)$, let us consider the system

$$\begin{aligned} x_1'(t) &= x_2(t), \\ x_2'(t) &= -1.5x_1(t) - 4x_2(t) - 0.02x_1^2(t - 0.1) + v(t), \end{aligned} \tag{5.5.15}$$

where $v(t)$ is the solution of (5.4.12) represented in Fig. 5.1.

The x_1- and x_2-coordinates of the solution of (5.5.15) corresponding to the initial conditions $x_1(t) = 0.18$, $x_2(t) = 0.01$, $t \in [-0.1, 0]$ are shown in Fig. 5.2. The figure supports the result of Theorem 5.1 such that (5.5.14) possesses an unpredictable solution, and it reveals that the dynamics of (5.5.15) is chaotic. Moreover, the trajectory of the same solution is depicted in Fig. 5.3, and this simulation also confirms the presence of chaos in system (5.5.15).

Example 2 We take into account the discrete system

$$\begin{aligned} x_{i+1} &= \frac{x_i}{2} - \frac{y_i}{7} + 3\psi_i^3 \\ y_{i+1} &= -\frac{x_i}{8} + \frac{y_i}{3} + \frac{x_i^{2/3}}{12} + 4\psi_i, \end{aligned} \tag{5.5.16}$$

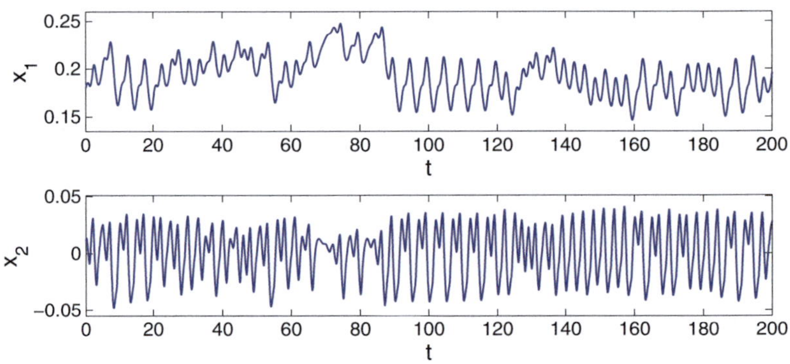

Fig. 5.2 Time series of the x_1- and x_2-coordinates of system (5.5.15). Chaotic behavior in both coordinates is observable in the figure

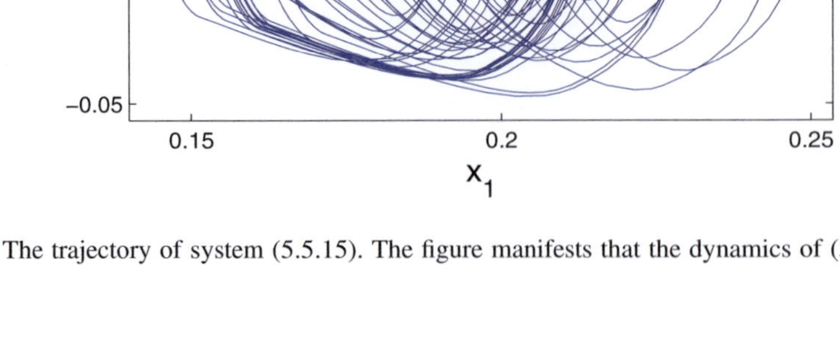

Fig. 5.3 The trajectory of system (5.5.15). The figure manifests that the dynamics of (5.5.15) is chaotic

where $\{\psi_i\}$ is an unpredictable solution of (5.4.9) with $\mu = 3.91$. Theorem 5.3 implies that the sequence $\{\overline{\psi}_i\}$, $i \in \mathbb{Z}$, defined by $\overline{\psi}_i = (3\psi_i^3,\ 4\psi_i) \in \mathbb{R}^2$ is also unpredictable.

In order to demonstrate the chaotic behavior of (5.5.16), we consider the system

$$\begin{aligned} x_{i+1} &= \frac{x_i}{2} - \frac{y_i}{7} + 3\lambda_i^3 \\ y_{i+1} &= -\frac{x_i}{8} + \frac{y_i}{3} + \frac{x_i^{2/3}}{12} + 4\lambda_i, \end{aligned} \tag{5.5.17}$$

where $\{\lambda_i\}$ is a solution of (5.4.9), again with $\mu = 3.91$. One can numerically verify that for each $\{\lambda_i\}$, the bounded solutions of (5.5.17) take place inside the compact region

$$\mathscr{D} = \{(x, y) \in \mathbb{R}^2 : \ 0.1 \le x \le 2.7,\ 1.7 \le y \le 5.1\}.$$

Therefore, the conditions $(C4)$–$(C6)$ are satisfied for system (5.5.16), and there exists a unique unpredictable solution of (5.5.16) in accordance with Theorem 5.2.

Figure 5.4 shows the first and second coordinates of the solution of system (5.5.17) with the initial data $\lambda_0 = 0.4$, $x_0 = 0.58$, and $y_0 = 1.95$. Moreover, we represent in Fig. 5.5 the two dimensional trajectory of the same solution. Both Figs. 5.4 and 5.5 support the result of Theorem 5.2 such that an unpredictable sequence takes place in the dynamics of the discrete system (5.5.16) and the behavior of the system is chaotic.

Example 3 In this example, we will demonstrate the appearance of irregular behavior near periodic orbits of discrete systems. For that purpose, let us consider the system

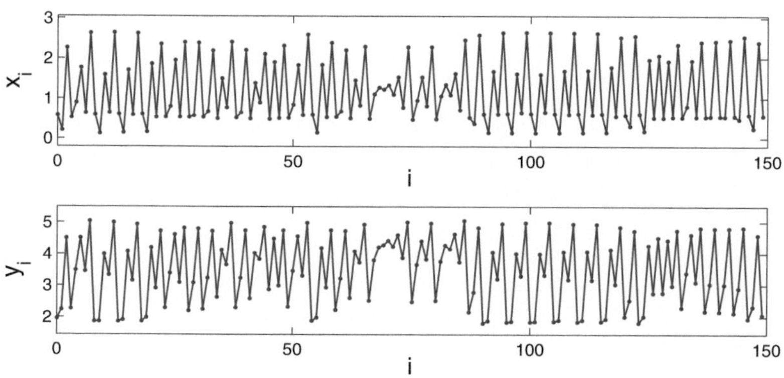

Fig. 5.4 The solution of (5.5.17) with the initial data $\lambda_0 = 0.4$, $x_0 = 0.58$, and $y_0 = 1.95$. The figure supports the result of Theorem 5.2 such that (5.5.16) possesses an unpredictable solution

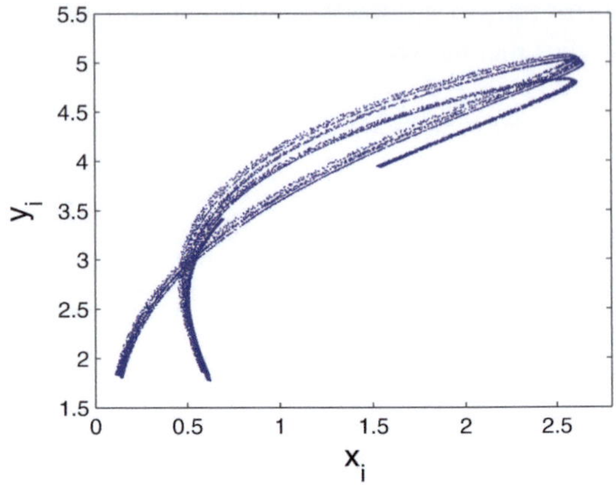

Fig. 5.5 The trajectory of the discrete system (5.5.17) corresponding to the initial data $\lambda_0 = 0.4$, $x_0 = 0.58$, and $y_0 = 1.95$. The figure indicates the chaotic behavior of system (5.5.16)

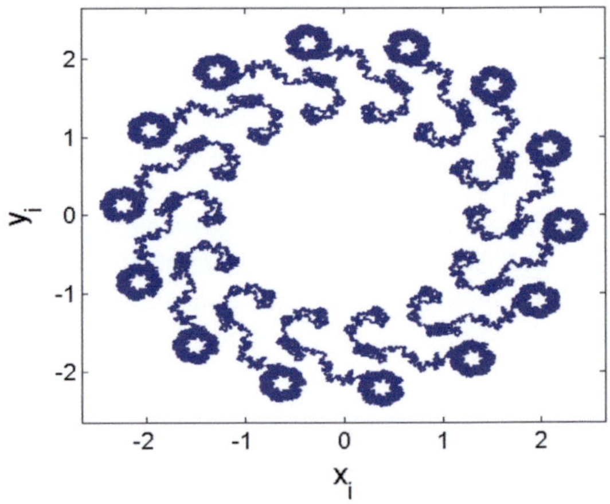

Fig. 5.6 The orbit of system (5.5.19) with $\lambda_0 = 0.4$, $x_0 = 1$, and $y_0 = 1$. The figure reveals that the orbit behaves chaotically near the 14-periodic orbit of (5.5.18)

$$\begin{aligned} x_{i+1} &= \cos(\omega_0)x_i + \sin(\omega_0)y_i \\ y_{i+1} &= -\sin(\omega_0)x_i + \cos(\omega_0)y_i, \end{aligned} \tag{5.5.18}$$

where ω_0 is a real parameter. It is shown in the book [15] that the system (5.5.18) admits a stable periodic orbit whenever the value $\omega_0/2\pi$ is rational. Taking $\mu = 3.86$ in the logistic map (5.4.9) and perturbing system (5.5.18) with solutions of (5.4.9), we set up the system

$$\begin{aligned} x_{i+1} &= \cos(\omega_0)x_i + \sin(\omega_0)y_i + 0.001\lambda_i \\ y_{i+1} &= -\sin(\omega_0)x_i + \cos(\omega_0)y_i + 0.001\lambda_i, \end{aligned} \tag{5.5.19}$$

where $\{\lambda_i\}$ is a solution of (5.4.9).

Let us use the value of $\omega_0 = \pi/7$ so that the non-perturbed system (5.5.18) possesses a one parameter family of stable 14-periodic orbits. We depict in Fig. 5.6 the trajectory of (5.5.19) corresponding to the initial data $\lambda_0 = 0.4$, $x_0 = 1$, and

$y_0 = 1$. The total number of iterations used in the simulation is 65×10^6. The utilized parameter value $\mu = 3.86$ of the logistic map (5.4.9) and the initial point $\lambda_0 = 0.4$ were analyzed for shadowing in paper [16]. It is seen in Fig. 5.6 that the applied perturbation makes system (5.5.19) behave chaotically near the 14-periodic orbit of (5.5.18). It is worth noting that Fig. 5.6 represents a single orbit. The fractal structure of the orbit is also observable in the simulation. Figure 5.6 manifests the appearance of chaos near the periodic orbit of (5.5.18).

5.6 A Hopfield Neural Network

This section is devoted to an application of our results to neural networks. Let us, first, consider the Hopfield neural network with delay and unpredictable input

$$\begin{aligned} x_1'(t) &= -1.7x_1(t) + 0.01\tanh(x_1(t-\tau_1)) - 0.02\tanh(x_2(t-\tau_1)) \\ &\quad + 0.04\tanh(x_3(t-\tau_1)) + 2.1\Theta(t) \\ x_2'(t) &= -3.5x_2(t) + 0.06\tanh(x_1(t-\tau_1)) + 0.03\tanh(x_2(t-\tau_1)) \\ &\quad + 0.08\tanh(x_3(t-\tau_1)) + 1.3\Theta(t) \\ x_3'(t) &= -2.8x_3(t) - 0.05\tanh(x_1(t-\tau_1)) - 0.06\tanh(x_2(t-\tau_1)) \\ &\quad + 0.03\tanh(x_3(t-\tau_1)) + 1.9\Theta(t), \end{aligned} \tag{5.6.20}$$

where $\tau_1 = 0.3$ and $\Theta(t)$ is the unpredictable function defined by (5.4.10).

The coefficients of the nonlinear terms in (5.6.20) are sufficiently small in absolute value such that the conditions of Theorem 5.1 hold, and accordingly, network (5.6.20) possesses a unique uniformly exponentially stable unpredictable solution $\widetilde{\Theta}(t) = \big(\widetilde{\Theta}_1(t), \widetilde{\Theta}_2(t), \widetilde{\Theta}_3(t)\big)$.

To demonstrate the chaotic dynamics of (5.6.20) numerically, in a similar way to the first example in this section, we utilize the network

$$\begin{aligned} x_1'(t) &= -1.7x_1(t) + 0.01\tanh(x_1(t-\tau_1)) - 0.02\tanh(x_2(t-\tau_1)) \\ &\quad + 0.04\tanh(x_3(t-\tau_1)) + 2.1v(t) \\ x_2'(t) &= -3.5x_2(t) + 0.06\tanh(x_1(t-\tau_1)) + 0.03\tanh(x_2(t-\tau_1)) \\ &\quad + 0.08\tanh(x_3(t-\tau_1)) + 1.3v(t) \\ x_3'(t) &= -2.8x_3(t) - 0.05\tanh(x_1(t-\tau_1)) - 0.06\tanh(x_2(t-\tau_1)) \\ &\quad + 0.03\tanh(x_3(t-\tau_1)) + 1.9v(t), \end{aligned} \tag{5.6.21}$$

where $v(t)$ is the solution of (5.4.12) shown in Fig. 5.1. We represent in Fig. 5.7 the x_2-coordinate of (5.6.21) using the initial data $x_1(t) = 0.34$, $x_2(t) = 0.12$, $x_3(t) = 0.19$, $t \in [-\tau_1, 0]$. Figure 5.7 reveals that the dynamics of the network (5.6.20) is Poincaré chaotic.

Next, we will discuss the extension of Poincaré chaos by Hopfield neural networks. For that purpose, we use the unpredictable output $\widetilde{\Theta}(t)$ of (5.6.20) as an external input for another Hopfield neural network and set up the system

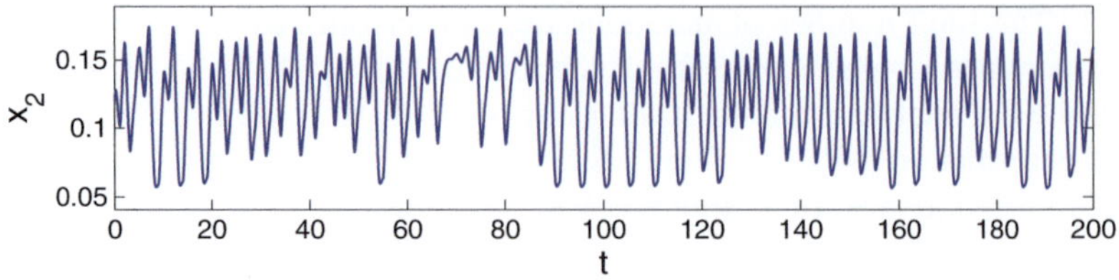

Fig. 5.7 The x_2-coordinate of the output of network (5.6.21) corresponding to the initial data $x_1(t) = 0.34$, $x_2(t) = 0.12$, $x_3(t) = 0.19$, $t \in [-\tau_1, 0]$. The simulation result demonstrates the presence of chaos

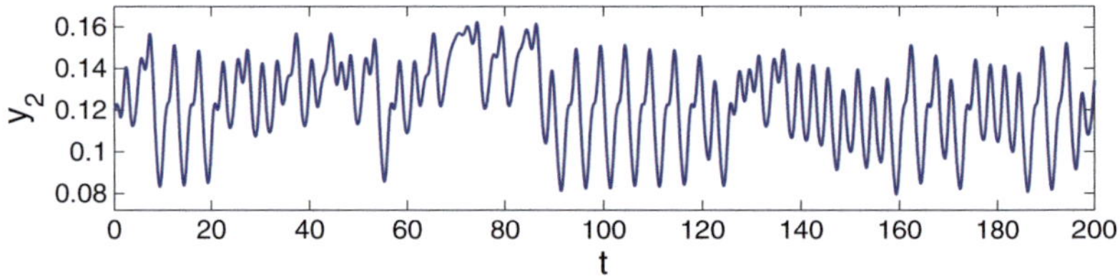

Fig. 5.8 The y_2-coordinate of the output of network (5.6.23) corresponding to the initial data $y_1(t) = 0.15$, $y_2(t) = 0.12$, $y_3(t) = 0.09$, $t \in [-\tau_2, 0]$. The figure confirms the extension of Poincaré chaos by the Hopfield neural network (5.6.22)

$$\begin{aligned} y_1'(t) &= -2.3y_1(t) + 0.05\tanh(y_1(t-\tau_2)) - 0.07\tanh(y_3(t-\tau_2)) + \widetilde{\Theta}_1(t),\\ y_2'(t) &= -y_2(t) + 0.02\tanh(y_1(t-\tau_2)) - 0.08\tanh(y_2(t-\tau_2))\\ &\quad + 0.09\tanh(y_3(t-\tau_2)) + \widetilde{\Theta}_2(t),\\ y_3'(t) &= -1.5y_3(t) + 0.04\tanh(y_2(t-\tau_2)) - 0.05\tanh(y_3(t-\tau_2)) + \widetilde{\Theta}_3(t), \end{aligned} \tag{5.6.22}$$

where $\tau_2 = 0.2$.

Using the result of Theorem 5.1 one more time, it can be confirmed that (5.6.22) admits a unique unpredictable output, which is exponentially stable. In order to simulate the extension of unpredictability, we consider the network

$$\begin{aligned} y_1'(t) &= -2.3y_1(t) + 0.05\tanh(y_1(t-\tau_2)) - 0.07\tanh(y_3(t-\tau_2)) + x_1(t),\\ y_2'(t) &= -y_2(t) + 0.02\tanh(y_1(t-\tau_2)) - 0.08\tanh(y_2(t-\tau_2))\\ &\quad + 0.09\tanh(y_3(t-\tau_2)) + x_2(t),\\ y_3'(t) &= -1.5y_3(t) + 0.04\tanh(y_2(t-\tau_2)) - 0.05\tanh(y_3(t-\tau_2)) + x_3(t), \end{aligned} \tag{5.6.23}$$

where $(x_1(t), x_2(t), x_3(t))$ is the output of (5.6.21) whose second coordinate is depicted in Fig. 5.7.

The time series of the y_2-coordinate of the Hopfield neural network (5.6.23) is shown in Fig. 5.8, where the chaotic behavior is observable. The initial data $y_1(t) = 0.15$, $y_2(t) = 0.12$, $y_3(t) = 0.09$, $t \in [-\tau_2, 0]$, are used in the simulation. Figure 5.8 confirms the extension of the unpredictable behavior of (5.6.20) by the network (5.6.22).

5.7 Notes

An unpredictable function was defined as an unpredictable point of the Bebutov dynamics, and the first theorems on the existence of unpredictable solutions were proved in [6, 7]. The metric of the Bebutov dynamics is not convenient for applications, since it is hard to verify. For this reason, in this chapter, we apply the topology of uniform convergence on compact sets to define unpredictable functions. The topology is metrizable and easy for applications with integral operators. Thus, one can accept that we lay a corner stone to the foundation of differential equations theory related to unpredictable solutions, and consequently, chaos. Therefore, in our opinion, a new field to analyze in the theory of differential equations has been discovered. Since many results of differential equations have their counterparts in discrete equations [21], one can suppose that theorems on the existence of unpredictable solutions can be proved for discrete equations. The present chapter is one to realize the both paradigms. The existence and uniqueness theorems for quasilinear delay and ordinary differential equations and difference equations have been proved, when the perturbation is an unpredictable function or sequence. This is visualized as Poincaré chaos in simulations.

We emphasize the meaning of our results for the development of theory of differential and discrete equations issuing from the general character of considered systems. In the next research, one can investigate the existence of unpredictable solutions, and consequently, chaos in discrete equations by applying well developed techniques such as averaging method, method of integral manifolds, method of asymptotic integrations, second Lyapunov method, and others [14].

We hope that the constructions of unpredictable functions and sequences suggested in our present research will be developed to more larger classes of functions, enlarging the applicability meaning of our results. This method of chaos appearance and of its consequent control cannot be underestimated in neuroscience [20, 26, 27]. Our approach suggests very effective applications and analysis method of generation and control of chaos in neural networks through a single function, an unpredictable one. This is why we can say that the motivation for our results is strong. In our research we provide unpredictable functions as external inputs which are not discontinuous but obtained from discontinuous functions by integration. This may enrich application variance for chaos analysis of neural network dynamics.

Our results can trigger further extension of the theory for discrete dynamical systems which can be defined as iterated maps. For this reason, we expect that the introduction of unpredictability in the discrete dynamics will be beneficial for new researches in hyperbolic dynamics, strange attractors, and ergodic theory [13, 14, 18, 19, 24, 32].

Unpredictable functions are compulsorily accompanied by Poincaré chaos, and this is considered in our previous papers [6, 7], too. It is significant that the unpredictable motion is still a Poisson stable one. On the basis of unpredictable functions one immediately considers a new type of chaos, Poincaré chaos, and consequently, the next question is how the chaos is related to previously known types of chaos. Developing this line of research, one will increase the practical role of the unpredictable functions and motions as much as of the previously known types of chaos. The results of this chapter are published in paper [8].

References

1. M.U. Akhmet, Devaney's chaos of a relay system. Commun. Nonlinear Sci. Numer. Simulat. **14**, 1486–1493 (2009)
2. M.U. Akhmet, Li–Yorke chaos in the system with impacts. J. Math. Anal. Appl. **351**, 804–810 (2009)
3. M.U. Akhmet, Creating a chaos in a system with relay. Int. J. Qualit. Th. Diff. Eqs. Appl. **3**, 3–7 (2009)
4. M.U. Akhmet, M.O. Fen, Replication of chaos. Commun. Nonlinear Sci. Numer. Simul. **18**, 2626–2666 (2013)
5. M. Akhmet, M.O. Fen, Unpredictable points and chaos. Commun. Nonlinear Sci. Numer. Simul. **40**, 1–5 (2016)
6. M. Akhmet, M.O. Fen, Existence of unpredictable solutions and chaos. Turk. J. Math. **41**, 254–266 (2017)
7. M. Akhmet, M.O. Fen, Poincaré chaos and unpredictable functions. Commun. Nonlinear Sci. Numer. Simul. **48**, 85–94 (2017)
8. M. Akhmet, M.O. Fen, Non-autonomous equations with unpredictable solutions. Commun. Nonlinear Sci. Numer. Simul. **59**, 657–670 (2018)
9. G.D. Birkhoff, *Dynamical Systems*, vol. 9 (Amer. Math. Soc., Colloquium Publications, Providence, 1927)
10. R.L. Devaney, *An Introduction to Chaotic Dynamical Systems* (Addison-Wesley, USA, 1989)
11. W.L. Ditto, K. Murali, S. Sinha, Chaos computing: ideas and implementations. Phil. Trans. R. Soc. A **366**, 653–664 (2008)
12. R.D. Driver, *Ordinary and Delay Differential Equations* (Springer, New York, 1977)
13. J. Guckenheimer, J. Moser, S.E. Newhouse, *Dynamical Systems* (Birkhäuser, Boston, 1980)
14. J. Guckenheimer, P.J. Holmes, *Nonlinear Oscillations, Dynamical Systems, and Bifurcations of Vector Fields* (Springer, New York, Heidelberg, Berlin, 1983)
15. J. Hale, H. Koçak, *Dynamics and Bifurcations* (Springer, New York, 1991)
16. S.M. Hammel, J.A. Yorke, C. Grebogi, Do numerical orbits of chaotic dynamical processes represent true orbits? J. Complexity **3**, 136–145 (1987)
17. J.J. Hopfield, Neurons with graded response have collective computational properties like those of two-state neurons. Proc. Natl. Acad. Sci. USA **81**, 3088–3092 (1984)
18. A. Katok, B. Hasselblatt, *Introduction to the Modern Theory of Dynamical Systems* (Cambridge University Press, Cambridge, 1997)
19. P.E. Kloeden, J. Ombach, Hyperbolic homeomorphisims and bishadowing. Ann. Polon. Math. **65**, 171–177 (1997)
20. H. Korn, P. Faure, Is there chaos in the brain? II. Experimental evidence and related models. C. R. Biologies **326**, 787–840 (2003)
21. V. Lakshmikantham, D. Trigiante, *Theory of Difference Equations: Numerical Methods and Applications* (Marcel Dekker, USA, 2002)
22. T.Y. Li, J.A. Yorke, Period three implies chaos. Am. Math. Monthly **82**, 985–992 (1975)
23. C.M. Marcus, R.M. Westervelt, Stability of analog neural networks with delay. Phys. Rev. A **39**, 347–359 (1989)
24. G. Palis, F. Takens, *Hyperbolicity and Sensitive Chaotic Dynamics at Homoclinic Bifurcations* (Cambridge University Press, Cambridge, 1995)

25. H. Poincaré, *Les méthodes nouvelles de la mécanique céleste*, Vol. 1, 2 (Gauthier-Villars, Paris, 1892)
26. M.I. Rabinovich, H.D.I. Abarbanel, The role of chaos in neural systems. Neuroscience **87**, 5–14 (1998)
27. S.N. Sarbadhikari, K. Chakrabarty, Chaos in the brain: a short review alluding to epilepsy, depression, exercise and lateralization. Medical Eng. Phys. **23**, 445–455 (2001)
28. G.R. Sell, *Topological Dynamics and Ordinary Differential Equations* (Van Nostrand Reinhold Company, London, 1971)
29. Y. Shi, G. Chen, Chaos of discrete dynamical systems in complete metric spaces. Chaos Solitons & Fractals **22**, 555–571 (2004)
30. Y. Shi, P. Yu, On chaos of the logistic maps. Dynam. Contin. Discrete Impuls. Syst. Ser. B **14**, 175–195 (2007)
31. S. Staingrube, M. Timme, F. Worgotter, P. Mannonpong, Self-organized adaptation of simple neural circuits enables complex robot behavior. Nature Physics **6**, 224–230 (2010)
32. S. Wiggins, *Global Bifurcation and Chaos: Analytical Methods* (Springer, New York, Berlin, 1988)

Chapter 6
Unpredictable Solutions of Hyperbolic Linear Equations

In this chapter, the existence and uniqueness of unpredictable solutions in the dynamics of non-homogeneous linear systems of differential and discrete equations are investigated. The hyperbolic cases are under discussion. The presence of unpredictable solutions confirms the existence of Poincaré chaos. Simulations illustrating the chaos are provided. The results of this chapter are published in paper [3].

6.1 Preliminaries

The concept of the unpredictable functions was introduced in the study [1]. The paper [2] was devoted to the investigation of sufficient conditions for the existence of unpredictable solutions of retarded quasilinear differential equations in the case that all eigenvalues of the matrix of coefficients admit negative real parts.

In this chapter, we investigate the existence and uniqueness of unpredictable solutions of linear differential and discrete equations in which unpredictable perturbations are used (Theorems 6.1 and 6.2). The results of the present chapter have two principal novelties compared to the previous results in the field. The first one is that we consider the hyperbolic cases such that the eigenvalues of the matrix of coefficients can admit positive real parts in the case of differential equations. The second one is that we propose a simpler and more comprehensible proof. Additionally, we consider new properties of unpredictable functions (Lemmas 6.1.1 and 6.1.2). An example of a piecewise constant unpredictable function is constructed. Moreover, continuous function as solution of a linear non-homogeneous scalar equation has been found, which approximates an unpredictable function asymptotically. These functions are used in numerical simulations which confirm the theoretical result of the chapter.

M. Akhmet et al., *Dynamics with Chaos and Fractals*, Nonlinear Systems and Complexity 29, https://doi.org/10.1007/978-3-030-35854-9_6

It was shown in papers [1, 2] that the existence of an unpredictable solution implies *Poincaré chaos* so that irregularity is observable in the solutions. In such a case, the chaos is present in the dynamics where unpredictable functions are considered as points moving by shifts of the time argument [9], and the irregularity of the solutions is a consequence of this chaotic dynamics. The reason for the appearance of Poincaré chaos relies on the fact that the topology of uniform convergence on compact sets of the real axis is metrizable [9], p. 29.

In the remaining parts of the chapter, we will make use of the usual Euclidean norm for vectors and the norm induced by the Euclidean norm for square matrices.

Definition 6.1 ([2]) A uniformly continuous and bounded function $\vartheta : \mathbb{R} \to \mathbb{R}^m$ is unpredictable if there exist positive numbers ϵ_0, δ and sequences $\{t_n\}$, $\{u_n\}$ both of which diverge to infinity such that $\|\vartheta(t+t_n) - \vartheta(t)\| \to 0$ as $n \to \infty$ uniformly on compact subsets of $\mathbb{R}$ and $\|\vartheta(t+t_n) - \vartheta(t)\| \geq \epsilon_0$ for each $t \in [u_n - \delta, u_n + \delta]$ and $n \in \mathbb{N}$.

Definition 6.2 ([2]) A bounded sequence $\{\kappa_i\}$, $i \in \mathbb{Z}$, in $\mathbb{R}^m$ is called unpredictable if there exist a positive number ϵ_0 and the sequences $\{\zeta_n\}$, $\{\eta_n\}$, $n \in \mathbb{N}$, of positive integers both of which diverge to infinity such that $\|\kappa_{i+\eta_n} - \kappa_i\| \to 0$ as $n \to \infty$ for each i in bounded intervals of integers and $\|\kappa_{\zeta_n+\eta_n} - \kappa_{\eta_n}\| \geq \epsilon_0$ for each $n \in \mathbb{N}$.

Definitions 6.1 and 6.2 are utilized in Sects. 6.2 and 6.3, respectively. Next, let us remind the definition of a Poisson stable function [9], p. 124, adapted to our case.

Definition 6.3 ([2]) A continuous and bounded function $\vartheta : \mathbb{R} \to \mathbb{R}^m$ is positively Poisson stable if there exists a sequence $\{t_n\}$, $t_n \to \infty$ as $n \to \infty$, such that $\|\vartheta(t+t_n) - \vartheta(t)\| \to 0$ as $n \to \infty$ uniformly on compact subsets of $\mathbb{R}$.

By comparing Definitions 6.1 and 6.3 one can see that any unpredictable function is Poisson stable. The Poisson stable function is a specification of Poisson stable point considered for dynamical systems in [9, p. 85], [8, p. 344, 345].

It is worth noting that in the literature a large number of results are obtained for periodic, quasi-periodic, and almost periodic solutions of differential equations due to the established mathematical methods and important applications. On the other hand, recurrent and Poisson stable solutions are also crucial for the theory of differential equations [4, 9]. The proposal can revive interests of specialists in differential equations theory for two reasons. The first one is related to the verification of the unpredictability which requests a more sophisticated technique than for recurrent and Poisson stable solutions. Thus the problem of the existence of unpredictable solutions is a challenging one. Unpredictable solutions can be investigated for various types of differential equations such as partial differential equations, evolution equations, and hybrid systems. Thus, a new approach for chaos extension in many types of dynamics is suggested in our study.

The following lemmas can be useful for applications of our results.

Lemma 6.1.1 ([3]) *Suppose that $\phi(t) : \mathbb{R} \to \mathbb{R}$ is an unpredictable function. Then the function $\phi^3(t)$ is unpredictable.*

Proof One can find numbers $\epsilon_0 > 0$, $\delta > 0$ and sequences $\{t_n\}$, $\{u_n\}$ both of which diverge to infinity such that $\|\phi(t + t_n) - \phi(t)\| \to 0$ as $n \to \infty$ uniformly on compact subsets of $\mathbb{R}$ and $\|\phi(t + t_n) - \phi(t)\| \geq \epsilon_0$ for each $t \in [u_n - \delta, u_n + \delta]$ and $n \in \mathbb{N}$. It is easy to check that $\|\phi^3(t + t_n) - \phi^3(t)\| \to 0$ as $n \to \infty$ uniformly on compact subsets of $\mathbb{R}$, since it follows from the uniform continuity of the cubic function on a compact set.

Fix a natural number n. Let us show that for $t \in [u_n - \delta, u_n + \delta]$ the inequality $\|\phi(t + t_n) - \phi(t)\| \geq \epsilon_0$ implies $\|\phi^3(t + t_n) - \phi^3(t)\| \geq \epsilon_0^3/4$.

We have that

$$|\phi^3(t + t_n) - \phi^3(t)| = \frac{1}{2}|\phi(t + t_n) - \phi(t)| \left[\phi^2(t + t_n) + \phi^2(t) + (\phi(t + t_n) + \phi(t))^2\right]$$
$$\geq \frac{1}{2}\left(\phi^2(t + t_n) + \phi^2(t)\right)\epsilon_0.$$

Consider the function $F(a, b) = a^2+b^2$ for $|a-b| \geq \epsilon_0$. The minimum of F occurs at the points (a, b) with $|a| = |b| = \epsilon_0/2$. Therefore, $|\phi^3(t + t_n) - \phi^3(t)| \geq \epsilon_0^3/4$ for $t \in [u_n - \delta, u_n + \delta]$. □

Lemma 6.1.2 ([3]) *If the function $\phi(t) : \mathbb{R} \to \mathbb{R}$ is unpredictable, then the function $\phi(t) + c$, where c is a constant, is also unpredictable.*

Proof There exist positive numbers ϵ_0, δ and sequences $\{t_n\}$, $\{u_n\}$ both of which diverge to infinity such that $\|\phi(t + t_n) - \phi(t)\| \to 0$ as $n \to \infty$ uniformly on compact subsets of $\mathbb{R}$ and $\|\phi(t+t_n)-\phi(t)\| \geq \epsilon_0$ for each $t \in [u_n-\delta, u_n+\delta]$ and $n \in \mathbb{N}$. Let us denote $\omega(t) = \phi(t) + c$. Then we have that $\|\omega(t + t_n) - \omega(t)\| = \|\phi(t + t_n) - \phi(t)\| \to 0$ as $n \to \infty$ uniformly on compact subsets of $\mathbb{R}$ and $\|\omega(t + t_n) - \omega(t)\| = \|\phi(t + t_n) - \phi(t)\| \geq \epsilon_0$ for each $t \in [u_n - \delta, u_n + \delta]$ and $n \in \mathbb{N}$. Therefore, the function $\phi(t) + c$ is unpredictable. □

6.2 Differential Equations with Unpredictable Solutions

Let us consider the system of linear differential equations

$$x'(t) = Ax(t) + g(t), \tag{6.2.1}$$

where $x \in \mathbb{R}^p$ and the function $g : \mathbb{R} \to \mathbb{R}^p$ is uniformly continuous and bounded. Moreover, we assume that all eigenvalues of the constant matrix $A \in \mathbb{R}^{p\times p}$ have nonzero real parts.

Assume that the following condition is valid:

(C1) $\Re e\lambda_i < 0$, $i = 1, 2, \ldots, r$, and $\Re e\lambda_i > 0$, $i = r+1, r+2, \ldots, p$, $1 \leq r < p$, where λ_i, $i = 1, \ldots, p$, are the eigenvalues of the matrix A and $\Re e\lambda_i$ denotes the real part of λ_i.

One can find a nonsingular matrix B such that the substitution $x = By$ transforms system (6.2.1) to the equation

$$y'(t) = B^{-1}ABy(t) + B^{-1}g(t), \tag{6.2.2}$$

with the block diagonal matrix of coefficients [5]. Therefore, we assume without loss of generality that the matrix A in system (6.2.1) is block diagonal such that $A = \mathrm{diag}(A_-, A_+)$, where the eigenvalues of the matrices A_- and A_+ possess negative and positive real parts, respectively. There exist numbers $K \geq 1$ and $\alpha > 0$ such that $\|e^{A_- t}\| \leq Ke^{-\alpha t}$ for $t \geq 0$ and $\|e^{A_+ t}\| \leq Ke^{\alpha t}$ for $t \leq 0$.

From Eq. (6.2.2), it implies that the following auxiliary assertion is needed.

Lemma 6.2.3 ([3]) *If the function $g(t)$ is unpredictable, then the function $f(t) = B^{-1}g(t)$ is also unpredictable.*

The proof of the lemma immediately follows from the inequalities $\|f(t+t_n) - f(t)\| \leq \|B^{-1}\|\|g(t+t_n) - g(t)\|$ and $\|f(t+t_n) - f(t)\| \geq \dfrac{1}{\|B\|}\|g(t+t_n) - g(t)\|$.

In what follows we will denote $g(t) = (g_-(t), g_+(t))$, where the vector-functions $g_-(t)$ and $g_+(t)$ are of dimensions r and $p - r$, respectively.

As it is known from the theory of differential equations [5], system (6.2.1) admits a unique bounded on $\mathbb{R}$ solution $\varphi(t) = (\varphi_-(t), \varphi_+(t))$,

$$\varphi_-(t) = \int_{-\infty}^{t} e^{A_-(t-s)}g_-(s)ds, \quad \varphi_+(t) = -\int_{t}^{\infty} e^{A_+(t-s)}g_+(s)ds, \tag{6.2.3}$$

if the function $g(t)$ is bounded on $\mathbb{R}$. One can confirm that $\sup_{t\in\mathbb{R}} \|\varphi(t)\| \leq \dfrac{2M_g K}{\alpha}$, where $M_g = \sup_{t\in\mathbb{R}} \|g(t)\|$. Moreover, $\varphi(t)$ is periodic, quasi-periodic, or almost periodic if the perturbation function $g(t)$ is, respectively, of the same type.

The following theorem is concerned with the unpredictable solution of system (6.2.1).

Theorem 6.1 ([3]) *Assume that $g(t)$ is an unpredictable function and condition $(C1)$ is valid. Then system (6.2.1) possesses a unique unpredictable solution. Additionally, if all eigenvalues of the matrix A have negative real parts, then the unpredictable solution is uniformly asymptotically stable.*

Proof The boundedness of $g(t)$ implies that system (6.2.1) admits a unique bounded solution $\varphi(t) = (\varphi_-(t), \varphi_+(t))$, which satisfies (6.2.3), and it is uniformly continuous since its derivative is bounded. Moreover, the bounded solution is uniformly asymptotically stable provided that all eigenvalues of the matrix A have negative real parts. Hence, it is sufficient to prove that $\varphi(t)$ is an unpredictable function.

The function $\varphi(t)$ is uniformly continuous since its derivative is bounded. According to the Poisson stability of $g(t)$, there exists a sequence $\{t_n\}$ with $t_n \to \infty$ as $n \to \infty$ such that $\|g(t+t_n) - g(t)\| \to 0$ uniformly on compact subsets of $\mathbb{R}$. One can easily find that

$$\begin{aligned}\|\varphi_-(t+t_n) - \varphi_-(t)\| &= \left\| \int_{-\infty}^{t} e^{A_-(t-s)}[g_-(s+t_n) - g_-(s)]ds \right\| \\ &\leq \int_{-\infty}^{t} Ke^{-\alpha(t-s)}\|g_-(s+t_n) - g_-(t)\|ds\end{aligned}$$

and

$$\begin{aligned}\|\varphi_+(t+t_n) - \varphi_+(t)\| &= \left\| \int_{t}^{\infty} e^{A_+(t-s)}[g_+(s+t_n) - g_+(s)]ds \right\| \\ &\leq \int_{t}^{\infty} Ke^{\alpha(t-s)}\|g_+(s+t_n) - g_+(t)\|ds.\end{aligned}$$

Fix an arbitrary positive number ϵ and a closed interval $[a, b]$, $-\infty < a < b < \infty$, of the real axis. We will show that for sufficiently large n it is true that $\|\varphi(t+t_n) - \varphi(t)\| < \epsilon$ on $[a, b]$. Let us choose numbers $c < a$, $d > b$, $\xi > 0$ such that $\frac{2M_g K}{\alpha}e^{-\alpha(a-c)} \leq \frac{\epsilon}{4}$, $\frac{2M_g K}{\alpha}e^{-\alpha(d-b)} \leq \frac{\epsilon}{4}$, and $\frac{K\xi}{\alpha} \leq \frac{\epsilon}{4}$.

Consider n sufficiently large such that $\|g(t+t_n) - g(t)\| < \xi$ for $t \in [c, d]$. Then we have for all $t \in [a, b]$ that

$$\begin{aligned}\|\varphi_-(t+t_n) - \varphi_-(t)\| &\leq \int_{-\infty}^{c} Ke^{-\alpha(t-s)}\|g_-(s+t_n) - g_-(s)\|ds \\ &\quad + \int_{c}^{t} Ke^{-\alpha(t-s)}\|g_-(s+t_n) - g_-(s)\|ds \\ &\leq \int_{-\infty}^{c} 2M_g Ke^{-\alpha(t-s)}ds + \int_{c}^{t} K\xi e^{-\alpha(t-s)}ds \\ &< \frac{2M_g K}{\alpha}e^{-\alpha(a-c)} + \frac{K\xi}{\alpha} \\ &\leq \frac{\epsilon}{2}\end{aligned}$$

and similarly one can show that

$$\begin{aligned}\|\varphi_+(t+t_n) - \varphi_+(t)\| &\leq \int_{t}^{d} Ke^{\alpha(t-s)}\|g_+(s+t_n) - g_+(s)\|ds \\ &\quad + \int_{d}^{\infty} Ke^{\alpha(t-s)}\|g_+(s+t_n) - g_+(s)\|ds\end{aligned}$$

$$\leq \int_t^d K\xi e^{\alpha(t-s)}ds + \int_d^\infty 2M_g K e^{\alpha(t-s)}ds$$
$$< \frac{K\xi}{\alpha} + \frac{2M_g K}{\alpha} e^{-\alpha(d-b)}$$
$$\leq \frac{\epsilon}{2}.$$

Thus, for sufficiently large n it is true that

$$\|\varphi(t+t_n) - \varphi(t)\| \leq \|\varphi_+(t+t_n) - \varphi_+(t)\| + \|\varphi_-(t+t_n) - \varphi_-(t)\| < \epsilon$$

for $t \in [a, b]$.

Next, we will show the existence of a sequence $\{\bar{u}_n\}$, $\bar{u}_n \to \infty$ as $n \to \infty$, and positive numbers $\bar{\epsilon}_0$, δ such that $\|\varphi(t+t_n) - \varphi(t_n)\| \geq \bar{\epsilon}_0$ for $t \in [\bar{u}_n - \delta, \bar{u}_n + \delta]$.

According to uniform continuity of $g(t)$, there exists a positive number $\bar{\delta}$, which does not depend on the sequences $\{t_n\}$ and $\{u_n\}$, such that the inequalities

$$\|g(t+t_n) - g(t_n + u_n)\| \leq \frac{\epsilon_0}{4\sqrt{p}}$$

and

$$\|g(t) - g(u_n)\| \leq \frac{\epsilon_0}{4\sqrt{p}}$$

are valid for every $t \in \left[u_n - \bar{\delta}, u_n + \bar{\delta}\right]$ and $n \in \mathbb{N}$.

Fix an arbitrary natural number n, and suppose that $g(t) = (g_1(t), g_2(t), \ldots, g_p(t))$, where each $g_k(t)$, $k = 1, 2, \ldots, p$, is a real valued function. It can be verified that there exists an integer j_n, $1 \leq j_n \leq p$, such that

$$|g_{j_n}(t_n + u_n) - g_{j_n}(u_n)| \geq \frac{\epsilon_0}{\sqrt{p}}.$$

Hence, we have

$$\begin{aligned} |g_{j_n}(t+t_n) - g_{j_n}(t)| &\geq |g_{j_n}(t_n+u_n) - g_{j_n}(u_n)| - |g_{j_n}(t+t_n) - g_{j_n}(t_n+u_n)| \\ &\quad - |g_{j_n}(t) - g_{j_n}(u_n)| \\ &\geq \frac{\epsilon_0}{2\sqrt{p}} \end{aligned} \tag{6.2.4}$$

for $t \in \left[u_n - \bar{\delta}, u_n + \bar{\delta}\right]$.

One can confirm that there exist numbers $s_1^n, s_2^n, \ldots, s_p^n$ in the interval $[u_n - \bar{\delta}, u_n + \bar{\delta}]$ such that the equation

$$\left\| \int_{u_n-\bar{\delta}}^{u_n+\bar{\delta}} (g(s+t_n)-g(s))ds \right\| = 2\bar{\delta}\Big[\sum_{i=1}^{p}(g_i(s_i^n+t_n)-g_i(s_i^n))^2\Big]^{1/2}$$

is valid. Using inequality (6.2.4) we obtain that

$$\left\| \int_{u_n-\bar{\delta}}^{u_n+\bar{\delta}} (g(s+t_n)-g(s))ds \right\| \geq 2\bar{\delta}|g_{j_n}(s_{j_n}^n+t_n)-g_{j_n}(s_{j_n}^n)| \geq \frac{\bar{\delta}\epsilon_0}{\sqrt{p}}.$$

By means of the equation

$$\begin{aligned}\varphi(t_n+u_n+\bar{\delta})-\varphi(u_n+\bar{\delta}) &= \varphi(t_n+u_n-\bar{\delta})-\varphi(u_n-\bar{\delta}) \\ &\quad + \int_{u_n-\bar{\delta}}^{u_n+\bar{\delta}} A[\varphi(s+t_n)-\varphi(s)]ds \\ &\quad + \int_{u_n-\bar{\delta}}^{u_n+\bar{\delta}} [g(s+t_n)-g(s)]ds,\end{aligned}$$

one can obtain that

$$\|\varphi(t_n+u_n+\bar{\delta})-\varphi(u_n+\bar{\delta})\| \geq \frac{\bar{\delta}\epsilon_0}{\sqrt{p}} - (1+2\bar{\delta}\|A\|) \sup_{t\in[u_n-\bar{\delta},u_n+\bar{\delta}]} \|\varphi(t+t_n)-\varphi(t)\|$$

Thus, $\sup_{t\in[u_n-\bar{\delta},u_n+\bar{\delta}]} \|\varphi(t+t_n)-\varphi(t)\| \geq \dfrac{\bar{\delta}\epsilon_0}{2(1+\bar{\delta}\|A\|)\sqrt{p}}.$

Now, suppose that $\sup_{t\in[u_n-\bar{\delta},u_n+\bar{\delta}]} \|\varphi(t+t_n)-\varphi(t)\| = \|\varphi(t_n+\bar{u}_n)-\varphi(\bar{u}_n)\|$ for some $\bar{u}_n \in [u_n-\bar{\delta}, u_n+\bar{\delta}]$, and let us denote

$$\bar{\epsilon}_0 = \frac{\bar{\delta}\epsilon_0}{4(1+\bar{\delta}\|A\|)\sqrt{p}}$$

and

$$\delta = \frac{\bar{\delta}\alpha\epsilon_0}{8M_g(1+\bar{\delta}\|A\|)\,(\alpha+2K\,\|A\|)\,\sqrt{p}}.$$

If $t \in [\bar{u}_n-\delta, \bar{u}_n+\delta]$, then we have

$$\begin{aligned}\|\varphi(t+t_n)-\varphi(t)\| &\geq \|\varphi(t_n+\bar{u}_n)-\varphi(\bar{u}_n)\| - \left|\int_{\bar{u}_n}^{t} \|A\|\varphi(s+t_n)-\varphi(s)\|ds\right| \\ &\quad - \left|\int_{\bar{u}_n}^{t} \|g(s+t_n)-g(s)\|ds\right|\end{aligned}$$

$$\geq \frac{\bar{\delta}\epsilon_0}{2(1+\bar{\delta}\|A\|)\sqrt{p}} - \frac{4\delta M_g K \|A\|}{\alpha} - 2\delta M_g$$
$$= \bar{\epsilon}_0.$$

Hence, $\|\varphi(t+t_n) - \varphi(t)\| \geq \bar{\epsilon}_0$ for each t from the intervals $[\bar{u}_n - \delta, \bar{u}_n + \delta]$, $n \in \mathbb{N}$. One can confirm that the sequence $\{\bar{u}_n\}$ diverges to infinity. Consequently, $\varphi(t)$ is the unique unpredictable solution of system (6.2.1). □

The next section is devoted to unpredictable solutions of linear discrete equations.

6.3 Discrete Equations with Unpredictable Solutions

Let us take into account the discrete equation

$$z_{i+1} = Dz_i + \phi_i, \tag{6.3.5}$$

where $i \in \mathbb{Z}$, $D \in \mathbb{R}^{q\times q}$ is a nonsingular matrix, and $\{\phi_i\}$ is a bounded sequence. In this section the following condition is needed:

(C2) $D = \text{diag}(D_-, D_+)$, where D_- and D_+ are, respectively, $k\times k$ and $(q-k)\times(q-k)$ matrices with $0 \leq k \leq q$ such that $\|D_-\| < 1$ and $\|D_+\| > 1$.

We will denote $\phi_i = (\phi_i^-, \phi_i^+)$, where ϕ_i^-, ϕ_i^+ are, respectively, k and $q-k$ dimensional.

According to the results of [7], Eq. (6.3.5) possesses a unique bounded solution $\psi_i = (\psi_i^-, \psi_i^+)$, $i \in \mathbb{Z}$, which satisfies the relations

$$\psi_i^- = \sum_{j=-\infty}^{i} D_-^{i-j} \phi_{j-1}^- \tag{6.3.6}$$

and

$$\psi_i^+ = -\sum_{j=i}^{\infty} D_+^{i-j-1} \phi_j^+. \tag{6.3.7}$$

One can verify for each $i \in \mathbb{Z}$ that $\|\psi_i^-\| \leq \dfrac{M_\phi}{1-\|D_-\|}$ and $\|\psi_i^+\| \leq \dfrac{M_\phi}{\|D_+\|-1}$, where $M_\phi = \sup_{i\in\mathbb{Z}} \|\phi_i\|$.

The following theorem is concerned with the existence of an unpredictable solution of Eq. (6.3.5).

Theorem 6.2 ([3]) *If $\{\phi_i\}$, $i \in \mathbb{Z}$, is an unpredictable sequence and the condition $(C2)$ is valid, then the Eq. (6.3.5) possesses a unique unpredictable solution.*

Proof Because the sequence $\{\phi_i\}$, $i \in \mathbb{Z}$, is unpredictable, according to Definition 6.2, there exist a positive number ϵ_0 and sequences $\{\zeta_n\}$, $\{\eta_n\}$, $n \in \mathbb{N}$, of positive integers both of which diverge to infinity such that $\|\phi_{i+\zeta_n} - \phi_i\| \to 0$ as $n \to \infty$ on each bounded intervals of integers and $\|\phi_{\zeta_n+\eta_n} - \phi_{\eta_n}\| \geq \epsilon_0$ for each $n \in \mathbb{N}$. We will show that the unique bounded solution $\psi_i = (\psi_i^-, \psi_i^+)$, $i \in \mathbb{Z}$, of (6.3.5) is unpredictable.

It is easy to check that

$$\psi^-_{i+\zeta_n} - \psi^-_i = \sum_{j=-\infty}^{i+\zeta_n} D_-^{i+\zeta_n-j}\phi^-_{j+\zeta-1} - \sum_{j=-\infty}^{i} D_-^{i-j}\phi^-_{j-1}$$

$$= \sum_{j=-\infty}^{i} D_-^{i-j}(\phi^-_{j+\zeta_n-1} - \phi^-_{j-1}),$$

and

$$\psi^+_{i+\zeta_n} - \psi^+_i = -\sum_{j=i+\zeta_n}^{\infty} D_+^{i+\zeta_n-j-1}\phi^+_j + \sum_{j=i}^{\infty} D_+^{i-j-1}\phi^+_j$$

$$= \sum_{j=i}^{\infty} D_+^{i-j-1}(\phi^+_j - \phi^+_{j+\zeta_n}).$$

Fix an arbitrary positive number ϵ and let a and b be integers with $a < b$. We will show that for sufficiently large n it is true that $\|\psi_{i+\zeta_n} - \psi_i\| < \epsilon$ for $a \leq i \leq b$. Let us choose integers $c < a$, $b < d$, $\xi > 0$, such that $\frac{1}{1-\|D_-\|}(2M_\phi\|D_-\|^{a-c}+\xi) < \frac{\epsilon}{2}$ and $\frac{1}{\|D_+\|-1}(\xi + 2M_\phi\|D_+\|^{d-b+1}) < \frac{\epsilon}{2}$.

Consider n sufficiently large such that $\|\psi_{i+\zeta_n} - \psi_i\| < \xi$ for $c \leq i \leq d$. Then we have for $a \leq i \leq b$ that

$$\|\psi^-_{i+\zeta_n} - \psi^-_i\| = \|\sum_{j=-\infty}^{c} D_-^{c-j}(\phi^-_{j+\zeta_n-1} - \phi^-_{j-1}) + \sum_{j=c}^{i} D_-^{i-j}(\phi^-_{j+\zeta_n-1} - \phi^-_{j-1})\|$$

$$\leq \frac{2M_\phi\|D_-\|^{a-c}}{1-\|D_-\|} + \frac{\xi}{1-\|D_-\|} < \frac{\epsilon}{2},$$

and similarly,

$$\|\psi^+_{i+\zeta_n} - \psi^+_i\| = \|\sum_{j=i}^{d} D_+^{d-j-1}(\phi^+_j - \phi^+_{j+\zeta_n}) + \sum_{j=d}^{\infty} D_+^{d-j}(\phi^+_j - \phi^+_{j+\zeta_n})\|$$

$$\leq \frac{\xi}{\|D_+\|-1} + \frac{2M_\phi \|D_+\|^{b-d+1}}{\|D_+\|-1} < \frac{\epsilon}{2}.$$

Thus, $\|\psi_{i+\zeta_n} - \psi_i\| \leq \|\psi^+_{i+\zeta_n} - \psi^+_i\| + \|\psi^-_{i+\zeta_n} - \psi^-_i\| < \epsilon$ for $a \leq i \leq b$, and hence, $\|\psi_{i+\zeta_n} - \psi_i\| \to 0$ as $n \to \infty$ on each bounded intervals of integers.

On the other hand, the equation

$$\psi_{\zeta_n+\eta_n+1} - \psi_{\eta_n+1} = D\left(\psi_{\zeta_n+\eta_n} - \psi_{\eta_n}\right) + \phi_{\zeta_n+\eta_n} - \phi_{\eta_n}$$

implies that

$$\|\psi_{\zeta_n+\eta_n+1} - \psi_{\eta_n+1}\| \geq \epsilon_0 - \|D\|\|\psi_{\zeta_n+\eta_n} - \psi_{\eta_n}\|.$$

Therefore, the inequality

$$(1+\|D\|)\max\{\|\psi_{\zeta_n+\eta_n+1} - \psi_{\eta_n+1}\|, \|\psi_{\zeta_n+\eta_n} - \psi_{\eta_n}\|\} \geq \epsilon_0$$

holds.

Let us define the sequence $\overline{\eta}_n$, $n \in \mathbb{N}$, such that $\overline{\eta}_n = \eta_n$ if $\|\psi_{\zeta_n+\eta_n} - \psi_{\eta_n}\| \geq \|\psi_{\zeta_n+\eta_n+1} - \psi_{\eta_n+1}\|$ and $\overline{\eta}_n = \eta_n + 1$, otherwise. Accordingly, we have that

$$\|\psi_{\zeta_n+\overline{\eta}_n} - \psi_{\overline{\eta}_n}\| \geq \frac{\epsilon_0}{1+\|D\|}$$

for each $n \in \mathbb{N}$. It is clear that $\overline{\eta}_n \to \infty$ as $n \to \infty$. Consequently, the bounded solution ψ_i, $i \in \mathbb{Z}$, of (6.3.5) is unpredictable. □

6.4 Examples

It was shown in paper [2] that the presence of an unpredictable function is inevitably accompanied with Poincaré chaos. Consequently, we can look for a confirmation of the results for unpredictability observing irregularity in simulations. The approach is effective for asymptotically stable unpredictable solutions, and it is just illustrative for hyperbolic systems with unstable solutions. In the latter case we rely on the fact that any solution becomes unpredictable ultimately.

In this section, first of all, we will show the construction of an unpredictable function using the dynamics of the logistic map in a similar way to paper [2].

Let us take into account the logistic map

$$\lambda_{i+1} = F_\mu(\lambda_i), \tag{6.4.8}$$

where $i \in \mathbb{Z}$ and $F_\mu(s) = \mu s(1-s)$. The interval $[0,1]$ is invariant under the iterations of (6.4.8) for $\mu \in (0,4]$ [6].

Using the results of paper [10], it was shown in Theorem 4.1 [1] that the logistic map (6.4.8) possesses an unpredictable solution for each $\mu \in [3+(2/3)^{1/2}, 4]$.

Let $\{\rho_i\}$, $i \in \mathbb{Z}$, be an unpredictable solution of the logistic map (6.4.8) with $\mu = 3.92$ inside the unit interval $[0,1]$, and consider the function

$$\Theta(t) = \int_{-\infty}^{t} e^{-5(t-s)/2}\Omega(s)ds, \tag{6.4.9}$$

where $\Omega(t)$ is a piecewise constant function defined on the real axis through the equation $\Omega(t) = \rho_i$ for $t \in [i, i+1)$, $i \in \mathbb{Z}$.

It is worth noting that $\Theta(t)$ is the unique globally exponentially stable solution of the differential equation

$$v'(t) = -\frac{5}{2}v(t) + \Omega(t).$$

Additionally, one can confirm that the function $\Theta(t)$ is bounded on the whole real axis such that

$$\sup_{t\in\mathbb{R}} |\Theta(t)| \le \frac{2}{5},$$

and it is uniformly continuous since its derivative is bounded.

Because the sequence $\{\rho_i\}$ is unpredictable, there exist a positive number ϵ_0 and sequences $\{\zeta_n\}$, $\{\eta_n\}$ both of which diverge to infinity such that $|\rho_{i+\zeta_n} - \rho_i| \to 0$ as $n \to \infty$ for each i in bounded intervals of integers and $|\rho_{\zeta_n+\eta_n} - \rho_{\eta_n}| \ge \epsilon_0$ for each $n \in \mathbb{N}$.

Fix an arbitrary positive number ϵ and arbitrary real numbers α, β with $\beta > \alpha$. Let N be a sufficiently large natural number satisfying

$$N \ge \frac{2}{5}\ln\left(\frac{6}{5\epsilon}\right).$$

There exists a natural number n_0 such that for each $n \ge n_0$ the inequality

$$|\rho_{i+\zeta_n} - \rho_i| < \frac{5\epsilon}{6}$$

holds for $i = \lfloor\alpha\rfloor - N, \lfloor\alpha\rfloor - N + 1, \ldots, \lfloor\beta\rfloor$, where $\lfloor\alpha\rfloor$ and $\lfloor\beta\rfloor$, respectively, denote the largest integers which are not greater than α and β. Accordingly, if $n \ge n_0$, then we have

$$|\Omega(t+\zeta_n)-\Omega(t)| < \frac{5\epsilon}{6} \tag{6.4.10}$$

for $t \in \big[\lfloor\alpha\rfloor - N, \lfloor\beta\rfloor + 1\big)$.

Fix a natural number $n \geq n_0$. Using the relations

$$\Theta(t) = e^{-5(t-\lfloor\alpha\rfloor+N)/2}\Theta\left(\lfloor\alpha\rfloor - N\right) + \int_{\lfloor\alpha\rfloor-N}^{t} e^{-5(t-s)/2}\Omega(s)ds$$

and

$$\Theta\left(t+\zeta_n\right) = e^{-5(t-\lfloor\alpha\rfloor+N)/2}\Theta\left(\lfloor\alpha\rfloor - N + \zeta_n\right) + \int_{\lfloor\alpha\rfloor-N}^{t} e^{-5(t-s)/2}\Omega(s+\zeta_n)ds$$

together with inequality (6.4.10), we obtain for $t \in \big[\lfloor\alpha\rfloor - N, \lfloor\beta\rfloor + 1\big)$ that

$$\begin{aligned}|\Theta(t+\zeta_n)-\Theta(t)| &\leq e^{-5(t-\lfloor\alpha\rfloor+N)/2}\left|\Theta(\lfloor\alpha\rfloor - N+\zeta_n)-\Theta(\lfloor\alpha\rfloor - N)\right| \\ &\quad + \int_{\lfloor\alpha\rfloor-N}^{t} e^{-5(t-s)/2}\left|\Omega(s+\zeta_n)-\Omega(s)\right| ds \\ &\leq \frac{4}{5}e^{-5(t-\lfloor\alpha\rfloor+N)/2} + \frac{\epsilon}{3}\left(1-e^{-5(t-\lfloor\alpha\rfloor+N)/2}\right).\end{aligned}$$

The last inequality implies that $|\Theta(t+\zeta_n)-\Theta(t)| < \epsilon$ for $t \in [\lfloor\alpha\rfloor, \lfloor\beta\rfloor + 1]$. Hence, $|\Theta(t+\zeta_n)-\Theta(t)| \to 0$ as $n \to \infty$ uniformly on the interval $[\alpha, \beta]$.

On the other hand, one can confirm for each $n \in \mathbb{N}$ that $|\Omega(t+\zeta_n)-\Omega(t)| \geq \epsilon_0$ for $t \in [\eta_n, \eta_n+1)$. For fixed $n \in \mathbb{N}$, using the equation

$$\begin{aligned}\Theta(t+\zeta_n)-\Theta(t) &= \Theta(\zeta_n+\eta_n)-\Theta(\eta_n) - \frac{5}{2}\int_{\eta_n}^{t}\left(\Theta(s+\zeta_n)-\Theta(s)\right)ds \\ &\quad + \int_{\eta_n}^{t}\left(\Omega(s+\zeta_n)-\Omega(s)\right)ds\end{aligned}$$

we attain that

$$\begin{aligned}|\Theta(\zeta_n+\eta_n+1)-\Theta(\eta_n+1)| \geq &\left|\int_{\eta_n}^{\eta_n+1}\left(\Omega(s+\zeta_n)-\Omega(s)\right)ds\right| \\ &- |\Theta(\zeta_n+\eta_n)-\Theta(\eta_n)| \\ &- \frac{5}{2}\left|\int_{\eta_n}^{\eta_n+1}\left(\Theta(s+\zeta_n)-\Theta(s)\right)ds\right|.\end{aligned}$$

Therefore, one can verify that

$$\sup_{t\in[\eta_n,\eta_n+1]} |\Theta(t+\zeta_n)-\Theta(t)| \ge \frac{2\epsilon_0}{9}.$$

Thus, there exists a sequence $\{u_n\}$ with $\eta_n \le u_n \le \eta_n + 1$, $n \in \mathbb{N}$, such that

$$|\Theta(u_n+\zeta_n)-\Theta(u_n)| \ge \frac{2\epsilon_0}{9}.$$

For $t \in [u_n - \delta, u_n + \delta]$, where $\delta = \epsilon_0/36$, we have

$$\begin{aligned}|\Theta(t+\zeta_n)-\Theta(t)| &\ge |\Theta(u_n+\zeta_n)-\Theta(u_n)| - \frac{5}{2}\left|\int_{u_n}^{t} |\Theta(s+\zeta_n)-\Theta(s)|\,ds\right| \\ &\quad - \left|\int_{u_n}^{t} |\Omega(s+\zeta_n)-\Omega(s)|\,ds\right| \\ &\ge \frac{\epsilon_0}{9}.\end{aligned}$$

It is clear that $u_n \to \infty$ as $n \to \infty$. Thus, the function $\Theta(t)$ is unpredictable.

Example 1 ([3]) Consider the system

$$\begin{aligned}x_1' &= -2x_1 + 2x_2 - 50\Theta(t)\\ x_2' &= x_1 - 3x_2 + 5\Theta^3(t),\end{aligned} \tag{6.4.11}$$

where $\Theta(t)$ is the unpredictable function defined by (6.4.9). The eigenvalues of the matrix of coefficients of system (6.4.11) are -2 and -0.5. One can confirm that the perturbation function $(-50\Theta(t), 5\Theta^3(t))$ is unpredictable in accordance with Lemma 6.1.1. By the main result of our paper, there is an asymptotically stable unpredictable solution $(\varphi_1(t), \varphi_2(t))$ of system (6.4.11). Consequently, any solution of the equation behaves irregularly ultimately. This is seen from the simulation of the solution with $x_1(0) = 0, 18$, $x_2(0) = 0, 01$ in Fig. 6.1.

The next example is devoted to a system of differential equations whose matrix of coefficients admits both positive and negative eigenvalues.

Example 2 ([3]) Let us take into account the system

$$\begin{aligned}y_1' &= -1000y_1 + 0.23y_2 + 120x_2^3(t) + 160\\ y_2' &= 6y_1 + 0.000002y_2 - 0.1x_1(t) + 20,\end{aligned} \tag{6.4.12}$$

where $(x_1(t), x_2(t))$ is the solution of (6.4.11) depicted in Fig. 6.1. The eigenvalues of the matrix of coefficients of system (6.4.12) are -1000 and 0.00138. The perturbation function $(120x_2^3(t) + 160, -0.1x_1(t) + 20)$ is unpredictable by Lemmas 6.1.1 and 6.1.2. According to the result of Theorem 6.1, system (6.4.12) possesses a unique unpredictable solution. The simulation results for system (6.4.12) corresponding to the initial conditions $y_1(0) = 0$ and $y_2(0) = 0.1$ are

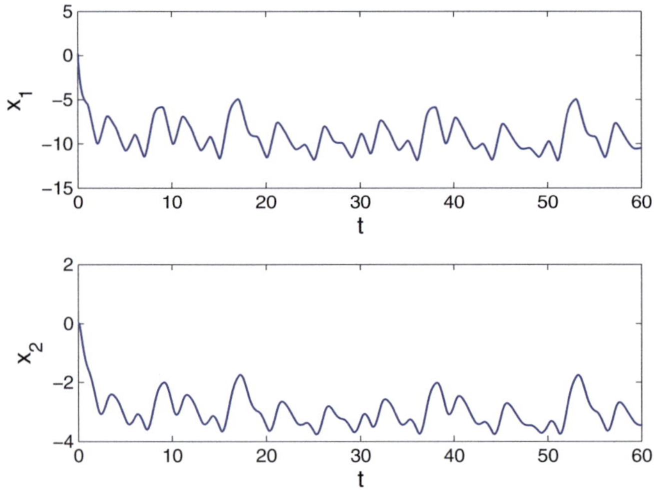

Fig. 6.1 The time series of the x_1 and x_2 coordinates of system (6.4.11) with the initial conditions $x_1(0) = 0, 18$, $x_2(0) = 0, 01$. The figure manifests the irregular behavior of the solution

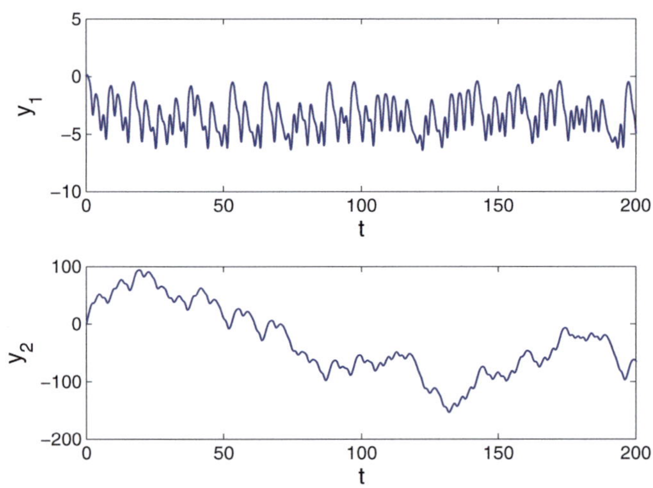

Fig. 6.2 The time series for the y_1 and y_2 coordinates of system (6.4.12) with the initial conditions $y_1(0) = 0$, $y_2(0) = 0.1$. The irregular behavior of the solution reveals the presence of an unpredictable solution in the dynamics of (6.4.12)

shown in Fig. 6.2. The time series of both y_1 and y_2 coordinates in the figure confirm the presence of irregularity in the dynamics of system (6.4.12).

References

1. M. Akhmet, M.O. Fen, Poincaré chaos and unpredictable functions. Commun. Nonlinear Sci. Numer. Simul. **48**, 85–94 (2017)
2. M. Akhmet, M.O. Fen, Non-autonomous equations with unpredictable solutions. Commun. Nonlinear Sci. Numer. Simul. **59**, 657–670 (2018)
3. M. Akhmet, M.O. Fen, M. Tleubergenova, A. Zhamanshin, Unpredictable solutions of linear differential and discrete equations. Turk. J. Math. **43**, 2377–2389 (2019)
4. P.R. Bener, Recurrent solutions to systems of ordinary differential equations. J. Differ. Equ. **5**, 271–282 (1969)
5. J.K. Hale, *Ordinary Differential Equations* (Krieger Publishing Company, Malabar, Florida, 1980)
6. J. Hale, H. Koçak, *Dynamics and Bifurcations* (Springer, New York, 1991)
7. V. Lakshmikantham, D. Trigiante, *Theory of Difference Equations: Numerical Methods and Applications* (Marcel Dekker, USA, 2002)
8. V.V. Nemytskii, V.V. Stepanov, *Qualitative Theory of Differential Equations* (Princeton University Press, Princeton, New Jersey, 1960)
9. G.R. Sell, *Topological Dynamics and Ordinary Differential Equations* (Van Nostrand Reinhold Company, London, 1971)
10. Y. Shi, P. Yu, On chaos of the logistic maps. Dynam. Contin. Discrete Impuls. Syst. Ser. B **14**, 175–195 (2007)

Chapter 7
Strongly Unpredictable Solutions

This chapter is devoted to strongly unpredictable solutions, that is, solutions whose all coordinates are unpredictable functions, of nonautonomous systems of differential equations. Moreover, systems with perturbations which are strongly unpredictable in the time variable are considered. The results of this chapter are published in paper [5].

7.1 Preliminaries

Considerable amount of researches in the theory of differential equations focus on oscillations owing to their importance and applications in science and industry. For that reason the investigation of periodic, quasiperiodic, and almost periodic solutions [6–9, 11] are crucial for the development of the theory and science.

A new type of oscillation, unpredictable function, has been introduced in the study [2]. It was shown in papers [2] and [3] that the existence of an unpredictable solution implies Poincaré chaos for a special dynamics in a functional space. Owing to the presence of the chaos, irregularity is observable in the oscillations. Consequently, the investigation of unpredictable solutions is as much beneficial as it is of chaos. The existence of unpredictable solutions in the dynamics of hyperbolic quasilinear systems whose matrix of coefficients possesses eigenvalues with negative and positive real parts was studied in paper [4]. In the present chapter we consider oscillations with all coordinates unpredictable. The novelty may increase the application meaning of the solutions, because irregularity is guaranteed in all dimensions of a dynamical process. Another significant novelty is that perturbations are assumed to be nonlinear functions of the time and space variables, and they are strongly unpredictable in the time variable.

Throughout the chapter, $\mathbb{R}$ and $\mathbb{N}$ will stand for the set of real and natural numbers, respectively. Additionally, the norm $\|u\|_1 = \sup_{t\in\mathbb{R}} \|u(t)\|$, where

M. Akhmet et al., *Dynamics with Chaos and Fractals*, Nonlinear Systems and Complexity 29, https://doi.org/10.1007/978-3-030-35854-9_7

$\|u\| = \max_{1\le i\le p} |u_i|$, $u = (u_1, \ldots, u_p), u_i \in \mathbb{R}, i = 1, 2, \ldots, p$, will be used. Correspondingly, for a square matrix $A = \{a_{ij}\}, i, j = 1, 2, \ldots, p$, the norm $\|A\| = \max_{i=1,\ldots,p} \sum_{j=1}^{p} |a_{ij}|$, will be utilized.

The following definition is the starting point of our research.

Definition 7.1 ([2]) A uniformly continuous and bounded function $v : \mathbb{R} \to \mathbb{R}^p$ is unpredictable if there exist positive numbers ϵ_0, σ and sequences $\{t_n\}$, $\{u_n\}$ both of which diverge to infinity such that $v(t + t_n) \to v(t)$ as $n \to \infty$ uniformly on compact subsets of $\mathbb{R}$ and $\|v(t+t_n) - v(t)\| \ge \epsilon_0$ for each $t \in [u_n - \sigma, u_n + \sigma]$ and $n \in \mathbb{N}$.

In papers [1, 3], an example of an unpredictable function was provided and it was shown that properties of unpredictable functions are convenient to be verified and they are easy for numerical simulations. Thus, existence of unpredictable solutions for the differential equation

$$x'(t) = Ax(t) + f(x) + g(t), \tag{7.1.1}$$

where $g(t)$ is unpredictable function, was investigated.

In the present investigation we extend Definition 7.1 to the class of functions with several independent variables. The following new definition will be of use.

Definition 7.2 ([5]) A continuous and bounded function $f(t, x) : \mathbb{R} \times G \to \mathbb{R}^p$, $f = (f_1, f_2, \ldots, f_p)$, $G \subset \mathbb{R}^p$ is a bounded domain, is unpredictable in t if it is uniformly continuous in t and there exist positive numbers ϵ_0, σ and sequences $\{t_n\}$, $\{u_n\}$ both of which diverge to infinity such that $\sup_G \|f(t+t_n, x) - f(t, x)\| \to 0$ as $n \to \infty$ uniformly on compact sets in $\mathbb{R}$ and $\inf_G \|f(t + t_n, x) - f(t, x)\| \ge \epsilon_0$ for $t \in [u_n - \sigma, u_n + \sigma]$ and $n \in \mathbb{N}$.

The present chapter contains two principal novelties. The first one is that strongly unpredictable solutions are considered instead of unpredictable ones. Second, we consider nonlinear perturbations, which are functions unpredictable in the time variable. Thus, in this chapter we have significantly enlarged the set of systems, which can be investigated for unpredictable solutions. To this end, we shall need the following two new notions, which are analogues to the last two definitions.

Definition 7.3 ([5]) A uniformly continuous and bounded function $v : \mathbb{R} \to \mathbb{R}^p$, $v = (v_1, \ldots, v_p)$, is strongly unpredictable if there exist positive numbers ϵ_0, σ and sequences t_n, u_n both of which diverge to infinity such that $v(t + t_n) \to v(t)$ as $n \to \infty$ uniformly on compact subsets of $\mathbb{R}$ and $|v_i(t + t_n) - v_i(t)| \ge \epsilon_0$ for each $t \in [u_n - \sigma, u_n + \sigma]$, $i = 1, 2, \ldots, p$ and $n \in \mathbb{N}$.

Definition 7.4 ([5]) A continuous and bounded function $f(t, x) : \mathbb{R} \times G \to \mathbb{R}^p$, $f = (f_1, f_2, \ldots, f_p)$, $G \subset \mathbb{R}^p$ is a bounded domain, is strongly unpredictable in t if it is uniformly continuous in t and there exist positive numbers ϵ_0, σ and sequences t_n, u_n both of which diverge to infinity such that $\sup_G \|f(t+t_n,x) - f(t,x)\| \to 0$ as $n \to \infty$ uniformly on compact sets in $\mathbb{R}$ and $\inf_{[u_n-\sigma,u_n+\sigma]\times G} |f_i(t+t_n,x) - f_i(t,x)| > \epsilon_0$ for all $i = 1, 2, \ldots, p$ and $n \in \mathbb{N}$.

Comparing Definitions 7.1 and 7.3 as well as Definitions 7.2 and 7.4 one can realize that an unpredictable function may admit some of coordinates which are not unpredictable, whereas each coordinate of a strongly unpredictable function is unpredictable. That is, the set of all strongly unpredictable functions is a subclass of unpredictable functions.

In this chapter, we take into account the system of quasilinear differential equations

$$x'(t) = Ax(t) + f(t,x), \tag{7.1.2}$$

where $t \in \mathbb{R}$, $x \in \mathbb{R}^p$, p is a fixed natural number, all eigenvalues of the constant matrix $A \in \mathbb{R}^{p\times p}$ have negative real parts, $f : \mathbb{R} \times G \to \mathbb{R}^p$, $f = (f_1, \ldots, f_p)$, $G = \{x \in \mathbb{R}^p, \|x\| < H\}$, where H is a positive number. It is true that there exist two real numbers $K \geq 1$ and $\gamma < 0$ such that $\|e^{At}\| \leq Ke^{\gamma t}$ for all $t \geq 0$.

One can see that the main difference between system (7.1.1) and system (7.1.2) is that the perturbation in the former one is less general than that of the latter one.

The following conditions will be needed in the paper:

(C1) the function *f(t, x)* is strongly unpredictable in the sense of Definition 7.4;
(C2) there exists a positive constant L such that the function $f(t,x)$ satisfies the inequality $\|f(t,x_1) - f(t,x_2)\| \leq L\,\|x_1 - x_2\|$ for all $t \in \mathbb{R}$, $x_1, x_2 \in G$; Definition 7.4 implies that there exists a positive number M such that $\sup_{\mathbb{R}\times G} \|f(t,x)\| = M < \infty$.
(C3) $\gamma < -\dfrac{KM}{H}$;
(C4) $\gamma < -KL$.

Our purpose is to prove that system (7.1.2) possesses a unique strongly unpredictable solution, provided that the function $f(t,x)$ is strongly unpredictable in t. Moreover, we prove that the solution is uniformly globally exponentially stable. Additionally, existence of an unpredictable solution, which is not strongly unpredictable, is considered for the system.

7.2 Main Results

Let U be the set of all uniformly continuous functions $\psi(t) = (\psi_1(t), \psi_2(t), \ldots, \psi_p(t))$, where for each $i = 1, 2, \ldots, p$, $\psi_i(t)$ is a real valued function defined on $\mathbb{R}$, such that $\|\psi\|_1 \leq H$, and $\psi(t+t_n) \to \psi(t)$ as $n \to \infty$ uniformly on each closed and bounded interval of the real axis, where the sequence $\{t_n\}$ is the same as for function $f(t,x)$ in system (7.1.2) in the case that condition $(C1)$ is valid.

According to the theory of differential equations [10], a function $\omega(t) = \big(\omega_1(t), \omega_2(t), \ldots, \omega_p(t)\big)$ bounded on the whole real axis, where for each $i = 1, 2, \ldots, p$, $\omega_i(t)$ is a real valued function defined on $\mathbb{R}$, is a solution of system (7.1.2) if and only if the integral equation

$$\omega(t) = \int_{-\infty}^{t} e^{A(t-s)} f(t, \omega(s))ds, \ t \in \mathbb{R}, \tag{7.2.3}$$

is satisfied.

Lemma 7.2.1 ([5]) *Suppose that conditions* $(C1)$–$(C4)$ *are valid, then the system (7.1.2) possesses a unique solution* $\omega(t) \in U$.

Proof Define an operator Π on U by the equation

$$\Pi\psi(t) = \int_{-\infty}^{t} e^{A(t-s)} f(s, \psi(s))ds, \ t \in \mathbb{R}. \tag{7.2.4}$$

Fix an arbitrary function $\psi(t)$ that belongs to U. We have that

$$\|\Pi\psi(t)\| \le \int_{-\infty}^{t} \|e^{A(t-s)}\| \|f(s, \psi(s))\|ds \le \frac{KM}{|\gamma|}$$

for all $t \in \mathbb{R}$. Thus, according to condition $(C3)$ we have that $\|\Pi\psi\|_1 \le H$.

Fix an arbitrary positive number ϵ and a closed interval $[a, b]$, $-\infty < a < b < \infty$, of the real axis. We will show that for sufficiently large n it is true that $\|\Pi\psi(t + t_n) - \Pi\psi(t)\| < \epsilon$ on $[a, b]$. Let us choose two numbers $c < a$, and $\xi > 0$ satisfying

$$\frac{2KM}{|\gamma|} e^{\gamma(a-c)} < \frac{\epsilon}{2} \tag{7.2.5}$$

and

$$\frac{K}{|\gamma|}\xi(L + 1)[1 - e^{\gamma(b-c)}] < \frac{\epsilon}{2}. \tag{7.2.6}$$

Consider n sufficiently large such that $\|f(t + t_n, x) - f(t, x)\| < \xi$ and $\|\psi(t + t_n) - \psi(t)\| < \xi$ for $t \in [c, b]$ and $x \in G$. Then, the inequality

$$\|\Pi\psi(t + t_n) - \Pi\psi(t)\| \le \int_{-\infty}^{c} \|e^{A(t-s)}\| \|f(s + t_n, \psi(s + t_n)) - f(s, \psi(s))\|ds +$$

$$\int_{c}^{t} \|e^{A(t-s)}\| \|f(s + t_n, \psi(s + t_n)) - f(s, \psi(s))\|ds \le \int_{-\infty}^{c} 2KMe^{\gamma(t-s)}ds +$$

$$\int_{c}^{t} \xi(L + 1)Ke^{\gamma(t-s)}ds \le \frac{2}{|\gamma|}KMe^{\gamma(a-c)} + \frac{K}{|\gamma|}\xi(L + 1)[1 - e^{\gamma(b-c)}] \tag{7.2.7}$$

is correct for all $t \in [a, b]$.

By inequalities (7.2.5) and (7.2.6) it is true that $\|\Pi\psi(t + t_n) - \Pi\psi(t)\| < \epsilon$ for $t \in [a, b]$, and, therefore, $\Pi\psi(t + t_n) \to \Pi\psi(t)$ uniformly as $n \to \infty$ on each closed and bounded interval of $\mathbb{R}$.

It is easy to verify that $\Pi\psi(t)$ is a uniformly continuous function. Thus, the set U is invariant for the operator Π.

We proceed to show that the operator $\Pi : U \to U$ is contractive. Let $u(t)$ and $v(t)$ be members of U. Then, we obtain that

$$\|\Pi u(t) - \Pi v(t)\| \leq \int_{-\infty}^{t} \|e^{A(t-s)}\| \|f(s, u(s)) - f(s, v(s))\| ds \leq$$

$$\int_{-\infty}^{t} Ke^{\gamma(t-s)} L\|u(s) - v(s)\| ds \leq \frac{KL}{|\gamma|}\|u(t) - v(t)\|_1$$

for all $t \in \mathbb{R}$. Therefore, the inequality $\|\Pi u - \Pi v\|_1 \leq \dfrac{KL}{|\gamma|}\|u - v\|_1$ holds, and according to condition $(C4)$ the operator $\Pi : U \to U$ is contractive. The next task is to show that the space U is complete. Consider a Cauchy sequence $\phi_k(t)$ in U, which converges to a limit function $\phi(t)$ on $\mathbb{R}$. Fix a closed and bounded interval $I \subset \mathbb{R}$. We have that

$$\begin{aligned} &\|\phi(t + t_p) - \phi(t)\| < \|\phi(t + t_n) - \phi_k(t + t_n)\| + \\ &\|\phi_k(t + t_n) - \phi_k(t)\| + \|\phi_k(t) - \phi(t)\|. \end{aligned} \tag{7.2.8}$$

Now, one can take sufficiently large n and k such that each term on right-hand side of (7.2.8) is smaller than $\frac{\epsilon}{3}$ for an arbitrary small ϵ and $t \in I$. The inequality implies that $\|\phi(t + t_n) - \phi(t)\| < \epsilon$ on I. That is the sequence $\phi(t + t_n)$ uniformly converges to $\phi(t)$ on I. Likewise, one can check that the limit function is bounded and uniformly continuous [10]. The completeness of U is proved. By contraction mapping theorem there exists the unique fixed point, $\omega(t) \in U$, of the operator Π, which is the unique bounded solution of the system (7.1.2). The lemma is proved.

□

Theorem 7.1 ([5]) *If conditions $(C1)$–$(C4)$ are fulfilled, then system (7.1.2) admits a unique uniformly globally exponentially stable strongly unpredictable solution.*

Proof According to Lemma 7.2.1, system (7.1.2) has a unique solution $\omega(t) \in U$. What is left is to verify that the unpredictability property is valid.

One can show that there exist a positive number κ and natural numbers l and k such that the following inequalities are valid:

$$\kappa < \sigma, \tag{7.2.9}$$

$$\kappa\Big(1 - \frac{2}{k}(L + \|A\|)\Big) \geq \frac{3}{l}, \tag{7.2.10}$$

$$\|\omega(t+s) - \omega(t)\| < \frac{\epsilon_0}{k}, \qquad t \in \mathbb{R}, |s| < \kappa. \tag{7.2.11}$$

Let the numbers l, k and κ as well as a number $n \in \mathbb{N}$, and $i = 1, 2, \ldots, p$, be fixed.

Using the relations

$$\omega_i(t) = \omega_i(u_n) + \int_{u_n}^{t} \sum_{j=1}^{p} a_{ij}\omega_j(s)ds + \int_{u_n}^{t} f_i(s, \omega(s))ds$$

and

$$\omega_i(t + t_n) = \omega_i(u_n + t_n) + \int_{u_n}^{t} \sum_{j=1}^{p} a_{ij}\omega_j(s + t_n)ds + \int_{u_n}^{t} f_i(s + t_n, \omega(s + t_n))ds$$

we obtain that

$$\omega_i(t + t_n) - \omega_i(t) = \omega_i(u_n + t_n) - \omega_i(u_n) + \int_{u_n}^{t} \sum_{j=1}^{p} a_{ij}[\omega_j(s + t_n) - \omega_j(s)]ds +$$

$$\int_{u_n}^{t} [f_i(s + t_n, \omega(s + t_n)) - f_i(s, \omega(s))]ds. \tag{7.2.12}$$

Denote $\Delta = |\omega_i(u_n + t_n) - \omega_i(u_n)|$ and consider two cases (i) $\Delta \geq \epsilon_0/l$, (ii) $\Delta < \epsilon_0/l$ such that the remaining proof naturally falls into two parts.

(i) One can find positive number $\kappa_1 < \kappa$ such that $\|\omega(t) - \omega(s)\| < \frac{\epsilon_0}{4l}$ provided $|t - s| < \kappa_1$. Therefore, the inequality

$$|\omega_i(t_n + t) - \omega_i(t)| \geq |\omega_i(t_n + u_n) - \omega_i(u_n)| - |\omega_i(u_n) - \omega_i(t)| -$$
$$|\omega_i(t_n + t) - \omega_i(t_n + u_n)| \geq \frac{\epsilon_0}{l} - \frac{\epsilon_0}{4l} - \frac{\epsilon_0}{4l} = \frac{\epsilon_0}{2l}$$

is valid if $t \in [u_n - \kappa_1, u_n + \kappa_1]$.

(ii) It is true that $|f_i(t + u_n, x) - f_i(u_n, x)| \geq \epsilon_0$, for all $\|x\| < H$, $t \in [u_n - \sigma, u_n + \sigma]$.

Then, using relations (7.2.9)–(7.2.11) we get that

$$|\omega_i(t_n+t)-\omega_i(t)| \geq \int_{u_n}^{t} |f_i(t_n+s,\omega(t_n+s))-f_i(s,\omega(t_n+s))|ds -$$

$$\int_{u_n}^{t} |f_i(s,\omega(t_n+s))-f_i(s,\omega(s))|ds - \int_{u_n}^{t} |\sum_{j=1}^{p} a_{ij}[\omega_j(s)-\omega_j(t_n+s)]|ds -$$

$$|\omega_j(t_n+u_n)-\omega_j(u_n)| \geq \frac{\kappa}{2}\epsilon_0 - \kappa\frac{1}{k}\epsilon_0(L+\|A\|) - \frac{\epsilon_0}{l} \geq \frac{\epsilon_0}{2l},$$

for $t \in [u_n+\frac{\kappa}{2}, u_n+\kappa]$.

Thus, one can conclude that $\omega(t)$ is a strongly unpredictable function.

The asymptotical stability of the solution $\omega(t)$ can be verified as stability of a bounded solution in [10]. The theorem is proved. □

We have considered the problem of existence and uniqueness of strongly unpredictable solutions. In what follows, we will search for quasilinear systems with unpredictable solutions, which are not strongly unpredictable. For this reason, assume that the following condition is valid.

(C5) The function *f(t, x)* is unpredictable in the sense of Definition 7.2.

Theorem 7.2 ([5]) *Suppose that the conditions (C2)–(C5) hold. Then system (7.1.2) admits a unique uniformly globally exponentially stable unpredictable solution.*

Proof One can easily see, proceeding in the way of the last theorem, that there exists a unique solution $\omega(t) \in U$ for system (7.1.2). The solution is globally uniformly asymptotically stable. What is left is to show that the unpredictability property is valid.

We have that

$$\omega(t+t_n)-\omega(t) = \omega(u_n+t_n)-\omega(u_n) + \int_{u_n}^{t} A[\omega(s+t_n)-\omega(s)]ds$$

$$+\int_{u_n}^{t} [f(t_n+s,\omega(s+t_n))-f(s,\omega(s))]ds,\ t\in\mathbb{R}. \tag{7.2.13}$$

One can find a positive number κ, natural numbers l, k and $j = 1,\ldots,p$, such that

$$\kappa < \sigma, \tag{7.2.14}$$

$$\|\omega(t+s)-\omega(t)\| < \frac{\epsilon_0}{k}, \quad t\in\mathbb{R}, |s|<\kappa, \tag{7.2.15}$$

$$|f_j(t_n+u_n+s,x)-f_j(u_n+s,x)| \geq \epsilon_0/2,\ \|x\|<H,\ |s|<\kappa, n\in\mathbb{N}, \tag{7.2.16}$$

$$\kappa\Big(1/2 - (\frac{1}{l} + \frac{2}{k})(L + \|A\|)\Big) > \frac{3}{2l}. \tag{7.2.17}$$

Denote $\Delta = \|\omega(t_n + u_n) - \omega(u_n)\|$ and consider two alternative cases (i) $\Delta \geq \epsilon_0/l$ and (ii) $\Delta < \epsilon_0/l$.

(i) For the case $\Delta \geq \epsilon_0/l$, fix additionally a positive number κ_1 sufficiently small for

$$\|\omega(t+s) - \omega(t)\| < \frac{\epsilon_0}{4l}, \quad t \in \mathbb{R}, |s| < \kappa_1.$$

Therefore,

$$\begin{aligned}&\|\omega(t_n + t) - \omega(t)\| \geq \|\omega(t_n + u_n) - \omega(u_n)\| - \|\omega(u_n) - \omega(t)\| - \\ &\|\omega(t_n + t) - \omega(t_n + u_n)\| \geq \frac{\epsilon_0}{l} - \frac{\epsilon_0}{4l} - \frac{\epsilon_0}{4l} = \frac{\epsilon}{2l},\end{aligned} \tag{7.2.18}$$

if $t \in [u_n - \kappa_1, u_n + \kappa_1]$ and $n \in \mathbb{N}$.

(ii) One can find that from (7.2.15) it follows that

$$\|\omega(t_n + t) - \omega(t)\| < \frac{\epsilon_0}{l} + \frac{\epsilon_0}{k} + \frac{\epsilon_0}{k} = \epsilon_0\Big(\frac{1}{l} + \frac{2}{k}\Big) \tag{7.2.19}$$

if $t \in [u_n, u_n + \kappa]$.

We obtain from (7.2.14)–(7.2.17) that

$$\begin{aligned}&|\omega_j(t_n + t) - \omega_j(t)| \geq \int_{u_n}^{t} |f_j(t_n + s, \omega(t_n + s)) - f_j(s, \omega(t_n + s))|ds - \\ &\int_{u_n}^{t} |f_j(s, \omega(t_n + s)) - f_j(s, \omega(s))|ds - \int_{u_n}^{t} |\sum_{j=1}^{p} a_{ji}[\omega_i(s) - \omega_i(t_n + s)]|ds - \\ &|\omega_i(t_n + u_n) - \omega_i(u_n)| \geq \frac{\kappa}{2}\epsilon_0 - \kappa L\epsilon_0\Big(\frac{1}{l} + \frac{2}{k}\Big) - \kappa\|A\|\epsilon_0\Big(\frac{1}{l} + \frac{2}{k}\Big) - \frac{\epsilon_0}{l} \geq \frac{\epsilon_0}{2l}\end{aligned}$$

for $t \in [u_n + \kappa/2, u_n + \kappa]$.

Thus, the solution $\omega(t)$ is unpredictable. The asymptotical stability of the solution $\omega(t)$ can be verified as stability of a bounded solution in [10]. The theorem is proved. □

7.3 Examples

First, we will construct two examples of unpredictable functions.

Example 1 Let $\{\psi_i\}, i \in \mathbb{Z}$, be an unpredictable solution [2] of the logistic map

$$\lambda_{i+1} = \mu\lambda_i(1-\lambda_i) \tag{7.3.20}$$

with $\mu = 3.92$. The sequence belongs to the unit interval $[0,1]$. There exist a positive number ϵ_0 and sequences $\{\zeta_n\}$, $\{\eta_n\}$ both of which diverge to infinity such that $|\psi_{i+\zeta_n} - \psi_i| \to 0$ as $n \to \infty$ for each i in bounded intervals of integers and $|\psi_{\zeta_n+\eta_n} - \psi_{\eta_n}| \geq \epsilon_0$ for each $n \in \mathbb{N}$.

Let us take into account the function $\Theta : \mathbb{R} \to \mathbb{R}$ defined by

$$\Theta(t) = \int_{-\infty}^{t} e^{-2(t-s)}\Omega(s)ds, \; t \in \mathbb{R}, \tag{7.3.21}$$

where $\Omega(t)$ is the piecewise constant function defined on the real axis through the equation $\Omega(t) = \psi_i$ for $t \in [i, i+1)$, $i \in \mathbb{Z}$.

One can confirm that $\Theta(t)$ is bounded on the whole real axis such that $\sup\limits_{t\in\mathbb{R}} |\Theta(t)| \leq 1/2$.

Next, we will show that $\Theta(t)$ is an unpredictable scalar function.

Consider a fixed bounded and closed interval $[\alpha, \beta]$, of the axis and a positive number ϵ. Without loss of generality one can assume that α and β are integers. Let us fix a positive number ξ and an integer $\gamma < \alpha$, which satisfy the following inequalities $e^{-2(\alpha-\gamma)} < \frac{\epsilon}{2}$ and $\xi[1-e^{-2(\beta-\gamma)}] < \epsilon$. Let n be a large natural number such that $|\Omega(t+\zeta_n) - \Omega(t)| < \xi$ on $[\gamma, \beta]$. Then for all $t \in [\alpha, \beta]$ we obtain that

$$|\Theta(t+\zeta_n) - \Theta(t)| = |\int_{-\infty}^{t} e^{-2(t-s)}(\Omega(s+\zeta_n) - \Omega(s))ds| =$$

$$|\int_{-\infty}^{\gamma} e^{-2(t-s)}(\Omega(s+\zeta_n) - \Omega(s))ds + \int_{\gamma}^{t} e^{-2(t-s)}(\Omega(s+\zeta_n) - \Omega(s))ds| \leq$$

$$\int_{-\infty}^{\gamma} e^{-2(t-s)}2ds + \int_{\gamma}^{\beta} e^{-2(t-s)}\xi ds \leq e^{-2(\alpha-\gamma)} + \frac{\xi}{2}[1-e^{-2(\beta-\gamma)}] < \frac{\epsilon}{2} + \frac{\epsilon}{2} = \epsilon.$$

Thus, $|\Theta(t+\zeta_n) - \Theta(t)| \to 0$ as $n \to \infty$ uniformly on the interval $[\alpha, \beta]$. Moreover, the following inequalities are valid, $|\Omega(t+\zeta_n) - \Omega(t)| = |\psi_{\zeta_n+\eta_n} - \psi_{\eta_n}| \geq \epsilon_0$, $t \in [\eta_n, \eta_n+1)$, $n \in \mathbb{N}$.

Let us fix the numbers κ and n and consider two alternative cases: (i) $|\Theta(\eta_n + \zeta_n) - \Theta(\eta_n)| < \frac{\epsilon_0}{8}$ and (ii) $|\Theta(\eta_n+\zeta_n) - \Theta(\eta_n)| \geq \frac{\epsilon_0}{8}$.

(i) Let $\kappa < 1$ be a positive number satisfying equation $e^{-2\kappa} = \frac{2}{3}$. Using the relation

$$\Theta(t+\zeta_n) - \Theta(t) = \Theta(\eta_n+\zeta_n) - \Theta(\eta_n) + \int_{\eta_n}^{t} e^{-2(t-s)}(\Omega(s+\zeta_n) - \Omega(s))ds \tag{7.3.22}$$

we obtain that

$$|\Theta(t+\zeta_n) - \Theta(t)| \geq \int_{\eta_n}^{t} e^{-2(t-s)} |\Omega(s+\zeta_n) - \Omega(s))| ds$$

$$-|\Theta(\eta_n+\zeta_n) - \Theta(\eta_n)| \geq$$

$$\int_{\eta_n}^{t} e^{-2(t-s)} \epsilon_0 ds - \frac{\epsilon_0}{8} \geq \frac{\epsilon_0}{2}(1-e^{-2\kappa}) - \frac{\epsilon_0}{8} = \frac{\epsilon_0}{24}$$

for $t \in [\eta_n + \kappa, \eta_n + 1)$.

(ii) Let $\kappa < 1$ be a positive number satisfying equation $1 - e^{-2\kappa} = \frac{\epsilon_0}{12}$. From the relation (7.3.22) we get

$$|\Theta(t+\zeta_n) - \Theta(t)| \geq |\Theta(\eta_n+\zeta_n)$$

$$-\Theta(\eta_n)| - |\int_{\eta_n}^{t} e^{-2(t-s)} (\Omega(s+\zeta_n) - \Omega(s)) ds|$$

$$\geq \frac{\epsilon_0}{8} - \int_{\eta_n}^{t} e^{-2(t-s)} 2ds \geq \frac{\epsilon_0}{8} - [1 - e^{-2\kappa}] = \frac{\epsilon_0}{24}$$

for $t \in [\eta_n, \eta_n + \kappa)$.

Thus, $\Theta(t)$ is an (strongly) unpredictable function.

Example 2 ([5]) Consider the function $g(t,x) = (arctan(x) + 2)\Theta(t)$ of two variables t and x, where $\Theta(t)$ is the function from defined by Eq. (7.3.21). The function $g(t,x)$ is continuously differentiable if $t \neq i, i \in \mathbb{Z}$, and it is bounded such that $\sup_{\mathbb{R}\times G} |g(t,x)| = \frac{\pi}{4} + 1$. Moreover, $\sup_{\mathbb{R}\times G} |\frac{\partial g(t,x)}{\partial x}| = 1/2, t \neq i, i \in \mathbb{Z}$.

Let us fix an arbitrary compact interval $I \subset \mathbb{R}$ and positive number ϵ. We have that $|\Theta(t+t_n) - \Theta(t)| < \epsilon$ for $t \in I$ sufficiently large n. Consequently,

$$|g(t+t_n,x) - g(t,x)| \leq |arctan(x) + 2||\Theta(t+t_n) - \Theta(t)| < (\frac{\pi}{2} + 2)\epsilon.$$

Hence, $g(t+t_n,x) \to g(t,x)$ as $n \to \infty$ uniformly for $(t,x) \in I \times G$.

On the other hand, it is true that $|\Theta(t+t_n) - \Theta(t)| \geq \bar{\epsilon}_0$ for all $t \in [u_n - \kappa, u_n + \kappa]$ and $n \in \mathbb{N}$. Therefore, we have

$$|g(t+t_n,x) - g(t,x)| = |arctan(x) + 2||\Theta(t+\zeta_n) - \Theta(t)| \geq (-\frac{\pi}{2} + 2)\bar{\epsilon}_0,$$

for every $(t,x) \in [u_n - \kappa, u_n + \kappa] \times G, n \in \mathbb{N}$. Thus, $g(t,x)$ is an unpredictable (strongly) in t function.

Example 3 ([5]) Let us consider the system of differential equations

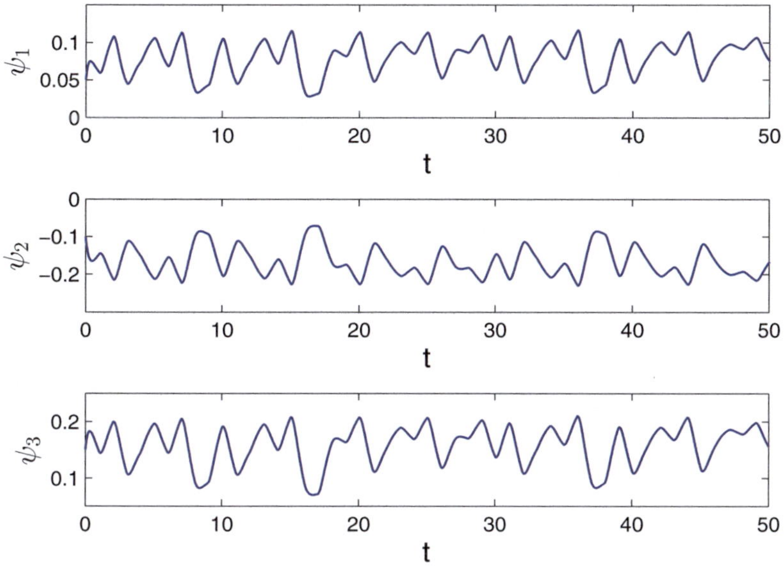

Fig. 7.1 The coordinates of the solution, $\psi(t)$, of system (7.3.23)

$$\begin{aligned} x_1' &= -3x_1 - x_2 - x_3 + 0.51g(t, x_3) \\ x_2' &= -x_1 - 3x_2 - x_3 - 0.62g(t, x_1) \\ x_3' &= x_1 + x_2 - x_3 + 0.51g(t, x_2), \end{aligned} \tag{7.3.23}$$

where $g(t, x)$ is the function from the previous example. The eigenvalues of the matrix of coefficients are -2 and -3. It can be verified that conditions $(C2)$–$(C5)$ are valid for system (7.3.23) with $\gamma = -2$, $K = 6$, and $L = 0.31$. According to Theorem 7.1, there exists the unique asymptotically stable unpredictable solution of system (7.3.23). The simulation results for the solution of (7.3.23) with initial data $\psi_1(0) = 0.05$, $\psi_2(0) = -0.1$, $\psi_3(0) = 0.15$ are depicted in Fig. 7.1. The figure confirms the irregularity of the dynamics.

References

1. M. Akhmet, M.O. Fen, Existence of unpredictable solutions and chaos. Turk. J. Math. **41**, 254–266 (2017)
2. M. Akhmet, M.O. Fen, Poincaré chaos and unpredictable functions. Commun. Nonlinear Sci. Numer. Simul. **48**, 85–94 (2017)
3. M. Akhmet, M.O. Fen, Non-autonomous equations with unpredictable solutions. Commun. Nonlinear Sci. Numer. Simul. **59**, 657–670 (2018)
4. M. Akhmet, M.O. Fen, M. Tleubergenova, A. Zhamanshin, Poincaré chaos for a hyperbolic quasilinear system. Miskolc Math. Notes **20**, 33–44 (2019)

5. M. Akhmet, M. Tleubergenova, A. Zhamanshin, Quasilinear differential equations with strongly unpredictable solutions. Carpathian J. Math. (in press)
6. H.A. Bohr, *Almost Periodic Functions* (Chelsea Publishing Company, 1947)
7. C. Corduneanu, *Almost Periodic Oscillations and Waves* (Springer, Berlin, 2009)
8. M. Farkas, *Periodic Motions* (Springer, New York, 1994)
9. A.M. Fink, *Almost Periodic Differential Equations* (Springer, New York, 1974)
10. P. Hartman, *Ordinary Differential Equations* (SIAM, New York, 2002)
11. Y. Hino, T. Naito, N. VanMinh, J.S. Shin, *Almost Periodic Solutions of Differential Equations in Banach Spaces* (CRC Press, 2001)

Chapter 8
Li–Yorke Chaos in Hybrid Systems on a Time Scale

In this chapter, we apply the method of replication of chaos introduced and developed in our studies [1–9, 11–13, 15–17, 19]. We rigorously prove the presence of chaos in dynamic equations on time scales by using the reduction technique to impulsive differential equations [20]. The results are based on the Li–Yorke definition of chaos. An illustrative example is presented by means of a Duffing equation on a time scale.

8.1 Introduction

The concept of chaos has come into prominence starting with the studies of Poincaré [23], Cartwright and Littlewood [31], Levinson [41], Lorenz [44], and Ueda [50]. Another subject that is also popular is the theory of time scales, which is first presented by Hilger [35]. Both concepts have many applications in various disciplines such as mechanics, electronics, neural networks, population models, and economics. See, for instance, [25, 26, 32, 33, 45, 46, 48, 49, 51] and the references therein.

Dynamic equations on time scales (DETS) have been extensively investigated in the literature [26, 38]. However, to the best of our knowledge, the presence of chaos has never been achieved in DETS. Motivated by the deficiency of mathematical methods for the investigation of chaos in such equations, we suggest the results of the present chapter.

The studies [2, 4, 5, 10–12, 14] were concerned with the extension of chaos in continuous-time systems that possess asymptotically stable and hyperbolic equilibria as well as orbitally stable limit cycles. It was found in these papers that the solutions admit the same type of chaos as the perturbations. The paper [2] deals with the general technique of dynamical synthesis, which was developed in [27–30].

M. Akhmet et al., *Dynamics with Chaos and Fractals*, Nonlinear Systems and Complexity 29, https://doi.org/10.1007/978-3-030-35854-9_8

In the present chapter, we develop the concept of Li–Yorke chaos for DETS and prove its existence rigorously. Our results are appropriate to obtain chaotic DETS with arbitrary high dimensions.

Throughout the chapter, we will denote by $\mathbb{R}$, $\mathbb{Z}$, and $\mathbb{N}$ the sets of real numbers, integers, and natural numbers, respectively. In this chapter, we consider the following equation:

$$y^{\Delta}(t) = Ay(t) + f(t, y(t)) + g(t, \zeta), \ t \in \mathbb{T}_0, \tag{8.1.1}$$

where A is a constant $n \times n$ real valued matrix, the function $f : \mathbb{T}_0 \times \mathbb{R}^n \to \mathbb{R}^n$ is rd-continuous and the function $g(t, \zeta)$ is defined through the equation $g(t, \zeta) = \zeta_k$ for $t \in [\theta_{2k-1}, \theta_{2k}]$, $k \in \mathbb{Z}$, such that $\zeta = \{\zeta_k\}$ is a sequence generated by the map

$$\zeta_{k+1} = F(\zeta_k), \tag{8.1.2}$$

where $\zeta_0 \in \Lambda$, $F : \Lambda \to \Lambda$ is a continuous function and Λ is a compact subset of $\mathbb{R}^n$. In Eq. (8.1.1) the time scale $\mathbb{T}_0$ is defined as $\mathbb{T}_0 = \bigcup_{k=-\infty}^{\infty} [\theta_{2k-1}, \theta_{2k}]$ in which $\{\theta_k\}$ is a strictly increasing sequence of real numbers such that $|\theta_k| \to \infty$ as $|k| \to \infty$ and $\sum_{-\infty}(\theta_{2k} - \theta_{2k-1}) = \infty$, $\sum^{\infty}(\theta_{2k} - \theta_{2k-1}) = \infty$.

In the present chapter, we investigate the existence of chaos in the dynamics of Eq. (8.1.1). The system under discussion is a hybrid one, since it combines the continuous dynamics on the time scale with the discrete equation used in the right-hand side of the system. We theoretically prove that chaos exists in (8.1.1) provided that the map (8.1.2) is chaotic. For that purpose, we make use of the reduction technique to impulsive differential equations, which was presented by Akhmet and Turan [20]. As far as we know, there is no paper on chaos in dynamics on time scales. The reason is that the dynamics is essentially non-autonomous and it is difficult to verify the ingredients of chaos for unspecified time scales. That is why we utilize the time scale introduced in the papers [20, 21] and the method of reduction of the dynamics to impulsive differential equations [20].

The rest of this chapter is organized as follows. In Sect. 8.2, some preliminary results as well as basic concepts about DETS are mentioned. Section 8.3 is devoted to the bounded solutions of (8.1.1). In Sect. 8.4, we give the description of the chaos of equation (8.1.1) and prove its presence rigorously. An example concerning Duffing equations on a time scale is presented in Sect. 8.5 to support the theoretical results. Finally, some concluding remarks are given in Sect. 8.6.

8.2 Preliminaries

The basic concepts on differential equations on time scales that are needed in this chapter are as follows [26, 38–40]. A time scale is a nonempty closed subset of $\mathbb{R}$. On a time scale $\mathbb{T}$, the forward and backward jump operators are defined as $\sigma(t) = \inf\{s \in \mathbb{T} : s > t\}$ and $\rho(t) = \sup\{s \in \mathbb{T} : s < t\}$, respectively. We say that a point

$t \in \mathbb{T}$ is right-scattered if $\sigma(t) > t$ and right-dense if $\sigma(t) = t$. In a similar way, if $\rho(t) < t$, then $t \in \mathbb{T}$ is called left-scattered, and otherwise it is called left-dense. Besides, a function $h : \mathbb{T} \times \mathbb{R}^n \to \mathbb{R}^n$ is called rd-continuous if it is continuous at each $(t, u) \in \mathbb{T} \times \mathbb{R}^n$ with right-dense t, and the limits $\lim_{(r,\nu)\to(t^-,u)} h(r, \nu) = h(t-, u)$ and $\lim_{\nu\to u} h(t, \nu) = h(t, u)$ exist at each (t, u) with left-dense t. At a right-scattered point $t \in \mathbb{T}$, the Δ-derivative of a continuous function ϑ is defined as $\vartheta^{\Delta}(t) = \dfrac{\vartheta(\sigma(t)) - \vartheta(t)}{\sigma(t) - t}$. On the other hand, at a right-dense point t, we have $\vartheta^{\Delta}(t) = \lim_{r\to t} \dfrac{\vartheta(t) - \vartheta(r)}{t - r}$ provided that the limit exists.

It is worth noting that on the time scale $\mathbb{T}_0$ used in system (8.1.1) the points $\theta_{2k-1}, k \in \mathbb{Z}$, are left-scattered and right-dense, and the points $\theta_{2k}, k \in \mathbb{Z}$, are right-scattered and left-dense. Moreover, $\sigma(\theta_{2k}) = \theta_{2k+1}$, $\rho(\theta_{2k+1}) = \theta_{2k}$, $k \in \mathbb{Z}$, and $\sigma(t) = \rho(t) = t$ for any $t \in \mathbb{T}_0$ except at the points θ_k, $k \in \mathbb{Z}$.

Suppose that the time scale $\mathbb{T}_0$ used in the description of Eq. (8.1.1) satisfies the ω-property. That is, there exists a number $\omega > 0$ such that $t + \omega \in \mathbb{T}_0$ whenever $t \in \mathbb{T}_0$. In this case, there exists a natural number p such that $\delta_{k+p} = \delta_k$ for all $k \in \mathbb{Z}$, where $\delta_k = \theta_{2k+1} - \theta_{2k}$ [20]. Suppose that p is the minimal among those numbers.

We assume without loss of generality that $\theta_{-1} < 0 < \theta_0$. Define on the set $\mathbb{T}_0' = \mathbb{T}_0 \setminus \bigcup_{k=-\infty}^{\infty} \{\theta_{2k-1}\}$ the ψ-substitution [20] as

$$\psi(t) = \begin{cases} t - \sum\limits_{0<\theta_{2k}<t} \delta_k, & t \geq 0, \\ t + \sum\limits_{t\leq\theta_{2k}<0} \delta_k, & t < 0. \end{cases} \tag{8.2.3}$$

The function $\psi(t)$ is one-to-one, $\psi(0) = 0$, $\psi(\mathbb{T}_0') = \mathbb{R}$ and $\lim_{t\to\infty,\, t\in\mathbb{T}_0'} \psi(t) = \infty$. According to the results of the paper [20], $d\psi(t)/dt = 1$, $t \in \mathbb{T}_0'$, and $d\psi^{-1}(s)/ds = 1$ provided that $s \neq s_k$, $k \in \mathbb{Z}$, where

$$\psi^{-1}(s) = \begin{cases} s + \sum\limits_{0<s_k<s} \delta_k, & s \geq 0, \\ s - \sum\limits_{s\leq s_k<0} \delta_k, & s < 0, \end{cases} \tag{8.2.4}$$

and the sequence $\{s_k\}$, $k \in \mathbb{Z}$, is defined through the equation $s_k = \psi(\theta_{2k})$. The function ψ^{-1} is piecewise continuous with discontinuities of the first kind at the points s_k, $k \in \mathbb{Z}$, such that $\psi^{-1}(s_k+) - \psi^{-1}(s_k) = \delta_k$, where $\psi^{-1}(s_k+) = \lim_{s\to s_k^+} \psi^{-1}(s)$, the sequence $\{s_k\}$ is $(\psi(\omega), p)$-periodic, i.e., $s_{k+p} = s_k + \psi(\omega)$ for all $k \in \mathbb{Z}$, and $\psi(t + \omega) = \psi(t) + \psi(\omega)$, $t \in \mathbb{T}_0'$. Moreover, if a function $h(t)$ is ω-periodic on $\mathbb{T}_0$, then $h(\psi^{-1}(s))$ is $\psi(\omega)$-periodic, and vice versa.

Let us denote by $\mathscr{C}_{rd}(\mathbb{T}_0)$ the set of all functions which are rd-continuous on $\mathbb{T}_0$, and let $\mathscr{C}_{rd}^1(\mathbb{T}_0) \subset \mathscr{C}_{rd}(\mathbb{T}_0)$ be the set of all continuously differentiable functions on $\mathbb{T}_0$, assuming that the functions have a one sided derivative at θ_k, $k \in \mathbb{Z}$. On the other hand, we say that a function defined on $\mathbb{R}$ is an element of the set $\mathscr{PC}_0$ if it is left-continuous on $\mathbb{R}$ and continuous on $\mathbb{R} \setminus \bigcup_{k=-\infty}^{\infty} \{s_k\}$, and it has discontinuities of the first kind at the points s_k, $k \in \mathbb{Z}$. Moreover, a function $h : \mathbb{R} \to \mathbb{R}^n$ belongs to the set $\mathscr{PC}_0^1$ if both h and h' are elements of $\mathscr{PC}_0$, where $h'(s_k) = \lim_{s \to s_k^-} \dfrac{h(s) - h(s_k)}{s - s_k}$, $k \in \mathbb{Z}$. It was shown by Akhmet and Turan [20] that a function $\vartheta(t)$ belongs to $\mathscr{C}_{rd}(\mathbb{T}_0)$ $\left(\mathscr{C}_{rd}^1(\mathbb{T}_0)\right)$ if and only if $\vartheta(\psi^{-1}(s))$ belongs to $\mathscr{PC}_0 \left(\mathscr{PC}_0^1\right)$.

In accordance with the equation $y^{\Delta}(\theta_{2k}) = \dfrac{y(\theta_{2k+1}) - y(\theta_{2k})}{\theta_{2k+1} - \theta_{2k}}$, $k \in \mathbb{Z}$, system (8.1.1) can be written as

$$\begin{aligned} y'(t) &= Ay(t) + f(t, y(t)) + g(t, \zeta), \; t \in \mathbb{T}_0, \\ y(\theta_{2k+1}) &= \delta_k Ay(\theta_{2k}) + f(\theta_{2k}, y(\theta_{2k}))\delta_k + \zeta_k \delta_k + y(\theta_{2k}). \end{aligned} \tag{8.2.5}$$

Applying the transformation $s = \psi(t)$ to (8.2.5) we obtain the following impulsive system:

$$\begin{aligned} x'(s) &= Ax(s) + f(\psi^{-1}(s), x(s)) + g(\psi^{-1}(s), \zeta), \; s \neq s_k, \\ \Delta x|_{s=s_k} &= \delta_k Ax(s_k) + f(\psi^{-1}(s_k), x(s_k))\delta_k + \zeta_k \delta_k, \end{aligned} \tag{8.2.6}$$

where $x(s) = y(\psi^{-1}(s))$, $\Delta x|_{s=s_k} = x(s_k+) - x(s_k)$, $k \in \mathbb{Z}$, and $x(s_k+) = \lim_{s \to s_k^+} x(s)$.

In what follows, we will make use of the usual Euclidean norm for vectors and the norm induced by the Euclidean norm for square matrices [36].

The following conditions are required throughout the paper.

(C1) $\det(I + \delta_k A) \neq 0$ for all $k \in \mathbb{Z}$, where I is the $n \times n$ identity matrix;
(C2) All eigenvalues of the matrix $e^{\psi(\omega)A} \Pi_{j=0}^{p-1} (I + \delta_j A)$ lie inside the unit circle;
(C3) There exists a positive number M_f such that $\sup_{t \in \mathbb{T}_0,\, y \in \mathbb{R}^n} \|f(t, y)\| \leq M_f$;
(C4) There exists a positive number L_f such that $\|f(t, y_1) - f(t, y_2)\| \leq L_f \|y_1 - y_2\|$ for all $t \in \mathbb{T}_0$ and $y_1, y_2 \in \mathbb{R}^n$.

Let us denote by $X(s, r)$ the transition matrix of the linear homogeneous system

$$\begin{aligned} x'(s) &= Ax(s), \; s \neq s_k, \\ \Delta x|_{s=s_k} &= \delta_k Ax(s_k). \end{aligned} \tag{8.2.7}$$

Under the conditions $(C1)$ and $(C2)$ there exist positive numbers N and λ such that $\|X(s, r)\| \leq Ne^{-\lambda(s-r)}$ for $s \geq r$ [9, 47].

The following conditions are also needed.

(C5) $NL_f\left(\frac{1}{\lambda}+\frac{p\bar{\delta}}{1-e^{-\lambda\psi(\omega)}}\right)<1$, where $\bar{\delta}=\max\limits_{0\le k\le p-1}\delta_k$;

(C6) $-\lambda+NL_f+\frac{p}{\psi(\omega)}\ln\left(1+NL_f\bar{\delta}\right)<0$;

(C7) $f(t+\omega,y)=f(t,y)$ for all $(t,y)\in\mathbb{T}_0\times\mathbb{R}^n$.

The next section is devoted to the bounded solutions of system (8.1.1).

8.3 Bounded Solutions

Under the conditions $(C1)$–$(C5)$, one can verify by using the results of [9, 47] that for a fixed sequence $\zeta=\{\zeta_k\}$, $k\in\mathbb{Z}$, there exists a unique bounded on $\mathbb{R}$ solution $\phi_\zeta(s)$ of (8.2.6), which satisfies the relation

$$\begin{aligned}\phi_\zeta(s)&=\int_{-\infty}^{s}X(s,r)\left[f\left(\psi^{-1}(r),\phi_\zeta(r)\right)+g\big(\psi^{-1}(r),\zeta\big)\right]dr\\&+\sum_{-\infty<s_k<s}X(s,s_k+)\left[f\big(\psi^{-1}(s_k),\phi_\zeta(s_k)\big)+\zeta_k\right]\delta_k.\end{aligned}\tag{8.3.8}$$

Moreover, $\sup\limits_{s\in\mathbb{R}}\left\|\phi_\zeta(s)\right\|\le K_0$, where

$$K_0=N(M_f+M_F)\Big(\frac{1}{\lambda}+\frac{p\bar{\delta}}{1-e^{-\lambda\psi(\omega)}}\Big)$$

and

$$M_F=\max_{\eta\in\Lambda}\|F(\eta)\|.$$

Therefore, for a fixed sequence $\zeta=\{\zeta_k\}$, the function $\varphi_\zeta(t)=\phi_\zeta(\psi(t))$ satisfying $\varphi_\zeta(\theta_{2k+1})=\phi_\zeta(s_k+)$, $k\in\mathbb{Z}$, is the unique solution of (8.2.5), and hence of (8.1.1), which is bounded on $\mathbb{T}_0$ such that $\sup\limits_{t\in\mathbb{T}_0}\left\|\varphi_\zeta(t)\right\|\le K_0$.

We say that the bounded solution $\varphi_\zeta(t)$ attracts a solution $y(t)$ of (8.1.1) if $\left\|y(t)-\varphi_\zeta(t)\right\|\to 0$ as $t\to\infty$, $t\in\mathbb{T}_0$. The attractiveness feature of the bounded solutions of (8.1.1) is mentioned in the next assertion.

Lemma 8.3.1 ([18]) *If the conditions* $(C1)$–$(C6)$ *are valid, then for a fixed sequence* ζ*, the bounded solution* $\varphi_\zeta(t)$ *attracts all other solutions of (8.1.1).*

Proof Consider an arbitrary solution $y(t)$, $y(t^0)=y_0$, of (8.1.1) for some $t^0\in\mathbb{T}_0$ and $y_0\in\mathbb{R}^n$. Assume without loss of generality that $t^0\ne\theta_{2k-1}$ for any $k\in\mathbb{Z}$. Let $s^0=\psi(t^0)$ and $x(s)=y(\psi^{-1}(s))$. The relation

$$
\begin{aligned}
x(s) - \phi_\zeta(s) &= X(s, s^0)(y_0 - \phi_\zeta(s^0)) \\
&+ \int_{s^0}^{s} \Big[f\big(\psi^{-1}(r), x(r)\big) - f\big(\psi^{-1}(r), \phi_\zeta(r)\big) \Big] dr \\
&+ \sum_{s^0 \le s_k < s} X(s, s_k+) \Big[f\big(\psi^{-1}(s_k), x(s_k)\big) - f\big(\psi^{-1}(s_k), \phi_\zeta(s_k)\big) \Big] \delta_k
\end{aligned}
$$

implies for $s \ge s^0$ that

$$
\begin{aligned}
\left\|x(s) - \phi_\zeta(s)\right\| \le N e^{-\lambda(s-s^0)} \left\| y_0 - \phi_\zeta(s^0) \right\| + \int_{s^0}^{s} N L_f e^{-\lambda(s-r)} \left\| x(r) - \phi_\zeta(r) \right\| dr \\
+ \sum_{s^0 \le s_k < s} N L_f \bar{\delta} e^{-\lambda(s-s_k)} \left\| x(s_k) - \phi_\zeta(s_k) \right\| .
\end{aligned}
$$

Applying the Gronwall–Bellman Lemma for piecewise continuous functions [9] to the last inequality, one can obtain for $s \ge s^0$ that

$$
\left\|x(s) - \phi_\zeta(s)\right\| \le N(1+NL_f\bar{\delta})^p \left\| y_0 - \phi_\zeta(s^0) \right\| e^{[-\lambda + NL_f + p\ln(1+NL_f\bar{\delta})/\psi(\omega)](s-s^0)}.
$$

Therefore, we have for $t \ge t^0$, $t \in \mathbb{T}_0$, that

$$
\begin{aligned}
\left\| y(t) - \varphi_\zeta(t) \right\| &\le N(1 + NL_f\bar{\delta})^p \left\| y_0 - \varphi_\zeta(t^0) \right\| \\
&\quad \times e^{[-\lambda + NL_f + p\ln(1+NL_f\bar{\delta})/\psi(\omega)](\psi(t) - \psi(t^0))}.
\end{aligned}
$$

Consequently, $\left\| y(t) - \varphi_\zeta(t) \right\| \to 0$ as $t \to \infty$, $t \in \mathbb{T}_0$. □

In the next section, we will deal with the presence of chaos in system (8.1.1).

8.4 The Chaotic Dynamics

The map (8.1.2) is called Li–Yorke chaotic on Λ if [22, 24, 37, 42, 43]: (i) For every natural number p_0, there exists a p_0-periodic point of F in Λ; (ii) There is an uncountable set $\mathscr{S} \subset \Lambda$, the scrambled set, containing no periodic points, such that for every $\zeta^1, \zeta^2 \in \mathscr{S}$ with $\zeta^1 \neq \zeta^2$, we have $\limsup_{k\to\infty} \left\| F^k(\zeta^1) - F^k(\zeta^2) \right\| > 0$ and $\liminf_{k\to\infty} \left\| F^k(\zeta^1) - F^k(\zeta^2) \right\| = 0$; (iii) For every $\zeta^1 \in \mathscr{S}$ and a periodic point $\zeta^2 \in \Lambda$, we have $\limsup_{k\to\infty} \left\| F^k(\zeta^1) - F^k(\zeta^2) \right\| > 0$.

Let us denote by Θ the set of all sequences $\zeta = \{\zeta_k\}$, $k \in \mathbb{Z}$, obtained by Eq. (8.1.2). A pair of sequences $\zeta = \{\zeta_k\}$, $\tilde{\zeta} = \left\{\tilde{\zeta}_k\right\} \in \Theta$ is proximal if $\liminf_{k\to\infty} \|\zeta_k - \tilde{\zeta}_k\| = 0$. Moreover, the pair is frequently separated if $\limsup_{k\to\infty} \|\zeta_k - \tilde{\zeta}_k\| > 0$.

We say that a pair $\varphi_\zeta(t)$, $\varphi_{\tilde{\zeta}}(t)$ of bounded solutions of (8.1.1) is proximal if for an arbitrary small real number $\epsilon > 0$ and an arbitrary large natural number E, there exists an integer m such that $\|\varphi_\zeta(t) - \varphi_{\tilde{\zeta}}(t)\| < \epsilon$ for all $t \in [\theta_{2m-1}, \theta_{2(m+E)}] \cap \mathbb{T}_0$. On the other hand, the pair $\varphi_\zeta(t)$, $\varphi_{\tilde{\zeta}}(t)$ is frequently (ϵ_0, Δ)-separated if there exist numbers $\epsilon_0 > 0$, $\Delta > 0$ and infinitely many disjoint intervals $J_q \subset \mathbb{T}_0$, $q \in \mathbb{N}$, each with a length no less than Δ, such that $\|\varphi_\zeta(t) - \varphi_{\tilde{\zeta}}(t)\| > \epsilon_0$ for each t from these intervals. Furthermore, a pair $\varphi_\zeta(t)$, $\varphi_{\tilde{\zeta}}(t)$ of solutions of (8.1.1) is called a Li–Yorke pair if it is proximal and frequently (ϵ_0, Δ)-separated for some positive numbers ϵ_0 and Δ.

Let $\mathscr{A}$ be the collection of all bounded solutions $\varphi_\zeta(t)$ of (8.1.1) such that $\zeta \in \Theta$. The description of Li–Yorke chaos for system (8.1.1) is as follows.

Definition 8.1 ([18]) System (8.1.1) is called Li–Yorke chaotic if:

(i) There exists an $m\omega$-periodic solution of (8.1.1) for each $m \in \mathbb{N}$;
(ii) There exists an uncountable set $\Sigma \subset \mathscr{A}$, the scrambled set, which does not contain any periodic solution, such that any pair of different solutions of (8.1.1) inside Σ is a Li–Yorke pair;
(iii) For any $\varphi_\zeta(t) \in \Sigma$ and any periodic solution $\varphi_{\hat{\zeta}}(t) \in \mathscr{A}$, the pair $\varphi_\zeta(t)$, $\varphi_{\hat{\zeta}}(t)$ is frequently (ϵ_0, Δ)-separated for some positive numbers ϵ_0 and Δ.

One can verify that the sequence $\{\kappa_k\}$ defined through the equation $\kappa_k = \theta_{2k} - \theta_{2k-1}$, $k \in \mathbb{Z}$, is p-periodic. In what follows, we will denote $\underline{\kappa} = \min_{0\le k\le p-1} \kappa_k$ and $\overline{\kappa} = \max_{0\le k\le p-1} \kappa_k$. Moreover, let $i((a_0, b_0))$ be the number of the terms of the sequence $\{s_k\}$ that belong to the interval (a_0, b_0), where $a_0, b_0 \in \mathbb{R}$ with $a_0 < b_0$. One can verify that $i((a_0, b_0)) \le p + \dfrac{p}{\psi(\omega)}(b_0 - a_0)$.

The next assertion is about the proximality feature of bounded solutions of Eq. (8.1.1).

Lemma 8.4.2 ([18]) *Suppose that the conditions* $(C1)$–$(C6)$ *are fulfilled. If a pair of sequences* $\zeta, \tilde{\zeta} \in \Theta$ *is proximal, then the same is true for the pair* $\varphi_\zeta(t), \varphi_{\tilde{\zeta}}(t) \in \mathscr{A}$.

Proof Set $R_1 = 2N(M_f + M_F)\Big(\dfrac{1}{\lambda} + \dfrac{p\bar{\delta}}{1 - e^{-\lambda\psi(\omega)}}\Big)$ and $\alpha = \lambda - NL_f - \dfrac{p}{\psi(\omega)} \ln(1 + NL_f\bar{\delta})$. Suppose that γ is a real number which satisfies the inequality

$$\gamma \ge 1 + N\Big(\frac{1}{\lambda} + \frac{\bar{\delta}p}{1 - e^{-\lambda\psi(\omega)}}\Big)\Big(1 + \frac{NL_f(1 + NL_f\bar{\delta})^p}{\alpha} + \frac{NL_f\bar{\delta}p(1 + NL_f\bar{\delta})^p}{1 - e^{-\alpha\psi(\omega)}}\Big).$$

Fix an arbitrary small number $\epsilon > 0$ and an arbitrary large natural number E such that

$$E \geq \frac{1}{\alpha \underline{\kappa}} \ln\Big(\frac{\gamma R_1(1+NL_f\bar{\delta})^p}{\epsilon}\Big).$$

We assume without loss of generality that $\epsilon \leq 2M_F$. Since the pair ζ, $\tilde{\zeta}$ is proximal, there exists an integer k_0 such that

$$\Big\| g\big(t,\zeta\big) - g\big(t,\tilde{\zeta}\big)\Big\| < \epsilon/\gamma$$

for $t \in [\theta_{2k_0-1}, \theta_{2(k_0+2E)}] \cap \mathbb{T}_0$. In this case,

$$\Big\| g\big(\psi^{-1}(s),\zeta\big) - g\big(\psi^{-1}(s),\tilde{\zeta}\big)\Big\| < \epsilon/\gamma$$

for $s \in (s_{k_0-1}, s_{k_0+2E}]$.

The bounded solutions $\phi_\zeta(s) = \varphi_\zeta(\psi^{-1}(s))$ and $\phi_{\tilde{\zeta}}(s) = \varphi_{\tilde{\zeta}}(\psi^{-1}(s))$ of (8.2.6) satisfy the relation

$$\begin{aligned}
\phi_\zeta(s) - \phi_{\tilde{\zeta}}(s) = &\int_{-\infty}^{s} X(s,r)\Big[f\big(\psi^{-1}(r),\phi_\zeta(r)\big) - f\big(\psi^{-1}(r),\phi_{\tilde{\zeta}}(r)\big)\\
&+g\big(\psi^{-1}(r),\zeta\big) - g\big(\psi^{-1}(r),\tilde{\zeta}\big)\Big]dr\\
&+ \sum_{-\infty<s_k<s} X(s,s_k+)\Big[f\big(\psi^{-1}(s_k),\phi_\zeta(s_k)\big) - f\big(\psi^{-1}(s_k),\phi_{\tilde{\zeta}}(s_k)\big) + \zeta_k - \tilde{\zeta}_k\Big]\delta_k.
\end{aligned}$$

Thus, for $s \in (s_{k_0-1}, s_{k_0+2E}]$, we have that

$$\begin{aligned}
\Big\|\phi_\zeta(s) - \phi_{\tilde{\zeta}}(s)\Big\| \leq\ & R_1 e^{-\lambda(s-s_{k_0-1})} + \frac{N\epsilon}{\gamma\lambda}\Big(1 - e^{-\lambda(s-s_{k_0-1})}\Big)\\
&+\frac{N\bar{\delta}p\epsilon}{\gamma(1-e^{-\lambda\psi(\omega)})}\Big(1 - e^{-\lambda(s-s_{k_0-1}+\psi(\omega))}\Big)\\
&+\int_{s_{k_0-1}}^{s} NL_f e^{-\lambda(s-r)}\Big\|\phi_\zeta(r) - \phi_{\tilde{\zeta}}(r)\Big\|dr\\
&+\sum_{s_{k_0-1}<s_k<s} NL_f\bar{\delta}e^{-\lambda(s-s_k)}\Big\|\phi_\zeta(s_k) - \phi_{\tilde{\zeta}}(s_k)\Big\|.
\end{aligned} \qquad (8.4.9)$$

Let us define the functions $u(s) = e^{\lambda s}\Big\|\phi_\zeta(s) - \phi_{\tilde{\zeta}}(s)\Big\|$ and $v(s) = \beta_1 + \beta_2 e^{\lambda s}$, where

$$\beta_1 = R_1 e^{\lambda s_{k_0-1}} - \frac{N\epsilon}{\gamma\lambda}e^{\lambda s_{k_0-1}} - \frac{N\bar{\delta}p\epsilon}{\gamma(1-e^{-\lambda\psi(\omega)})}e^{\lambda(s_{k_0-1}-\psi(\omega))}$$

and

$$\beta_2 = \frac{N\epsilon}{\gamma}\left(\frac{1}{\lambda} + \frac{\bar{\delta}p}{1-e^{-\lambda\psi(\omega)}}\right).$$

One can confirm by means of (8.4.9) that

$$u(s) \le v(s) + \int_{s_{k_0-1}}^{s} NL_f u(r)dr + \sum_{s_{k_0-1}<s_k<s} NL_f\bar{\delta}u(s_k).$$

It can be shown by applying the analogue of the Gronwall's Lemma for piecewise continuous functions that

$$\begin{aligned} u(s) \le v(s) &+ \int_{s_{k_0-1}}^{s} NL_f(1+NL_f\bar{\delta})^{i((r,s))}e^{NL_f(s-r)}v(r)dr \\ &+ \sum_{s_{k_0-1}<s_k<s} NL_f\bar{\delta}(1+NL_f\bar{\delta})^{i((s_k,s))}e^{NL_f(s-s_k)}v(s_k). \end{aligned}$$

Accordingly, the inequality

$$\begin{aligned} u(s) \le\ & \beta_1(1+NL_f\bar{\delta})^p e^{(\lambda-\alpha)(s-s_{k_0-1})} + \beta_2 e^{\lambda s} \\ &+ \frac{NL_f\beta_2(1+NL_f\bar{\delta})^p}{\alpha}e^{\lambda s}\left(1-e^{-\alpha(s-s_{k_0-1})}\right) \\ &+ \frac{NL_f\bar{\delta}p\beta_2(1+NL_f\bar{\delta})^p}{1-e^{-\alpha\psi(\omega)}}e^{\lambda s}\left(1-e^{-\alpha(s-s_{k_0-1}+\psi(\omega))}\right) \end{aligned}$$

is valid. Therefore,

$$\begin{aligned} &\left\|\phi_\zeta(s) - \phi_{\tilde{\zeta}}(s)\right\| < R_1(1+NL_f\bar{\delta})^p e^{-\alpha(s-s_{k_0-1})} \\ &+ \frac{N\epsilon}{\gamma}\left(\frac{1}{\lambda} + \frac{\bar{\delta}p}{1-e^{-\lambda\psi(\omega)}}\right)\left(1 + \frac{NL_f(1+NL_f\bar{\delta})^p}{\alpha} + \frac{NL_f\bar{\delta}p(1+NL_f\bar{\delta})^p}{1-e^{-\alpha\psi(\omega)}}\right) \end{aligned}$$

for $s \in (s_{k_0-1}, s_{k_0+2E}]$.

Suppose that s belongs to the interval $(s_{k_0-1+E}, s_{k_0+2E}]$. Because the number E is sufficiently large such that $E \ge \dfrac{1}{\alpha\underline{\kappa}} \ln\left(\dfrac{\gamma R_1(1+NL_f\bar{\delta})^p}{\epsilon}\right)$ and $s - s_{k_0-1} > E\underline{\kappa}$, we have

$$R_1(1+NL_f\bar{\delta})^p e^{-\alpha(s-s_{k_0-1})} < \frac{\epsilon}{\gamma}.$$

Hence,

$$
\begin{aligned}
&\left\|\phi_{\zeta}(s)-\phi_{\tilde{\zeta}}(s)\right\| \\
&< \frac{\epsilon}{\gamma}+\frac{N\epsilon}{\gamma}\left(\frac{1}{\lambda}+\frac{\bar{\delta}p}{1-e^{-\lambda\psi(\omega)}}\right)\left(1+\frac{NL_f(1+NL_f\bar{\delta})^p}{\alpha}+\frac{NL_f\bar{\delta}p(1+NL_f\bar{\delta})^p}{1-e^{-\alpha\psi(\omega)}}\right) \\
&\leq \epsilon.
\end{aligned}
$$

The last inequality yields $\left\|\varphi_{\zeta}(t)-\varphi_{\tilde{\zeta}}(t)\right\| < \epsilon$ for $t \in [\theta_{2(k_0+E)-1}, \theta_{2(k_0+2E)}] \cap \mathbb{T}_0$. Consequently, the couple $\varphi_{\zeta}(t), \varphi_{\tilde{\zeta}}(t)$ is proximal. □

The frequent separation feature of the bounded solutions of (8.1.1) is presented in the next lemma.

Lemma 8.4.3 ([18]) *Under the conditions $(C1)$–$(C5)$, if a pair of sequences $\zeta, \tilde{\zeta} \in \Theta$ is frequently separated, then the pair of solutions $\varphi_{\zeta}(t), \varphi_{\tilde{\zeta}}(t) \in \mathscr{A}$ is frequently (ϵ_0, Δ)-separated for some positive numbers ϵ_0 and Δ.*

Proof Because the pair of sequences $\zeta, \tilde{\zeta}$ is frequently separated, there exists a positive number $\bar{\epsilon}_0$ and a sequence $\{k_q\}$ of integers satisfying $k_q \to \infty$ as $q \to \infty$ such that $\left\|\zeta_{k_q}-\tilde{\zeta}_{k_q}\right\| > \bar{\epsilon}_0$ for each $q \in \mathbb{N}$.

Let us fix a natural number q. For $s \in (s_{k_q-1}, s_{k_q}]$, the solutions $\phi_{\zeta}(s) = \varphi_{\zeta}(\psi^{-1}(s))$ and $\phi_{\tilde{\zeta}}(s) = \varphi_{\tilde{\zeta}}(\psi^{-1}(s))$ of (8.2.6) satisfy the relations

$$
\phi_{\zeta}(s) = \phi_{\zeta}(s_{k_q-1}+) + \int_{s_{k_q-1}}^{s} [A\phi_{\zeta}(r) + f(\psi^{-1}(r), \phi_{\zeta}(r)) + \zeta_{k_q}]dr
$$

and

$$
\phi_{\tilde{\zeta}}(s) = \phi_{\tilde{\zeta}}(s_{k_q-1}+) + \int_{s_{k_q-1}}^{s} [A\phi_{\tilde{\zeta}}(r) + f(\psi^{-1}(r), \phi_{\tilde{\zeta}}(r)) + \tilde{\zeta}_{k_q}]dr,
$$

respectively. Therefore, one can obtain that

$$
\begin{aligned}
\left\|\phi_{\zeta}(s_{k_q})-\phi_{\tilde{\zeta}}(s_{k_q})\right\| &> \bar{\epsilon}_0\underline{\kappa} - \left\|\phi_{\zeta}(s_{k_q-1}+)-\phi_{\tilde{\zeta}}(s_{k_q-1}+)\right\| \\
&\quad - \int_{s_{k_q-1}}^{s_{k_q}} \left(\|A\|+L_f\right)\left\|\phi_{\zeta}(r)-\phi_{\tilde{\zeta}}(r)\right\| dr \\
&\geq \bar{\epsilon}_0\underline{\kappa} - [1+\left(\|A\|+L_f\right)\overline{\kappa}] \sup_{s\in(s_{k_q-1}, s_{k_q}]} \left\|\phi_{\zeta}(s)-\phi_{\tilde{\zeta}}(s)\right\|.
\end{aligned}
$$

The last inequality implies that

$$\sup_{s\in(s_{k_q-1},s_{k_q}]} \left\|\phi_\zeta(s)-\phi_{\tilde{\zeta}}(s)\right\| > \frac{\bar{\epsilon}_0\underline{\kappa}}{2+\big(\|A\|+L_f\big)\overline{\kappa}}.$$

Define the number

$$\Delta = \min\left\{\frac{\underline{\kappa}}{2}, \frac{\bar{\epsilon}_0\underline{\kappa}}{4[2+(\|A\|+L_f)\overline{\kappa}](K_0\,\|A\|+M_f+M_F)}\right\}.$$

At first, suppose that $\sup_{s\in(s_{k_q-1},s_{k_q}]} \left\|\phi_\zeta(s)-\phi_{\tilde{\zeta}}(s)\right\| = \left\|\phi_\zeta(\eta)-\phi_{\tilde{\zeta}}(\eta)\right\|$ for some $\eta\in(s_{k_q-1},s_{k_q}]$, and let

$$\nu_q = \begin{cases} \eta, & \text{if } \eta \le (s_{k_q-1}+s_{k_q})/2, \\ \eta-\Delta, & \text{if } \eta > (s_{k_q-1}+s_{k_q})/2. \end{cases}$$

It can be verified for $s\in\widetilde{J}_q=[\nu_q,\nu_q+\Delta]$ that

$$\begin{aligned} &\left\|\phi_\zeta(s)-\phi_{\tilde{\zeta}}(s)\right\| \ge \left\|\phi_\zeta(\eta)-\phi_{\tilde{\zeta}}(\eta)\right\| - \left|\int_\eta^s \|A\|\left\|\phi_\zeta(r)-\phi_{\tilde{\zeta}}(r)\right\|dr\right| \\ &-\left|\int_\eta^s \left\|f(\psi^{-1}(r),\phi_\zeta(r))-f(\psi^{-1}(r),\phi_{\tilde{\zeta}}(r))\right\|dr\right| - \left|\int_\eta^s \left\|\eta_{k_q}-\tilde{\eta}_{k_q}\right\|dr\right| \\ &> \frac{\bar{\epsilon}_0\underline{\kappa}}{2[2+(\|A\|+L_f)\overline{\kappa}]}. \end{aligned}$$

On the other hand, the inequality $\left\|\phi_\zeta(s)-\phi_{\tilde{\zeta}}(s)\right\| > \dfrac{\bar{\epsilon}_0\underline{\kappa}}{2[2+(\|A\|+L_f)\overline{\kappa}]}$ is true also for $s\in\widetilde{J}_q=(s_{k_q-1},s_{k_q-1}+\Delta]$ in the case that $\sup_{s\in(s_{k_q-1},s_{k_q}]}\left\|\phi_\zeta(s)-\phi_{\tilde{\zeta}}(s)\right\| = \left\|\phi_\zeta(s_{k_q-1}+)-\phi_{\tilde{\zeta}}(s_{k_q-1}+)\right\|$.

Thus, $\left\|\varphi_\zeta(t)-\varphi_{\tilde{\zeta}}(t)\right\| > \epsilon_0$ for each t from the intervals J_q, $q\in\mathbb{N}$, where $\epsilon_0 = \dfrac{\bar{\epsilon}_0\underline{\kappa}}{2[2+(\|A\|+L_f)\overline{\kappa}]}$ and $J_q=\psi^{-1}(\widetilde{J}_q)$. Consequently, the pair $\varphi_\zeta(t),\varphi_{\tilde{\zeta}}(t)\in\mathscr{A}$ is frequently (ϵ_0,Δ)-separated. □

The main result of the chapter is mentioned in the following theorem.

Theorem 8.1 ([18]) *Assume that the conditions* $(C1)$–$(C7)$ *are fulfilled. If the map (8.1.2) is Li–Yorke chaotic on* Λ, *then system (8.1.1) chaotic in the sense of Definition 8.1.*

Proof Suppose that $\zeta=\{\zeta_k\}$ is a p_0-periodic solution of (8.1.2) for some $p_0\in\mathbb{N}$. In this case, the function $g(t,\zeta)$, which is used in the right-hand side of Eq. (8.1.1), is $m\omega$-periodic, where $m = lcm\{p_0,p\}/p$. Making use of the conditions $(C5)$

and $(C7)$, one can verify that the bounded solution $\varphi_\zeta(t)$ of (8.1.1) is $m\omega$-periodic. Therefore, (8.1.1) possesses an $m\omega$-periodic solution for each $m \in \mathbb{N}$.

Let us denote by Σ the set consisting of bounded solutions $\varphi_\zeta(t)$ of (8.1.1) for which the initial value ζ_0 of the sequence $\zeta = \{\zeta_k\}$ belongs to the scrambled set $\mathscr{S}$ of the map (8.1.2). Because the set $\mathscr{S}$ is uncountable, Σ is also uncountable. Moreover, Σ does not contain any periodic solutions, since no periodic points of F take place inside $\mathscr{S}$.

According to the Lemmas 8.4.2 and 8.4.3, any pair of different solutions inside Σ is a Li–Yorke pair, i.e., Σ is a scrambled set. Besides, Lemma 8.4.3 implies that for any solution $\varphi_\zeta(t) \in \Sigma$ and any periodic solution $\varphi_{\hat{\zeta}}(t) \in \mathscr{A}$, the pair $\varphi_\zeta(t)$, $\varphi_{\hat{\zeta}}(t)$ is frequently (ϵ_0, Δ)-separated for some positive numbers ϵ_0 and Δ. Consequently, system (8.1.1) is Li–Yorke chaotic. □

In the next section, a Duffing equation on a time scale will be utilized to illustrate the theoretical results.

8.5 An Example

Let $\mathbb{T}_0$ be the time scale defined by $\mathbb{T}_0 = \bigcup_{k=-\infty}^{\infty}[\theta_{2k-1}, \theta_{2k}]$, where $\theta_k = 3k + \frac{1}{2}(1 + (-1)^k)$, $k \in \mathbb{Z}$. We consider the forced Duffing equation,

$$y^{\Delta\Delta}(t) + 5y^{\Delta}(t) + \frac{35}{2}y(t) + 0.02y^3(t) = 0.1\cos\left(\frac{\pi}{3}t\right) + g(t, \zeta), \quad (8.5.10)$$

where $t \in \mathbb{T}_0$. The function $g(t, \zeta)$ is defined through the equation $g(t, \zeta) = \zeta_k$ for $t \in [\theta_{2k-1}, \theta_{2k}]$, $k \in \mathbb{Z}$, in which the sequence $\zeta = \{\zeta_k\}$, $\zeta_0 \in [0, 1]$, is generated by the logistic map

$$\zeta_{k+1} = 3.9\zeta_k(1 - \zeta_k). \quad (8.5.11)$$

The time scale $\mathbb{T}_0$ satisfies the ω-property with $\omega = 6$, and one can confirm that $\psi(\omega) = 4$ and $\delta_k = 2$ for all $k \in \mathbb{Z}$, where $\delta_k = \theta_{2k+1} - \theta_{2k}$. According to the results of the paper [42], the map (8.5.11) possesses Li–Yorke chaos. It is worth noting that the unit interval [0, 1] is invariant under the iterations of the map [34].

By using the variables $y_1 = y$ and $y_2 = y^{\Delta}$, Eq. (8.5.10) can be reduced to the system

$$\begin{aligned} y_1^{\Delta}(t) &= y_2(t), \\ y_2^{\Delta}(t) &= -17.5y_1(t) - 5y_2(t) - 0.02y_1^3(t) + 0.1\cos\left(\frac{\pi}{3}t\right) + g(t, \zeta), \end{aligned} \quad (8.5.12)$$

which is in the form of (8.1.1), where

$$A = \begin{pmatrix} 0 & 1 \\ -17.5 & -5 \end{pmatrix}$$

and

$$f(t, y_1, y_2) = \begin{pmatrix} 0 \\ -0.02y_1^3 + 0.1\cos\left(\frac{\pi}{3}t\right) \end{pmatrix}.$$

One can show that

$$e^{At} = e^{-\frac{5}{2}t} Q \begin{pmatrix} \cos\left(\frac{3\sqrt{5}}{2}t\right) & -\sin\left(\frac{3\sqrt{5}}{2}t\right) \\ \sin\left(\frac{3\sqrt{5}}{2}t\right) & \cos\left(\frac{3\sqrt{5}}{2}t\right) \end{pmatrix} Q^{-1},$$

where

$$Q = \begin{pmatrix} 0 & 1 \\ \dfrac{3\sqrt{5}}{2} & -\dfrac{5}{2} \end{pmatrix},$$

and the eigenvalues of the matrix $e^{4A}(I + 2A)$ are inside the unit circle, where I is the 2×2 identity matrix.

Because the coefficient of the nonlinear term $y_1^3(t)$ in (8.5.12) is sufficiently small in absolute value, it can be numerically verified for $\zeta_0 \in [0, 1]$ that the bounded solutions of system (8.5.12) lie inside the region

$$\mathscr{D} = \left\{(y_1, y_2) \in \mathbb{R}^2 : -0.01 \le y_1 \le 0.07, -0.12 \le y_2 \le 0.07\right\}.$$

Therefore, it is reasonable to consider the dynamics of (8.5.12) inside $\mathscr{D}$.

The conditions $(C5)$ and $(C6)$ hold for (8.5.12) with $N = 193$, $\lambda = 1.6$, $p = 1$, $\bar{\delta} = 2$, and $L_f = 0.000294$. In accordance with Theorem 8.1, system (8.5.12) is Li–Yorke chaotic. It is worth noting that the chaoticity of the logistic map (8.5.11) gives rise to the presence of chaos in (8.5.12). Moreover, Lemma 8.3.1 implies that for a fixed solution $\zeta = \{\zeta_k\}$ of (8.5.11) the unique bounded solution of (8.5.12) attracts all other solutions of the system.

Let us use the solution $\zeta = \{\zeta_k\}$ of (8.5.11) with $\zeta_0 = 0.19$ in system (8.5.12). We depict in Fig. 8.1 the y_1-coordinate of the solution of (8.5.12) corresponding to the initial data $y_1(0) = 0.019$ and $y_2(0) = -0.004$. Figure 8.1 supports the result of Theorem 8.1 such that system (8.5.12) possesses chaos. Moreover, the trajectory of the same solution in the $y_1 - y_2$ plane is represented in Fig. 8.2, which reveals the existence of a chaotic attractor in the dynamics of (8.5.12).

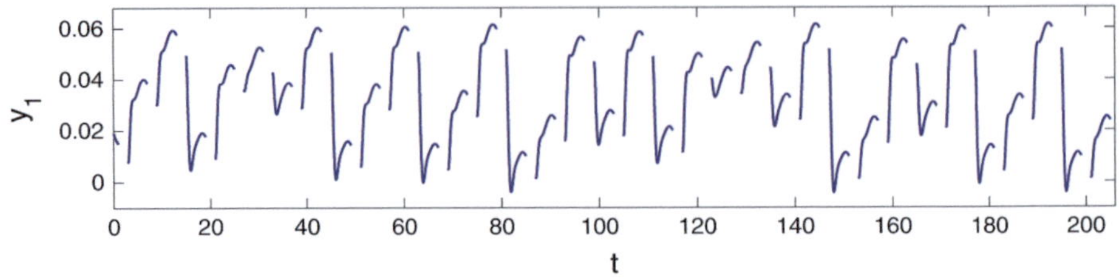

Fig. 8.1 The chaotic behavior in the solution of system (8.5.12)

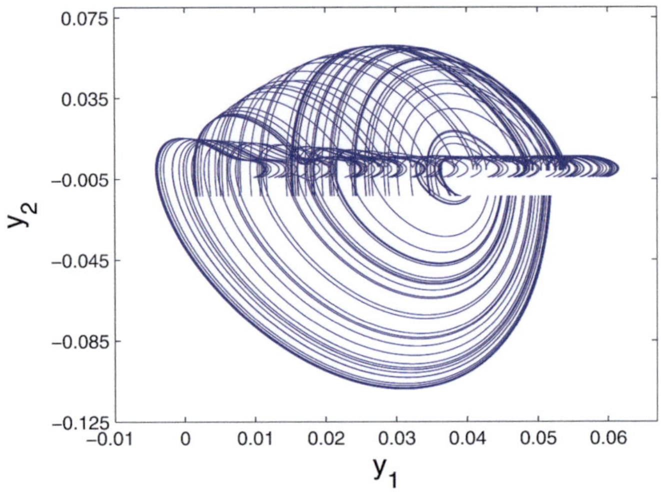

Fig. 8.2 The chaotic trajectory of system (8.5.12)

8.6 Notes

The replication of chaos technique is applied to prove rigorously the existence of chaos in dynamic equations on time scales, where the right-hand side of the equations depends on a chaotic map. The reduction technique to impulsive differential equations presented in the paper [20] is used in our investigations. A mathematical description of chaos in the sense of Li–Yorke is provided for DETS, and the ingredients of the Li–Yorke chaos, proximality and frequent separation, are theoretically proved. The results can be used to obtain chaotic mechanical systems and electrical circuits on time scales without any restriction in the dimension. The results of this chapter are published in paper [18].

References

1. M.U. Akhmet, Hyperbolic sets of impact systems, Dyn. Contin. Discrete Impuls. Syst. Ser. A Math. Anal. 15 (Suppl. S1), 1–2, in *Proceedings of the 5th International Conference on Impulsive and Hybrid Dynamical Systems and Applications* (Watan Press, Beijing, 2008)
2. M.U. Akhmet, Dynamical synthesis of quasi-minimal sets. Int. J. Bifurcat. Chaos **19**, 2423–2427 (2009)
3. M.U. Akhmet, Shadowing and dynamical synthesis. Int. J. Bifurcat. Chaos **19**, 3339–3346 (2009)
4. M.U. Akhmet, Devaney's chaos of a relay system. Commun. Nonlinear Sci. Numer. Simulat. **14**, 1486–1493 (2009)
5. M.U. Akhmet, Li-Yorke chaos in the system with impacts. J. Math. Anal. Appl. **351**, 804–810 (2009)
6. M.U. Akhmet, Creating a chaos in a system with relay. Int. J. Qualit. Th. Diff. Eqs. Appl. **3**, 3–7 (2009)
7. M.U. Akhmet, The complex dynamics of the cardiovascular system. Nonlinear Analysis **71**, e1922–e1931 (2009)
8. M.U. Akhmet, Homoclinical structure of the chaotic attractor. Commun. Nonlinear Sci. Numer. Simulat. **15**, 819–822 (2010)
9. M.U. Akhmet, *Principles of Discontinuous Dynamical Systems* (Springer, New York, 2010)
10. M.U. Akhmet, M.O. Fen, Chaotic period-doubling and OGY control for the forced Duffing equation. Commun. Nonlinear Sci. Numer. Simul. **17**, 1929–1946 (2012)
11. M.U. Akhmet, M.O. Fen, Replication of chaos. Commun. Nonlinear Sci. Numer. Simul. **18**, 2626–2666 (2013)
12. M.U. Akhmet, M.O. Fen, Chaos generation in hyperbolic systems. Discontinuity Nonlinearity Complexity **1**, 353–365 (2012)
13. M.U. Akhmet, M.O. Fen, Shunting inhibitory cellular neural networks with chaotic external inputs. Chaos **23**, 023112 (2013)
14. M.U. Akhmet, M.O. Fen, Entrainment by chaos. J. Nonlinear Sci. **24**, 411–439 (2014)
15. M. Akhmet, M.O. Fen, Chaotification of impulsive systems by perturbations. Int. J. Bifurcat. Chaos **24**, 1450078 (2014)
16. M.U. Akhmet, M.O. Fen, Replication of discrete chaos. Chaotic Model. Simul. (CMSIM) **2**, 129–140 (2014)
17. M. Akhmet, M.O. Fen, Attraction of Li-Yorke chaos by retarded SICNNs. Neurocomputing **147**, 330–342 (2015)
18. M. Akhmet, M.O. Fen, Li-Yorke chaos in hybrid systems on a time scale. Int. J. Bifurcat. Chaos **25**, 1540024 (2015)
19. M. Akhmet, M.O. Fen, *Replication of Chaos in Neural Networks, Economics and Physics* (Higher Education Press, Beijing; Springer, Heidelberg, 2016)
20. M.U. Akhmet, M. Turan, The differential equations on time scales through impulsive differential equations. Nonlinear Analysis **65**, 2043–2060 (2006)
21. M.U. Akhmet, M. Turan, Differential equations on variable time scales. Nonlinear Analysis **70**, 1175–1192 (2009)
22. E. Akin, S. Kolyada, Li-Yorke sensitivity. Nonlinearity **16**, 1421–1433 (2003)
23. K.G. Andersson, Poincaré's discovery of homoclinic points. Arch. Hist. Exact Sci. **48**, 133–147 (1994)
24. B. Aulbach, B. Kieninger, On three definitions of chaos. Nonlinear Dyn. Syst. Theory **1**, 23–37 (2001)
25. R. Barrio, M.A. Martinez, S. Serrano, A. Shilnikov, Macro- and micro-chaotic structures in the Hindmarsh-Rose model of bursting neurons. Chaos **24**, 023128 (2014)
26. M. Bohner, A. Peterson, *Dynamic Equations on Time Scales: An Introduction with Applications* (Birkhäuser, Boston, 2001)

27. R. Brown, L. Chua, Dynamical synthesis of Poincaré maps. Int. J. Bifurcation Chaos **3**, 1235–1267 (1993)
28. R. Brown, L. Chua, From almost periodic to chaotic: the fundamental map. Int. J. Bifurcation Chaos **6**, 1111–1125 (1996)
29. R. Brown, L. Chua, Chaos: generating complexity from simplicity. Int. J. Bifurcation Chaos **7**, 2427–2436 (1997)
30. R. Brown, R. Berezdivin, L. Chua, Chaos and complexity. Int. J. Bifurcation Chaos **11**, 19–26 (2001)
31. M. Cartwright, J. Littlewood, On nonlinear differential equations of the second order I: The equation $\ddot{y} - k(1 - y^2)'y + y = bkcos(\lambda t + a)$, k large. J. Lond. Math. Soc. **20**, 180–189 (1945)
32. M. Fečkan, *Bifurcation and Chaos in Discontinuous and Continuous Systems* (Springer, Heidelberg, 2011)
33. C. Grebogi, J.A. Yorke, *The Impact of Chaos on Science and Society* (United Nations University Press, Tokyo, 1997)
34. J. Hale, H. Koçak, *Dynamics and Bifurcations* (Springer, New York, 1991)
35. S. Hilger, Ein Maßkettenkalkül mit Anwendung auf Zentrumsmanningfaltigkeiten, PhD thesis, Universität Würzburg, 1988
36. R.A. Horn, C.R. Johnson, *Matrix Analysis* (Cambridge University Press, USA, 1992)
37. S.F. Kolyada, Li-Yorke sensitivity and other concepts of chaos. Ukrainian Math. J. **56**, 1242–1257 (2004)
38. V. Lakshmikantham, S. Sivasundaram, B. Kaymakcalan, *Dynamic Systems on Measure Chains* (Kluwer Academic Publishers, The Netherlands, 1996)
39. V. Lakshmikantham, A.S. Vatsala, Hybrid systems on time scales. J. Comput. Appl. Math. **141**, 227–235 (2002)
40. V. Lakshmikantham, J.V. Devi, Hybrid systems with time scales and impulses. Nonlinear Analysis **65**, 2147–2152 (2006)
41. N. Levinson, A second order differential equation with singular solutions. Ann. Math. **50**, 127–153 (1949)
42. T.Y. Li, J.A. Yorke, Period three implies chaos. Am. Math. Monthly **82**, 985–992 (1975)
43. S. Li, ω-chaos and topological entropy. Trans. Am. Math. Soc. **339**, 243–249 (1993)
44. E.N. Lorenz, Deterministic nonperiodic flow. J. Atmos. Sci. **20**, 130–141 (1963)
45. A.C.J. Luo, *Toward Analytical Chaos in Nonlinear Systems* (Wiley, UK, 2014)
46. B.A.M. Owens, M.T. Stahl, N.J. Corron, J.N. Blakely, L. Illing, Exactly solvable chaos in an electromechanical oscillator. Chaos **23**, 033109 (2013)
47. A.M. Samoilenko, N.A. Perestyuk, *Impulsive Differential Equations* (World Scientific, Singapore, 1995)
48. K. Thamilmaran, M. Lakshmanan, A. Venkatesan, Hyperchaos in a modified canonical Chua's circuit. Int. J. Bifurcation Chaos **14**, 221–243 (2004)
49. C.C. Tisdell, A. Zaidi, Basic qualitative and quantitative results for solutions to nonlinear, dynamic equations on time scales with an application to economic modelling. Nonlinear Analysis **68**, 3504–3524 (2008)
50. Y. Ueda, Random phenomena resulting from non-linearity in the system described by Duffing's equation. Trans. Inst. Electr. Eng. Jpn. **98A**, 167–173 (1978)
51. J. Zhang, M. Fan, H. Zhu, Periodic solution of single population models on time scales. Math. Comput. Model. **52**, 515–521 (2010)

Chapter 9
Homoclinic and Heteroclinic Motions in Economic Models

In this chapter, the technique of replication of chaos is utilized to prove the presence of homoclinic and heteroclinic motions in the dynamics of economic models perturbed with exogenous shocks. An illustrative example based on the Kaldor model of the aggregate economy is presented.

9.1 Introduction

Poincarés discovery of homoclinic orbits in the three body problem is one of the most significant discoveries in the theory of dynamical systems [20, 26]. A criterion for the presence of chaos is the existence of a structurally stable Poincaré homoclinic orbit [32, 39, 40]. Likewise homoclinic orbits, heteroclinic orbits also take part in the investigation of chaotic dynamics [23, 24].

The existence of homoclinic and heteroclinic motions in economic models has been extensively investigated in the literature [1, 2, 31, 37, 42]. The paper [1] deals with the occurrence of homoclinic and heteroclinic connections in a nonlinear overlapping generations (OLG) model with credit market imperfection and endogenous labor supply. By means of a singular perturbation method, the presence of transverse homoclinic points to the golden rule steady state in a two-dimensional Diamond-type OLG model was shown in [42]. In paper [2], the presence of homoclinic tangles associated with saddle points or saddle cycles of different period was demonstrated for a particular version of discrete-time Kaldor business cycle model. Homoclinic bifurcations in a class of models representing heterogeneous agents with adaptively rational rules were investigated within the scope of the paper [31]. Moreover, the existence of a homoclinic bifurcation was explored in [37] by using numerical simulations.

M. Akhmet et al., *Dynamics with Chaos and Fractals*, Nonlinear Systems and Complexity 29, https://doi.org/10.1007/978-3-030-35854-9_9

Chaos in the sense of Devaney [26] as well as the one obtained through period-doubling cascade [27] were investigated in our study [9] for economic models perturbed with exogenous shocks. It was shown in the paper [10] that exogenous shocks can cause economic models to exhibit chaotic business cycles. Other chaos generation techniques in systems of differential equations can be found in [4–6, 11–16, 18, 19]. In the present chapter, we theoretically prove that exogenous shocks are capable of generating homoclinic and heteroclinic motions in the continuous-time dynamics of economic systems, and we numerically demonstrate the presence of such motions in the Kaldor model of the aggregate economy. The usage of exogenous shocks in the formation of homoclinic and heteroclinic motions is the main novelty.

Exogenous shocks in a macroeconomic model of a country can occur in two types [9]. The first one is the generation of shocks that are either completely outside of human control or are shaped in some worldwide marketplace. One can think of the economic fluctuations caused by weather phenomena, commodity prices that are determined in the world markets, and the futures prices of wheat, sugar, corn, soybean, coffee, as well as oil product prices [25, 38, 41] as examples of the first type. The second type of exogenous shocks can be generated outside the economic system, but endogenous to some other system that is linked with the former through financial, trade and information flows. Exports to the foreign country may be viewed as an exogenous shock to the domestic economic system in the case that the real output in a foreign economy affects the level of demand by this economy for the exports of the home country, and exports to the foreign economy influence the economic activity at home [9].

The generation of homoclinic and heteroclinic motions in systems of ordinary differential equations by means of discontinuous perturbations was first considered in [7]. According to the results of [7], homoclinic solutions take place in the chaotic attractor of the relay system, which was introduced in [4]. On the other hand, by taking advantage of the moments of impulses, similar results were obtained for impulsive differential equations in [3].

The presence of homoclinic and heteroclinic motions in impulsive systems in the case that the systems are under the influence of a discrete map is considered in the study [29] and the existence of such motions in hybrid systems on a time scale is discussed in paper [28]. Moreover, the paper [30] is concerned with the presence of homoclinic and heteroclinic outputs in the dynamics of retarded shunting inhibitory cellular neural networks with rectangular input currents.

The rest of the chapter is organized as follows. In Sect. 9.2, we introduce the economic model that will be investigated. Section 9.3 is devoted to the theoretical results about the existence of homoclinic and heteroclinic solutions in the model. An example concerning the Kaldor model of the aggregate economy is presented in Sect. 9.4 in order to support the theoretical results. Finally, some concluding remarks are given in Sect. 9.5.

9.2 The Model

Throughout the chapter $\mathbb{R}$ and $\mathbb{Z}$ will stand for the sets of real numbers and integers, respectively. Moreover, the usual Euclidean norm for vectors and the norm induced by the Euclidean norm for square matrices [35] will be utilized.

In this chapter, we take into account economic models perturbed with pulse functions. A function $p : \mathbb{R} \to \mathbb{R}^n$ is called a pulse function if for each integer i there is $p_i \in \mathbb{R}^n$ such that $p(t) = p_i$ either for $t \in (\theta_i, \theta_{i+1}]$ or for $t \in [\theta_i, \theta_{i+1})$, where $\{\theta_i\}_{i\in\mathbb{Z}}$ is a strictly increasing sequence of real numbers such that $|\theta_i| \to \infty$ as $|i| \to \infty$.

Consider the following economic model:

$$\dot{v} = H(v), \tag{9.2.1}$$

where the function $H : \mathbb{R}^n \to \mathbb{R}^n$ is continuously differentiable in its arguments and $v : \mathbb{R} \to \mathbb{R}^n$ is a function of time t. We assume that (9.2.1) possesses a steady state at $v = v_0$.

Let $[\tau]$ denote the largest integer that is not greater than $\tau \in \mathbb{R}$, and fix a positive number h. We perturb (9.2.1) with the pulse function $\widetilde{d}_{[t/h]}$, and set up the model

$$\dot{v} = H(v) + \widetilde{d}_{[t/h]}, \tag{9.2.2}$$

where $t \in \mathbb{R}$, $\widetilde{d}_i = (g(d_i), 0, 0, \ldots, 0) \in \mathbb{R}^n$ for each $i \in \mathbb{Z}$, the function $g : \Lambda \to \mathbb{R}$ is continuous, $\Lambda \subset \mathbb{R}$ is a bounded interval, and the sequence $\{d_i\}_{i\in\mathbb{Z}}$, $d_0 \in \Lambda$, is a solution of the discrete equation

$$d_{i+1} = F(d_i), \tag{9.2.3}$$

where $F : \Lambda \to \Lambda$ is a continuous function. It is worth noting that $\widetilde{d}_{[t/h]} = \widetilde{d}_i$ for $t \in [ih, (i+1)h)$.

Because of the economic reasons mentioned in Sect. 9.4, we consider the perturbation $\widetilde{d}_{[t/h]}$ with only one nonzero coordinate, and the more general case can be investigated in a similar way.

Exogenous shocks with variable values have many applications from the economic point of view. Economic time series such as commodity prices, productivity indices, and international trade indicators are the examples of exogenous shocks, and they are usually gauged by economists at regular discrete intervals, no matter how disaggregated (year, month, day, minute, second). Moreover, the government budget that is determined once a year, earnings of a farm that sells its produce in accordance with the seasons, and a firm's capital equipment that changes with periodical investment can be considered as other examples [9]. Our main purpose is to rigorously prove that homoclinic and heteroclinic motions exist in the dynamics of (9.2.2) by taking advantage of the exogenous shocks.

If we transform the state variables $x = v - v_0$ in (9.2.2), then near the equilibrium point the linearized model takes the form

$$\dot{x} = Ax + f(x) + \widetilde{d}_{[t/h]}, \tag{9.2.4}$$

where A is an $n \times n$ constant real valued matrix and $f : \mathbb{R}^n \to \mathbb{R}^n$ is a function such that $f(0) = 0$. We suppose that all eigenvalues the matrix A have negative real parts.

The chaotic dynamics of the model under investigation was studied in the paper [9]. More precisely, it was theoretically proved in [9] that system (9.2.4) possesses chaos in the sense of Devaney [26] and through period-doubling cascade [27] provided that the same is true for the map (9.2.3). In the next section, we will show that homoclinic as well as heteroclinic motions take place in the continuous-time dynamics of (9.2.4) in the case that the map (9.2.3) possesses homoclinic and heteroclinic orbits.

9.3 Homoclinic and Heteroclinic Motions

According to the assumption that the eigenvalues of the matrix A in (9.2.4) have all negative real parts, there exist positive numbers N and ω such that $\left\|e^{At}\right\| \leq Ne^{-\omega t}$ for all $t \geq 0$.

The following conditions are required.

(C1) There exist positive numbers M_f and M_g such that

$$\sup_{x \in \mathbb{R}^n} \|f(x)\| \leq M_f$$

and

$$\sup_{z \in \Lambda} |g(z)| \leq M_g;$$

(C2) There exists a positive number $L_f < \omega/N$ such that

$$\|f(x_1) - f(x_2)\| \leq L_f \|x_1 - x_2\|$$

for all $x_1, x_2 \in \mathbb{R}^n$;

(C3) There exists a positive number L_g such that

$$|g(z_1) - g(z_2)| \leq L_g |z_1 - z_2|$$

for all $z_1, z_2 \in \Lambda$.

Let $\mathscr{D}$ be the set of all sequences $d = \{d_i\}_{i\in\mathbb{Z}}$ generated by the map (9.2.3). Under the conditions $(C1)$ and $(C2)$, for a given sequence $d = \{d_i\}_{i\in\mathbb{Z}} \in \mathscr{D}$, system (9.2.4) possesses a unique solution $\phi_d(t)$ which is bounded on $\mathbb{R}$ [33]. According to the results of [8], the bounded solution $\phi_d(t)$ satisfies the relation

$$\phi_d(t) = \int_{-\infty}^{t} e^{A(t-s)}(f(\phi_d(s)) + \widetilde{d}_{[s/h]})ds.$$

Denote by $\mathscr{B}$ the set of all bounded solutions $\phi_d(t)$, $d \in \mathscr{D}$, of (9.2.4). It can be verified that $\sup_{t\in\mathbb{R}} \|\phi_d(t)\| \leq \dfrac{N(M_f + M_g)}{\omega}$ for each $\phi_d(t) \in \mathscr{B}$. For a given sequence $d \in \mathscr{D}$, if $x_d(t, x_0)$ is the solution of (9.2.4) satisfying $x_d(0, x_0) = x_0$, then we have

$$\|x_d(t, x_0) - \phi_d(t)\| \leq N \|x_0 - \phi_d(0)\| e^{(NL_f-\omega)t}$$

for all $t \geq 0$. Thus, $\|x_d(t, x_0) - \phi_d(t)\| \to 0$ as $t \to \infty$ in accordance with condition $(C2)$, i.e., the bounded solution $\phi_d(t)$ attracts all other solutions of (9.2.4) for a fixed $d \in \mathscr{D}$.

Now, let us continue with the definitions of the stable and unstable sets as well as the hyperbolicity for system (9.2.2) and the map (9.2.3). These definitions are adapted from the paper [7].

The stable set of a sequence $d = \{d_i\}_{i\in\mathbb{Z}} \in \mathscr{D}$ is defined as

$$W^s(d) = \left\{c = \{c_i\}_{i\in\mathbb{Z}} \in \mathscr{D} : \ \|c_i - d_i\| \to 0 \text{ as } i \to \infty\right\},$$

and the unstable set of d is

$$W^u(d) = \left\{c = \{c_i\}_{i\in\mathbb{Z}} \in \mathscr{D} : \ \|c_i - d_i\| \to 0 \text{ as } i \to -\infty\right\}.$$

The set $\mathscr{D}$ is called hyperbolic if for each $d \in \mathscr{D}$ the stable and unstable sets of d contain at least one element different from d. A sequence $c \in \mathscr{D}$ is homoclinic to another sequence $d \in \mathscr{D}$ if $c \in W^s(d) \cap W^u(d)$. Moreover, $c \in \mathscr{D}$ is heteroclinic to the sequences $d^1, d^2 \in \mathscr{D}$, $c \neq d^1$, $c \neq d^2$, if $c \in W^s(d^1) \cap W^u(d^2)$.

On the other hand, a bounded solution $\phi_c(t) \in \mathscr{B}$ belongs to the stable set $W^s(\phi_d(t))$ of $\phi_d(t) \in \mathscr{B}$ if $\|\phi_c(t) - \phi_d(t)\| \to 0$ as $t \to \infty$. Besides, $\phi_c(t)$ is an element of the unstable set $W^u(\phi_d(t))$ of $\phi_d(t)$ provided that $\|\phi_c(t) - \phi_d(t)\| \to 0$ as $t \to -\infty$.

We say that the set $\mathscr{B}$ is hyperbolic if for each $\phi_d(t) \in \mathscr{B}$ the sets $W^s(\phi_d(t))$ and $W^u(\phi_d(t))$ contain at least one element different from $\phi_d(t)$. A solution $\phi_c(t) \in \mathscr{B}$ is homoclinic to another solution $\phi_d(t) \in \mathscr{B}$ if $\phi_c(t) \in W^s(\phi_d(t)) \cap W^u(\phi_d(t))$, and $\phi_c(t) \in \mathscr{B}$ is heteroclinic to the bounded solutions $\phi_{d^1}(t)$, $\phi_{d^2}(t) \in \mathscr{B}$, $\phi_c(t) \neq \phi_{d^1}(t)$, $\phi_c(t) \neq \phi_{d^2}(t)$, if $\phi_c(t) \in W^s(\phi_{d^1}(t)) \cap W^u(\phi_{d^2}(t))$.

In the next lemma, we deal with the connection between the stable sets of the solutions of (9.2.3) and (9.2.4).

Lemma 9.3.1 ([17]) *Suppose that the conditions* $(C1)$–$(C3)$ *hold, and let* $c = \{c_i\}_{i\in\mathbb{Z}}$ *and* $d = \{d_i\}_{i\in\mathbb{Z}}$ *be elements of* $\mathscr{D}$. *If* $c \in W^s(d)$, *then* $\phi_c(t) \in W^s(\phi_d(t))$.

Proof Fix an arbitrary positive number ϵ, and let γ be a number such that

$$\gamma \geq 1 + \frac{NL_g}{\omega - NL_f}.$$

Since $c \in W^s(d)$, there exists an integer j_0 such that $\left|c_{j_0} - d_{j_0}\right| < \dfrac{\epsilon}{\gamma}$ for all $i \geq j_0$. In this case, we have $\left\|\widetilde{c}_{[t/h]} - \widetilde{d}_{[t/h]}\right\| < \dfrac{L_g\epsilon}{\gamma}$ for $t \geq j_0h$.

By using the relation

$$\phi_c(t) - \phi_d(t) = \int_{-\infty}^{t} e^{A(t-s)}\left[f(\phi_c(s)) + \widetilde{c}_{[s/h]} - f(\phi_d(s)) - \widetilde{d}_{[s/h]}\right]ds,$$

one can obtain for $t \geq j_0h$ that

$$\begin{aligned}\|\phi_c(t) - \phi_d(t)\| \leq{}& \frac{2N(M_f + M_g)}{\omega}e^{-\omega(t-j_0h)} + \frac{NL_g\epsilon}{\omega\gamma}\left(1 - e^{-\omega(t-j_0h)}\right)\\ &+ \int_{j_0h}^{t} NL_f e^{-\omega(t-s)}\, \|\phi_c(s) - \phi_d(s)\|\, ds.\end{aligned}$$

If we denote $u(t) = e^{\omega t}\, \|\phi_c(t) - \phi_d(t)\|$ and $\alpha = \left(\dfrac{2N(M_f + M_g)}{\omega} - \dfrac{NL_g\epsilon}{\omega\gamma}\right)e^{\omega j_0h}$, then we attain

$$u(t) \leq \alpha + \frac{NL_g\epsilon}{\omega\gamma}e^{\omega t} + \int_{j_0h}^{t} NL_f u(s)ds.$$

It can be verified by applying the Gronwall's Lemma [33] that

$$u(t) \leq \alpha + \frac{NL_g\epsilon}{\omega\gamma}e^{\omega t} + \int_{j_0h}^{t} NL_f\left(\alpha + \frac{NL_g\epsilon}{\omega\gamma}e^{\omega s}\right)e^{NL_f(t-s)}ds.$$

Therefore,

$$u(t) \leq \frac{NL_g\epsilon}{\omega\gamma}e^{\omega t} + \alpha e^{NL_f(t-j_0h)} + \frac{N^2L_fL_g\epsilon}{\omega(\omega - NL_f)\gamma}e^{\omega t}\left(1 - e^{-(\omega - NL_f)(t-j_0h)}\right).$$

The last inequality yields

$$\|\phi_c(t) - \phi_d(t)\| < \frac{NL_g\epsilon}{(\omega - NL_f)\gamma} + \frac{2N(M_f + M_g)}{\omega}e^{(NL_f-\omega)(t-j_0h)},\ t \geq j_0h.$$

Now, let R be a sufficiently large positive number such that

$$\frac{2N(M_f + M_g)}{\omega} e^{(NL_f - \omega)R} \leq \frac{\epsilon}{\gamma}.$$

For $t \geq R + j_0 h$, we have that

$$\|\phi_c(t) - \phi_d(t)\| < \frac{\epsilon}{\gamma}\left(1 + \frac{NL_g}{\omega - NL_f}\right) \leq \epsilon.$$

Consequently, $\phi_c(t)$ belongs to the stable set $W^s(\phi_d(t))$ of $\phi_d(t) \in \mathscr{B}$. □

The following assertion is concerned with the unstable sets of the solutions of (9.2.3) and (9.2.4).

Lemma 9.3.2 ([17]) *Suppose that the conditions* $(C1)$*–*$(C3)$ *hold, and let* $c = \{c_i\}_{i\in\mathbb{Z}}$ *and* $d = \{d_i\}_{i\in\mathbb{Z}}$ *be elements of* $\mathscr{D}$*. If* $c \in W^u(d)$*, then* $\phi_c(t) \in W^u(\phi_d(t))$*.*

Proof Fix an arbitrary positive number ϵ, and let γ be a number such that $\gamma > \dfrac{NL_g}{\omega - NL_f}$. One can find an integer j_0 such that $|c_i - d_i| < \dfrac{\epsilon}{\gamma}$ for all $i \leq j_0$ since the sequence c belongs to the unstable set $W^u(d)$ of $d \in \mathscr{D}$. Hence,

$$\left\|\widetilde{c}_{[t/h]} - \widetilde{d}_{[t/h]}\right\| < \frac{L_g \epsilon}{\gamma}$$

for $t < (j_0 + 1)h$.

Making use of the equation

$$\phi_c(t) - \phi_d(t) = \int_{-\infty}^{t} e^{A(t-s)}\left[f(\phi_c(s)) + \widetilde{c}_{[s/h]} - f(\phi_d(s)) - \widetilde{d}_{[s/h]}\right]ds,$$

it can be obtained for $t < (j_0 + 1)h$ that

$$\|\phi_c(t) - \phi_d(t)\| < \frac{NL_g\epsilon}{\omega\gamma} + \int_{-\infty}^{t} NL_f e^{-\omega(t-s)} \|\phi_c(s) - \phi_d(s)\|\, ds.$$

Therefore,

$$\sup_{t<(j_0+1)h} \|\phi_c(t) - \phi_d(t)\| \leq \frac{NL_g\epsilon}{(\omega - NL_f)\gamma} < \epsilon.$$

Consequently, $\|\phi_c(t) - \phi_d(t)\| \to 0$ as $t \to -\infty$ so that $\phi_c(t) \in W^u(\phi_d(t))$. □

The following theorem, which is the main result in this chapter, can be proved by using Lemmas 9.3.1 and 9.3.2.

Theorem 9.1 ([17]) *Under the conditions* $(C1)$–$(C3)$, *the following assertions are valid.*

(i) If $c \in \mathscr{D}$ *is homoclinic to* $d \in \mathscr{D}$, *then* $\phi_c(t) \in \mathscr{B}$ *is homoclinic to* $\phi_d(t) \in \mathscr{B}$;
(ii) If $c \in \mathscr{D}$ *is heteroclinic to* d^1, $d^2 \in \mathscr{D}$, *then* $\phi_c(t) \in \mathscr{B}$ *is heteroclinic to* $\phi_{d^1}(t)$, $\phi_{d^2}(t) \in \mathscr{B}$;
(iii) If $\mathscr{D}$ *is hyperbolic, then the same is true for* $\mathscr{B}$.

An example concerning the Kaldor model of the aggregate economy will be taken into account in the next section.

9.4 An Example

Consider the following Kaldor model [36, 43]:

$$\begin{aligned} \dot{Y} &= \alpha[I(Y,K) - S(Y,K)] \\ \dot{K} &= I(Y,K) - \delta K, \end{aligned} \tag{9.4.5}$$

where $\delta \in (0,1)$ is the constant depreciation rate, $\alpha > 0$ is the adjustment coefficient, Y is income, K is capital stock, I is gross investment, and S is savings.

Let us use $I(Y,K) = Y - aY^3 + bK$ and $S(Y,K) = sY$ in (9.4.5) so that the system takes the form

$$\begin{aligned} \dot{Y} &= \alpha[(1-s)Y - aY^3 + bK] \\ \dot{K} &= Y - aY^3 + (b-\delta)K, \end{aligned} \tag{9.4.6}$$

where the constant parameters satisfy $a > 0$, $b < 0$, and $0 < s < 1$. In the case that $s(b-\delta)+\delta > 0$, system (9.4.6) admits the following steady state with positive coordinates:

$$Y^* = \sqrt{\frac{s(b-\delta)+\delta}{a\delta}}, \quad K^* = \frac{s}{\delta}\sqrt{\frac{s(b-\delta)+\delta}{a\delta}}.$$

The income Y of a given country is subject to many possible exogenous disturbances, such as productivity shocks and global economic fluctuations, while the capital stock K can be viewed as a mechanical relation between investment and capital stock, where there is little room for exogenous influences. Therefore, we modify (9.4.6) by using perturbation only in the equation for income Y, and constitute the system

$$\begin{aligned} \dot{Y} &= \alpha[(1-s)Y - aY^3 + bK] + g(d_{[t]}) \\ \dot{K} &= Y - aY^3 + (b-\delta)K, \end{aligned} \tag{9.4.7}$$

where the function g is defined as $g(z) = 0.02(z - 0.4 \sin z)$. The sequence $\{d_i\}_{i \in \mathbb{Z}}$, $d_0 \in [0, 1]$, is a solution of the logistic map

$$d_{i+1} = F(d_i), \tag{9.4.8}$$

where $F(\sigma) = 3.8\sigma(1 - \sigma)$. Notice that $d_{[t]} = d_i$ for $i \le t < i + 1$, $i \in \mathbb{Z}$. The unit interval $[0, 1]$ is invariant under the iterations of (9.4.8) [34]. Moreover, the inverses of the function F on the intervals $[0, 1/2]$ and $[1/2, 1]$ are $G_1(\sigma) = \frac{1}{2}\left(1 - \sqrt{1 - \frac{4\sigma}{3.8}}\right)$ and $G_2(\sigma) = \frac{1}{2}\left(1 + \sqrt{1 - \frac{4\sigma}{3.8}}\right)$, respectively. For the applications of the logistic map the reader is referred to [21].

In what follows, we will make use of the values $\alpha = 1$, $a = 0.01$, $b = -1/12$, $s = 1/3$, and $\delta = 1/6$. Using the transformation $y = Y - Y^*$, $k = K - K^*$ in (9.4.7), where $Y^* = 5\sqrt{2}$, $K^* = 10\sqrt{2}$, we obtain the system

$$\begin{aligned} \dot{y} &= -\frac{5}{6}y - \frac{1}{12}k - 0.01y^3 - \frac{3}{10\sqrt{2}}y^2 + g(d_{[t]}) \\ \dot{k} &= -\frac{1}{2}y - \frac{1}{4}k - 0.01y^3 - \frac{3}{10\sqrt{2}}y^2. \end{aligned} \tag{9.4.9}$$

System (9.4.9) is in the form of (9.2.4), where

$$A = \begin{pmatrix} -\frac{5}{6} & -\frac{1}{12} \\ -\frac{1}{2} & -\frac{1}{4} \end{pmatrix}, \quad f(y, k) = \begin{pmatrix} -0.01y^3 - \frac{3}{10\sqrt{2}}y^2 \\ -0.01y^3 - \frac{3}{10\sqrt{2}}y^2 \end{pmatrix}.$$

The eigenvalues of the matrix A are $\frac{-13 + \sqrt{73}}{24}$ and $\frac{-13 - \sqrt{73}}{24}$. One can confirm that $e^{At} = Pe^{Dt}P^{-1}$, where

$$P = \begin{pmatrix} \frac{7 - \sqrt{73}}{12} & 1 \\ 1 & \frac{-7 + \sqrt{73}}{2} \end{pmatrix}, \quad D = \begin{pmatrix} \frac{-13 + \sqrt{73}}{24} & 0 \\ 0 & \frac{-13 - \sqrt{73}}{24} \end{pmatrix}.$$

Thus, $\left\|e^{At}\right\| \le Ne^{-\omega t}$ for $t \ge 0$, where $N = \|P\| \left\|P^{-1}\right\| \approx 1.83005$ and $\omega = \frac{13 - \sqrt{73}}{24}$.

It can be numerically verified that the solutions of (9.4.9) which are bounded on the entire real axis lie inside the compact region

$$\mathscr{U} = \left\{(y, k) \in \mathbb{R}^2 : 0.008 \le y \le 0.016, \ -0.022 \le k \le -0.028\right\}.$$

Therefore, it is reasonable to consider the conditions $(C1)$ and $(C2)$ for the function $f(y,k)$ on the region $\mathscr{U}$ so that $(C2)$ is valid with $L_f = 0.0097$. On the other hand, $(C3)$ holds with $L_g = 0.0157$.

Since the inequality $|g(z_1) - g(z_2)| \geq 0.012\,|z_1 - z_2|$, $z_1, z_2 \in [0, 1]$, is also valid in addition to $(C1)$–$(C3)$, system (9.4.9) (and hence (9.4.7)) admits the chaos through period-doubling cascade according to the results of [9]. In order to demonstrate the presence of chaos, let us consider the solution of (9.4.9) with $y(0) = 0.014$, $k(0) = -0.025$, and $d_0 = 0.18$. The trajectory of the solution is depicted in Fig. 9.1, which reveals that (9.4.9) is chaotic.

Now, we will show the presence of homoclinic and heteroclinic motions in the dynamics of (9.4.9). It was mentioned in the paper [22] that the orbit

$$c = \left\{\ldots, G_2^3(c_0), G_2^2(c_0), G_2(c_0), c_0, F(c_0), F^2(c_0), F^3(c_0), \ldots\right\},$$

where $c_0 = 1/3.8$, is homoclinic to the fixed point $d^* = 2.8/3.8$ of (9.4.8). Let $\phi_c(t)$ and $\phi_{d^*}(t)$ be bounded on $\mathbb{R}$ solutions of (9.4.9) corresponding to c and d^*, respectively. The solution $\phi_c(t)$ is homoclinic to $\phi_{d^*}(t)$ in accordance with Theorem 9.1. Figure 9.2 shows the graphs of $\phi_c(t)$ and $\phi_{d^*}(t)$ in blue and red colors, respectively. Both the y and k coordinates of the solutions are represented in the figure. The simulation results confirm that $\|\phi_c(t) - \phi_{d^*}(t)\| \to 0$ as $t \to \pm\infty$, that is, $\phi_c(t)$ is homoclinic to $\phi_{d^*}(t)$.

Next, we take into account the orbit

$$\overline{c} = \left\{\ldots, G_1^3(\overline{c}_0), G_1^2(\overline{c}_0), G_1(\overline{c}_0), \overline{c}_0, F(\overline{c}_0), F^2(\overline{c}_0), F^3(\overline{c}_0), \ldots\right\},$$

where $\overline{c}_0 = 1/3.8$. According to [22], the orbit $\overline{c}$ is heteroclinic to the fixed points $d^* = 2.8/3.8$ and $d^{**} = 0$ of (9.4.8). One can conclude by using Theorem 9.1 that $\phi_{\overline{c}}(t)$ is heteroclinic to $\phi_{d^*}(t)$ and $\phi_{d^{**}}(t)$, where $\phi_{\overline{c}}(t)$, $\phi_{d^*}(t)$, and $\phi_{d^{**}}(t)$ are,

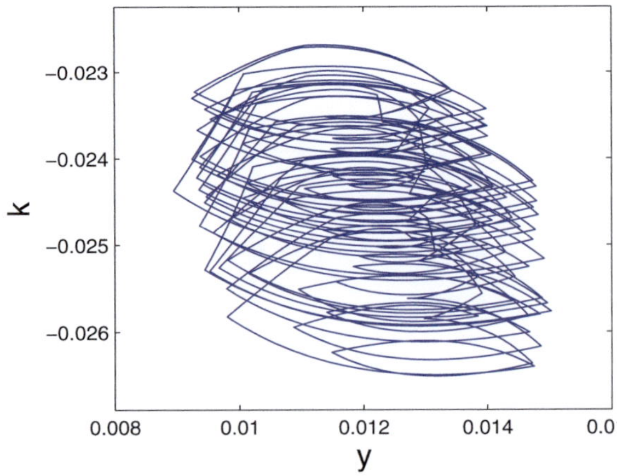

Fig. 9.1 The trajectory of system (9.4.7) corresponding to the initial data $y(0) = 0.014$ and $k(0) = -0.025$. The solution of (9.4.8) with $d_0 = 0.18$ is used in the simulation

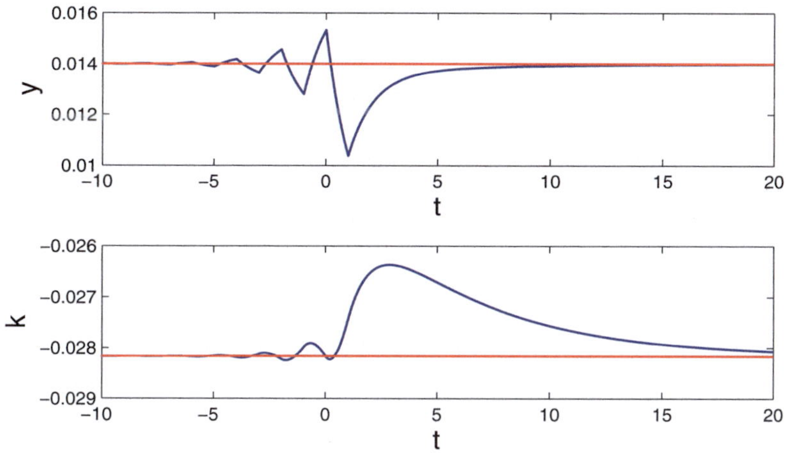

Fig. 9.2 The homoclinic solution of (9.4.9). The bounded solutions $\phi_c(t)$ and $\phi_{d^*}(t)$ are depicted in blue and red colors, respectively. It seen in the figure that $\phi_c(t)$ is homoclinic to $\phi_{d^*}(t)$

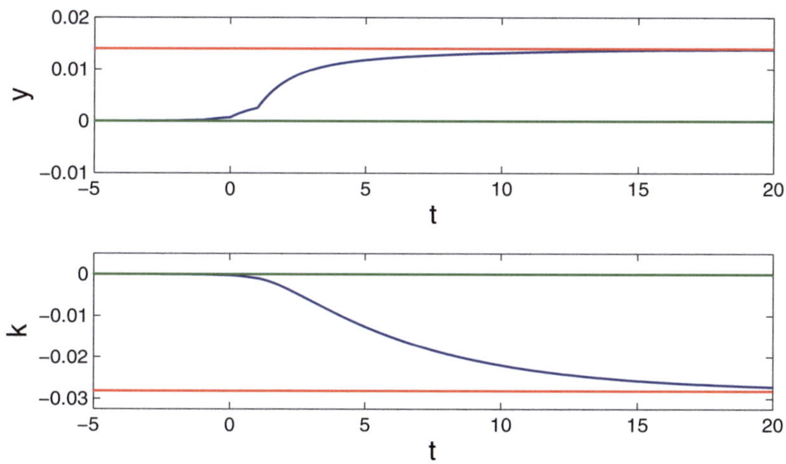

Fig. 9.3 The heteroclinic solution of (9.4.9). The bounded solutions $\phi_{\overline{c}}(t)$, $\phi_{d^*}(t)$, and $\phi_{d^{**}}(t)$ are depicted in blue, red, and green colors, respectively. The simulation demonstrates that $\phi_{\overline{c}}(t)$ is heteroclinic to $\phi_{d^*}(t)$, $\phi_{d^{**}}(t)$

respectively, bounded on $\mathbb{R}$ solutions of (9.4.9) corresponding to $\overline{c}$, d^*, and d^{**}. Figure 9.3 represents the y and k coordinates of $\phi_{\overline{c}}(t)$, $\phi_{d^*}(t)$, and $\phi_{d^{**}}(t)$ in blue, red, and green colors, respectively. The figure supports Theorem 9.1 such that $\phi_{\overline{c}}(t)$ is heteroclinic to $\phi_{d^*}(t)$, $\phi_{d^{**}}(t)$.

9.5 Notes

Homoclinic and heteroclinic orbits are crucial in the theory of dynamical systems since their presence is related to the existence of chaos. We provide a theoretical approach for the generation of homoclinic and heteroclinic motions in the continuous-time dynamics of economic models influenced by exogenous shocks. The main novelty of this chapter is the formation of such motions *exogenously*. The example concerning the Kaldor model of the aggregate economy presented in Sect. 9.4 supports the theoretical results such that exogenous shocks in economic models can lead to the occurrence of homoclinic and heteroclinic motions. The presented technique can be useful for the investigation of homoclinic and heteroclinic bifurcations in economic systems influenced by exogenous shocks. The results of this chapter are published in paper [17].

References

1. A. Agliari, G. Vachadze, Homoclinic and heteroclinic bifurcations in an overlapping generations model with credit market imperfection. Comput. Econ. **38** 241–260 (2011)
2. A. Agliari, R. Dieci, L. Gardini, Homoclinic tangles in a Kaldor-like business cycle model. J. Econ. Behav. Organ. **62**, 324–347 (2007)
3. M.U. Akhmet, Hyperbolic sets of impact systems, Dyn. Contin. Discrete Impuls. Syst. Ser. A Math. Anal. 15 (Suppl. S1), 1–2, in *Proceedings of the 5th International Conference on Impulsive and Hybrid Dynamical Systems and Applications* (Watan Press, Beijing, 2008)
4. M.U. Akhmet, Devaney's chaos of a relay system. Commun. Nonlinear Sci. Numer. Simulat. **14**, 1486–1493 (2009)
5. M.U. Akhmet, Li-Yorke chaos in the system with impacts. J. Math. Anal. Appl. **351**, 804–810 (2009)
6. M.U. Akhmet, The complex dynamics of the cardiovascular system. Nonlinear Analysis **71**, e1922–e1931 (2009)
7. M.U. Akhmet, Homoclinical structure of the chaotic attractor. Commun. Nonlinear Sci. Numer. Simulat. **15**, 819–822 (2010)
8. M.U. Akhmet, *Nonlinear Hybrid Continuous/Discrete-time Models* (Atlantis Press, Paris, Amsterdam, 2011)
9. M. Akhmet, Z. Akhmetova, M.O. Fen, Chaos in economic models with exogenous shocks. J. Econ. Behav. Organ. **106**, 95–108 (2014)
10. M. Akhmet, Z. Akhmetova, M.O. Fen, Exogenous versus endogenous for chaotic business cycles. Discontinuity Nonlinearity Complexity **5**(2), 101–119 (2016)
11. M.U. Akhmet, M.O. Fen, Replication of chaos. Commun. Nonlinear Sci. Numer. Simul. **18**, 2626–2666 (2013)
12. M.U. Akhmet, M.O. Fen, Chaos generation in hyperbolic systems. Discontinuity Nonlinearity Complexity **1**, 353–365 (2012)
13. M.U. Akhmet, M.O. Fen, Shunting inhibitory cellular neural networks with chaotic external inputs. Chaos **23**, 023112 (2013)
14. M.U. Akhmet, M.O. Fen, Entrainment by chaos. J. Nonlinear Sci. **24**, 411–439 (2014)
15. M.U. Akhmet, M.O. Fen, Replication of discrete chaos. Chaotic Model. Simul. (CMSIM) **2**, 129–140 (2014)
16. M. Akhmet, M.O. Fen, Attraction of Li-Yorke chaos by retarded SICNNs. Neurocomputing **147**, 330–342 (2015)

17. M. Akhmet, M.O. Fen, Homoclinic and heteroclinic motions in economic models with exogenous shocks. Appl. Math. Nonlinear Sci. **1**, 1–10 (2016)
18. M. Akhmet, M.O. Fen, *Replication of Chaos in Neural Networks, Economics and Physics* (Higher Education Press, Beijing; Springer, Heidelberg, 2016)
19. M. Akhmet, I. Rafatov, M.O. Fen, Extension of spatiotemporal chaos in glow discharge-semiconductor systems. Chaos **24**, 043127 (2014)
20. K.G. Andersson, Poincaré's discovery of homoclinic points. Arch. Hist. Exact Sci. **48**, 133–147 (1994)
21. M. Ausloos, M. Dirickx, *The Logistic Map and the Route to Chaos: From the Beginnings to Modern Applications* (Springer, Berlin, 2010)
22. V. Avrutin, B. Schenke, L. Gardini, Calculation of homoclinic and heteroclinic orbits in 1D maps. Commun. Nonlinear Sci. Numer Simul. **22**, 1201–1214 (2015)
23. A.L. Bertozzi, Heteroclinic orbits and chaotic dynamics in planar fluid flows. SIAM J. Math. Anal. **19**, 1271–1294 (1988)
24. R. Chacon, J.D. Bejarano, Homoclinic and heteroclinic chaos in a triple-well oscillator. J. Sound Vib. **186**, 269–278 (1995)
25. G.P. Decoster, W.C. Labys, D.W. Mitchell, Evidence of chaos in commodity futures prices. J. Futures Markets **12**, 291–305 (1992)
26. R.L. Devaney, *An Introduction to Chaotic Dynamical Systems* (Addison-Wesley, USA, 1989)
27. M.J. Feigenbaum, Universal behavior in nonlinear systems. Los Alamos Sci./Summer **1**, 4–27 (1980)
28. M.O. Fen, Homoclinic and heteroclinic motions in hybrid systems on a time scale. Proc. Dyn. Syst. Appl. **7**, 90–95 (2016)
29. M.O. Fen, F. Tokmak Fen, Homoclinic and heteroclinic motions in hybrid systems with impacts. Math. Slovaca **67**, 1179–1188 (2017)
30. M.O. Fen, F. Tokmak Fen, Homoclinical structure of retarded SICNNs with rectangular input currents. Neural Process. Lett. **49**, 521–538 (2019)
31. I. Foroni, L. Gardini, Homoclinic bifurcations in heterogeneous market models. Chaos Solitons & Fractals **15**, 743–760 (2003)
32. S.V. Gonchenko, L.P. Shil'nikov, D.V. Turaev, Dynamical phenomena in systems with structurally unstable Poincaré homoclinic orbits. Chaos **6**, 15–31 (1996)
33. J.K. Hale, *Ordinary Differential Equations* (Krieger Publishing Company, Malabar, Florida, 1980)
34. J. Hale, H. Koçak, *Dynamics and Bifurcations* (Springer, New York, 1991)
35. R.A. Horn, C.R. Johnson, *Matrix Analysis* (Cambridge University Press, USA, 1992)
36. H.-W. Lorenz, *Nonlinear Dynamical Economics and Chaotic Motion* (Springer, Berlin, Heidelberg, 1993)
37. A.K. Naimzada, G. Ricchiuti, Dynamic effects of increasing heterogeneity in financial markets. Chaos Solitons & Fractals **41**, 1764–1772 (2009)
38. E. Panas, V. Ninni, Are oil markets chaotic? A non-linear dynamic analysis. Energy Economics **22**, 549–568 (2000)
39. L.P. Shil'nikov, On a Poincaré-Birkhoff problem. Math. USSR-Sbornik **3**, 353–371 (1967)
40. S. Smale, Diffeomorphisms with many periodic points, in *Differential and Combinatorial Topology*, ed. by S.S. Cairns (Princeton University Press, Princeton, 1965), pp. 63–80
41. A. Wei, R.M. Leuthold, *Long Agricultural Futures Prices: ARCH, Long Memory or Chaos Processes? OFOR Paper 98–03* (University of Illinois at Urbana-Champaign, Urbana, 1998)
42. M. Yokoo, Chaotic dynamics in a two-dimensional overlapping generations model. J. Econ. Dyn. Control **24**, 909–934 (2000)
43. W.-B. Zhang, *Differential Equations, Bifurcations, and Chaos in Economics* (World Scientific, Singapore, 2005)

Chapter 10
Global Weather and Climate in the Light of El Niño-Southern Oscillation

In this chapter, we study the chaotic behavior of hydrosphere and its influence on global weather and climate. We give mathematical arguments for the sea surface temperature (SST) to be unpredictable over the global ocean. The impact of SST variability on global climate is clear during global climate patterns, which involve large-scale ocean-atmosphere fluctuations similar to the El Niño-Southern Oscillation (ENSO). Sensitivity (unpredictability) is the core ingredient of chaos. Several researches suggested that the ENSO might be chaotic. It was Vallis [65, 66] who revealed unpredictability of ENSO by reducing his model to the Lorenz equations. Interactions of ENSO and other global climate patterns may transmit chaos. We discuss the unpredictability as a global phenomenon through extension of chaos "horizontally" and "vertically" in coupled Vallis ENSO models, Lorenz systems, and advection equations by using theoretical as well as numerical analyses. To perform theoretical research, we apply the recent results on replication of chaos [1, 3] and unpredictable solutions of differential equations [6, 7], while for numerical analysis, we combine results on unpredictable solutions with numerical analysis of chaos in the advection equation. The main results presented in this chapter are published in [8].

10.1 Introduction and Preliminaries

The famous Lorenz equations give birth to the weather related observations. One of them is the unpredictability of weather in long period of time, which is a meteorological concept, and another one is that small changes of the climate and even weather at present may cause catastrophes for the human life in future. Issuing from this, we have taken into account the following three features of the Lorenz system, to emphasize the actuality of this chapter. Firstly, it is a regional model. Secondly, for some values of its parameters the equations are non-chaotic. Finally,

M. Akhmet et al., *Dynamics with Chaos and Fractals*, Nonlinear Systems and Complexity 29, https://doi.org/10.1007/978-3-030-35854-9_10

the model is of the atmosphere, but not of the hydrosphere. Therefore, one has to make additional investigations to reveal that the unpredictability of weather is a *global* phenomenon, and climatic catastrophes can be caused by physical processes at *any point* on the surface of the globe. The present chapter is concerned with all of the three factors issuing from the ocean surface dynamics of El Niño-Southern Oscillation type, and results of our former research.

10.1.1 Unpredictability of Weather and Deterministic Chaos

Global climate change has gained the attention of scientists and policymakers. The reason for that lies in its remarkable impact on human life on the Earth [48]. Climate change affects and controls many social, economic, and political human activities. It was an essential motive of human migration throughout history.

Weather is defined by the condition of the atmosphere at a specific place and time measured in terms of temperature, humidity, air pressure, wind, and precipitation, whereas climate can be viewed as the average of weather of a large area over a long period of time [42]. Some definitions of climate expand to include the conditions of not only the atmosphere, but also the rest components of the climate system: hydrosphere, cryosphere, lithosphere, biosphere, and, according to Vernadsky, noösphere [22].

During the last few decades, big efforts have been made to develop weather and climate change forecasting models. Due to the chaotic nature of weather, the forecasting range of weather prediction models is limited to only a few days. Climate models are more complicated than ordinary weather forecasting models, since they need to include additional factors of climate system that are not important in the weather forecast [52]. Understanding the concepts of chaos is an important step toward better comprehension of the natural variability of the climate system on different time scales. This involves the determination of the reasons and sources standing behind the presence of chaos in weather and climate models. Any progress made in this path will be helpful to adjust the conception of climate change and find solutions for climate control.

Chaos can be defined as aperiodic long-term behavior in a deterministic system that exhibits sensitive dependence on initial conditions [29, 62]. Predictability consists of constructing a relationship between cause and effect by which we can predict and estimate the future behavior of a physical property. Unpredictability means the failure of such empirical or theoretical relationships to predict due to the presence of noise term(s) and intrinsic irregularity of the physical property itself. Mathematically unpredictability is considered as a result of the sensitive dependence on initial conditions, which is an essential feature of Devaney chaos [20]. Recently, it is theoretically proved that a special kind of Poisson stable trajectory, called an unpredictable trajectory, gives rise to the existence of Poincaré chaos [4–6].

Unpredictability in the dynamics of weather forecast models was firstly observed by E. N. Lorenz. He developed a heat convection model consisting of twelve equations describing the relationship between weather variables such as temperature and pressure. Lorenz surprisingly found that his system was extremely sensitive to initial conditions. Later, in his famous paper [34], he simplified another heat convection model to a three-equation model that has the same sensitivity property [45]. This model is defined by the following nonlinear system of ordinary differential equations:

$$\begin{aligned} \frac{dx}{dt} &= -\sigma x + \sigma y, \\ \frac{dy}{dt} &= r x - xz - y, \\ \frac{dz}{dt} &= xy - b z, \end{aligned} \tag{10.1.1}$$

where the variable x represents the velocity of the convection motion, the variable y is proportional to the temperature difference between the ascending and descending currents, and the variable z is proportional to the deviation of the vertical temperature profile from linearity, whereas the constants σ, r, and b are positive physical parameters. Model (10.1.1) describes the thermal convection of a fluid heated from below between two layers. With certain values of these parameters, Lorenz system possesses intrinsic chaos and produces the so-called Lorenz butterfly attractor.

The dynamical nature of weather and climate requires a deeper understanding of the interaction and feedback mechanisms between the climate system components as well as the individual one between different regions of a certain component. To study the behavior of one component of the climate system at a specific region, a single model defined on particular spatial and temporal scales is acceptable and the accuracy of the outputs mainly depends on the number of climate variables and parameters included in the model. Due to the inevitable simplifications usually adopted for the construction of such models, the outputs include potential errors even for limited time scale. Super-modeling is a recently proposed technique that has been applied to reduce model error and improve prediction. The strategy is based on interconnection of different climate models that synchronize on a common solution, referred to as the supermodel solution [31, 55, 57]. Here, the synchronization in connected models plays an important role for compensating errors in order to achieve an optimal prediction. In this chapter, different models represent the dynamics of neighbor regions in the same component and different components of the climate system are coupled to investigate the global role of chaos in weather and climate through the inter-ocean and ocean–atmosphere interactions. The paper [2] was concerned with the extension of chaos through Lorenz systems. It was demonstrated in [2] that Lorenz systems can be unidirectionally coupled such that the drive system influences the response system, which is non-chaotic in the absence of driving, in such a way that the latter also possesses chaos. Additionally, it was showed that the synchronization does not take place in the dynamics of these types of coupled system. A possible connection of these results to the global weather dynamics was also provided in that study.

10.1.2 Ocean–Atmosphere Interaction and Its Effects on Global Weather

Coupled ocean-atmosphere models are the most fundamental tool for understanding the natural processes that affect climate. These models have been widely applied to interpret and predict global climate phenomena such as ENSO [58]. In meteorology and climate science, SST is considered as a very important factor in ocean–atmosphere interaction, where it plays a basic role in determining the magnitude and direction of the current velocity, as well as the ocean surface wind speed. It is difficult to give a precise definition of SST due to the complexity of the heat transfer operations in the mixed layer of upper ocean. In general, however, it can be defined as the bulk temperature of the oceanic mixed layer with a depth varies from 1 m to 20 m depending on the measurement method used [10]. The importance of SST stems from the fact that the world's oceans cover over 70% of the whole surface of the globe. This large contact area gives way to an active ocean–atmosphere interaction and sometimes becomes a fertile place for complex feedbacks between the ocean and atmosphere that drive an irregular climate change.

The most important example of the interactions and feedbacks between the ocean and the atmosphere is ENSO which is defined as a global coupled ocean-atmosphere phenomenon occurs irregularly in the Pacific Ocean about every 2 to 7 years [63]. This phenomenon is accompanied by undesirable changes in weather across the tropical Pacific and losses in agricultural and fishing industries especially in South America. The El Niño mechanism can be briefly summarized as follows: During normal conditions in the equatorial Pacific, trade winds blow from east to west driving the warm surface current in the same direction. As a consequence of this, warm water accumulates in the Western Pacific around Southeast Asia and Northern Australia. On the opposite side of the ocean around Central and South America, the warm water, pushed to the west, is replaced by upwelling cold deep water. During El Niño conditions, the trade winds are much weaker than normal. Because of this and due to SST difference, warm water flows back towards the Western Pacific. This situation involves large changes in air pressure and rainfall patterns in the tropical Pacific. The cool phase of this phenomenon is called La Nina, which is an intensification of the normal situation. The term "Southern Oscillation" is usually used to refer to the difference of the sea-level pressure (SLP) between Tahiti and Darwin, Australia. Bjerknes [13] conclude that El Niño and the Southern Oscillation are merely two different results of the same phenomenon. These phases of the phenomenon are scientifically called El Niño-Southern Oscillation or shortly ENSO. From the above mechanism we can note that the ENSO dynamics is a perfect example of self-excited oscillating systems.

The ramifications of El Niño are not restricted to the Pacific basin alone, but have widespread effects which severely disrupt global weather patterns. In the last few decades scientists developed theories about the climatic engine which produced El Niño, and they are trying to explain how that engine interacts with the great machine of global climate. Although remarkable progress has been made in monitoring and forecasting the onset of El Niño, it is still challenging to predict its intensity and the impact of the event on global weather. Study of ENSO is considered as a key to understanding climate change, it is a significant stride toward the meteorology's ultimate goal, "accurate prediction and control of world weather."

Besides the ENSO, there are several other atmospheric patterns that occur in different regions of the Earth. These phenomena are interacting in very complicated ways. Many researchers paid attention to the mutual influence of these phenomena and investigated if there is any co-occurrence relationship or interaction between them.

The most similar atmosphere-ocean coupled phenomenon to ENSO is the Indian Ocean Dipole (IOD), which occur in the tropical Indian Ocean, and it is sometimes called the Indian Niño. IOD has normal (neutral), negative, and positive phases. During neutral phase, Pacific warm water, driven by the Pacific trade winds, crosses between South Asia and Australia and flows toward the Indian Ocean. Because of the westerly wind, the warm water accumulates in the eastern basin of Indian Ocean. In the negative IOD phase with the coincidence of strength of the westerly wind, warmer water concentrates near Indonesia and Australia, and causes a heavy rainfall weather in these regions and cooler SST and droughts in the opposite side of the Indian Ocean basin around the eastern coast of Africa. The positive phase is the reversal mode of the negative phase, i.e., what happened in the east side will happen in west side and vice versa.

From the above we can see that there is a symmetry between the IOD and ENSO mechanisms. Indeed, SST data shows that the Indian Ocean warming appears as a near mirror image of ENSO in the Pacific [17]. In addition, the IOD is likely to have a link with ENSO events, where a positive (negative) IOD often occurs during El Niño (La Nina) [21, 69]. Luo et al. [37] investigated the ENSO-IOD interactions, and they suggest that IOD may significantly enhance ENSO and its onset forecast, and vice versa. Several other researchers like [12, 47] studied the relationship and interaction between ENSO and IOD. It should be noted here that (as in all these studies) the SST considered as the major variable, indicator and index for these events.

Other important atmosphere-ocean coupled phenomena like Pacific Decadal Oscillation (PDO), Atlantic Multidecadal Oscillation (AMO), Southern Annular Mode (SAM), Tropical Atlantic Variability (TAV), North Atlantic Oscillation (NAO), Arctic Oscillation/Northern Annular Mode (AO/NAM), Madden–Julian Oscillation (MJO), Pacific/North American pattern (PNA), Quasi-Biennial Oscillation (QBO), and Western Pacific pattern (WP) have significant influences on weather and climate variability throughout the world. Similar to the relationship between ENSO and IOD, various studies show expected relationships between these phenomena and mutual effects on their predictability. Figure 10.1 shows the places of occurrence of the major atmospheric patterns and Table 10.1 gives brief

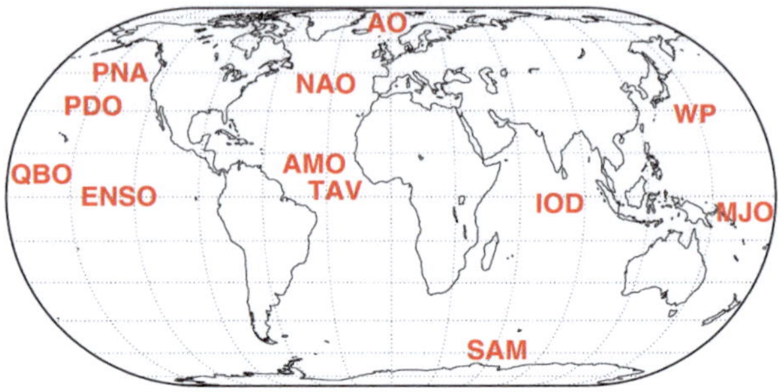

Fig. 10.1 The major global climate patterns

Table 10.1 The major climate variability systems

Term	Descriptions	Main Index	Timescale
ENSO	An irregularly periodical variation in sea surface temperatures over the tropical eastern Pacific Ocean	SST	3–7 years
QBO	An oscillation of the equatorial zonal wind in the tropical stratosphere	SLP	26–30 months
PDO	A low-frequency pattern similar to ENSO occurs primarily in the Northeast Pacific near North America	SST	20–30 years
PNA	An atmospheric pressure pattern driven by the relationship between the warm ocean water near Hawaii and the cool one near the Aleutian Islands of Alaska	SLP	7–8 days
AO/NAM	Defined by westerly winds changes driven by temperature contrasts between the tropics and northern polar areas	SLP	1–9 months
NAO	Large scale of pressure varies in opposite directions in the North Atlantic near Iceland in the north and the Azores in the south	SLP	9–10 days
TAV	Like ENSO, but it exhibits a north-south low frequency oscillation of the SST gradient across the equatorial Atlantic Ocean	SST	10–15 years
AMO	A mode of natural variability occurring in the North Atlantic Ocean and affects the SST on different modes on multidecadal timescales	SST	55–80 years
SAM	Defined by westerly winds changes driven by temperature contrasts between the tropics and southern polar areas	SLP	30–70 days
IOD	An irregular oscillation of sea surface temperatures in equatorial areas of the Indian Ocean	SST	2–5 years
WP	A low-frequency variability characterized by north-south dipolar anomalies in pressure over the far East and western North Pacific	SLP	7–8 days
MJO	An equatorial traveling pattern of anomalous rainfall located in the tropical Pacific and Indian Oceans	SLP	40–50 days

descriptions of them [32, 46, 67]. These pattern modes have different degrees of effect on SST. In Table 10.1, we see that the patterns that remarkably influence the ocean temperature are indexed by SST, whereas those that are most correlated with air pressure, the main indexes of them are based on SLP.

10.1.3 El Niño Chaotic Dynamics

The SST behavior associated with ENSO indicates irregular fluctuations. The ENSO indicator NINO3.4 index, for example, is one of the most commonly used indices, where the SST anomaly averaged over the region bounded by 5°N–5°S, 170°–120°W [16]. Figure 10.2 shows the oscillatory behavior of SST in the NINO3.4 region. Data from the Hadley Centre Sea-Ice and SST dataset HadISST1 [44] is used to generate the figure. This behavior encourages many scientists to answer the question: "Is ENSO a self-sustained chaotic oscillation or a damped one, requiring external stochastic forcing to be excited?" [56]. There are different hypotheses for the source of chaos in ENSO. According to Neelin and Latif [39], deterministic chaos within the nonlinear dynamics of coupled system, uncoupled atmospheric weather noise, and secular variation in the climatic state are the possible source of ENSO irregularity. Tiperman et al. [64] concluded that the chaotic behavior of ENSO is caused by the irregular jumping of the ocean-atmosphere system among different nonlinear resonances. Several studies like [11, 41] support this assumption and attributed the irregularity and the unpredictability of ENSO to influence of stochastic forcing generated by weather noise. Other studies like [38, 71] infer that ENSO is intrinsically chaotic, which means that the irregularity and the loss of predictability are independent of the chaotic nature of weather.

Practically, investigating chaos in ENSO needs long time series of data, which make the task quite difficult experimentally. Vallis [65, 66] developed a conceptual model of ENSO and suggested that the ENSO oscillation exhibits a chaotic behavior. Vallis used finite difference formulation to reduce two-dimensional versions of advection and continuity equations to a set of ordinary differential equations. In addition, he assumed that the zonal current is driven by the surface wind, which is in turn proportional to the temperature difference across the ocean. The model is described by the set of equations

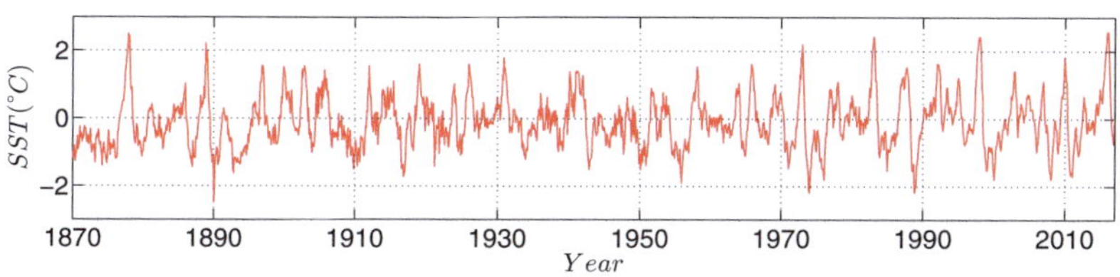

Fig. 10.2 Sea surface temperature anomalies of NINO3.4 region. The data utilized in the figure is from the Hadley Centre Sea-Ice and SST dataset HadISST1

$$\begin{aligned}
\frac{du}{dt} &= \beta\,(T_e - T_w) - \lambda\,(u - u^*),\\
\frac{dT_w}{dt} &= \frac{u}{2\Delta x}\,(\bar{T} - T_e) - \alpha\,(T_w - T^*),\\
\frac{dT_e}{dt} &= \frac{u}{2\Delta x}\,(T_w - \bar{T}) - \alpha\,(T_e - T^*),
\end{aligned} \tag{10.1.2}$$

where u represents the zonal velocity, T_w and T_e are the SST in the eastern and western ocean, respectively, $\bar{T}$ is the deep ocean temperature, T^* is the steady state temperature of ocean, u^* represents the effect of the mean trade winds, Δx is the width of the ocean basin, and α, β, and λ are constants.

By nondimensionalizing Eqs. (10.1.2) and forming the sum and difference of the two temperature equations, one can see that these equations have the same structure as the Lorenz system (10.1.1). Vallis utilized the fact that the Lorenz system, with specific parameters, is intrinsically chaotic, and showed that a chaotic behavior of the sum and difference of the west and east SST can be obtained.

ENSO, as mentioned above, occurs as a result of the interaction of the ocean and atmosphere. Therefore, modeling of ENSO would be a good instrument to research unpredictability not only in the atmosphere but also in the hydrosphere. Nevertheless, ENSO provides the arguments that unpredictability is also proper for seawater parameters which possibly can be reduced to a single one, the SST, if one excludes flow characteristics. Vallis saved in the model only hydrosphere variables ignoring the variation of atmosphere parameters when he considers chaos problem. In our opinion, however, the model is appreciated as a pioneer one, and furthermore, it implies chaos presence in the Pacific Ocean water. Hopefully, in the future, ENSO with both atmosphere and hydrosphere characteristics being variable will be modeled, but this time we focus on chaotic effects of ENSO by utilizing the Vallis model.

10.1.4 Sea Surface Temperature Advection Equation

The temporal and spatial evolution of the SST is governed by a first order quasilinear partial differential equation, the advection equation. If we denote the SST by T, the temperature advection equation of mixed layer of fixed depth can be written in the form [35, 68]

$$\frac{\partial T}{\partial t} + u\frac{\partial T}{\partial x} + v\frac{\partial T}{\partial y} + w\frac{\partial T}{\partial z} = f(t, x, y, z, T), \tag{10.1.3}$$

where u, v, w are the zonal, meridional, and vertical components of current velocity, respectively. These velocities theoretically must satisfy the continuity equation

$$\frac{\partial}{\partial x}(\rho u) + \frac{\partial}{\partial y}(\rho v) + \frac{\partial}{\partial z}(\rho w) = -\frac{\partial \rho}{\partial t}, \tag{10.1.4}$$

where ρ is the seawater density.

The inhomogeneous (forcing) term f on the right-hand side of Eq. (10.1.3) consists of the shortwave flux, the evaporative heat flux, the combined long-wave back-radiation and sensible heat flux and heat flux due to vertical mixing [30]. This term can be described by [26, 28, 60]

$$f \approx \frac{1}{h\rho C_p}\frac{\partial q}{\partial z} + D, \tag{10.1.5}$$

where h is the mixed layer depth, C_p is the heat capacity of seawater, q is radiative and diffusive heat flux, and D is the thermal damping (the numerical diffusion operator).

The spatial and temporal domain of Eq. (10.1.3) depends on the region and the nature of the phenomenon under study. For studying ENSO or IOD, for instance, there are various regions for monitoring SST. NINO3.4 is one of the most commonly used indices for ENSO. Dipole mode index (DMI) is usually used for IOD, and it depends on the difference in average SST anomalies between the western 50°E–70°E, 10°N–10°S and the eastern 90°E–110°E, 0°–10°S boxes [49]. The mixed layer depth h varies with season and depends on the vertical heat flux through the upper layers of the ocean. The average of mixed layer depth is about 30 m [36]. Different studies of ocean-atmosphere coupled models considered different regions of various sizes. Zebiak and Cane [71], for example, developed a model of ENSO. They considered a rectangular model extending from 124°E to 80°W and 29°N to 29°S, with constant mixed layer depth of 50 m and 90 years simulation.

From the above we find that the domain of Eq. (10.1.3) depends on the purpose of the study. To study ENSO, for instance, we would cover a big region of Pacific Ocean basin, and if we choose the origin of coordinates to be at 160°E on the equator, we can write the domain of (10.1.3) as follows:

$$t \geq 0, \quad 0 \leq x \leq 9000\,\text{km}, \quad -3000\,\text{km} \leq y \leq 3000\,\text{km}, \quad -100\,\text{m} \leq z \leq 0.$$

The inhomogeneous term in Eq. (10.1.3), which includes mixing processes of heat transfer, plays the main role for chaotic dynamics. In addition to this term, a chaotic behavior in ocean current velocity terms may also produce an unpredictable behavior in SST. These causes of unpredictability are proved analytically and numerically by perturbing these terms by unpredictable functions. In this chapter, we treat Eq. (10.1.3) mathematically without paying attention to the dimensions of the physical quantities. The important thing to us is the possibility of presence of chaos in this advection equation endogenously or be acquired from other equation or system. The advection equation, in addition to the Vallis model and the Lorenz system, will be used to demonstrate the extension of unpredictability "horizontally" through the global ocean and "vertically" between ocean and atmosphere.

10.1.5 Unpredictability and Poincaré Chaos

There are different types and definitions of chaos. Devaney [20] and Li–Yorke [33] chaos are the most frequently used types, which are characterized by transitivity, sensitivity, frequent separation, and proximality. Another common type is the period-doubling cascade, which is a sort of route to chaos through local bifurcation [23, 50, 53]. In the papers [4, 6], Poincaré chaos was developed by introducing the theory of unpredictable point and unpredictable function, which are built on the concepts of Poisson stable point and function. We define unpredictable point as follows. Let (X, d) be a metric space and $\pi : \mathbb{T} \times X \to X$ be a flow on X, where $\mathbb{T}$ refer to either the set of real numbers or the set of integers. We assume that the triple (π, X, d) defines a dynamical system.

Definition 10.1.1 ([4]) A point $p \in X$ and the trajectory through it are unpredictable if there exist a positive number ϵ (the unpredictability constant) and sequences $\{t_n\}$, $\{\tau_n\}$ both of which diverge to infinity such that $\lim_{n\to\infty} \pi(t_n, p) = p$ and $d[\pi(t_n + \tau_n, p), \pi(\tau_n, p)] \geq \epsilon$ for each $n \in \mathbb{N}$.

Definition 10.1.1 describes unpredictability as individual sensitivity for a motion, i.e., it is formulated for a single trajectory. The Poincaré chaos is distinguished by Poisson stable motions instead of periodic ones. Existence of infinitely many unpredictable Poisson stable trajectories that lie in a compact set meet all requirements of chaos. Based on this, chaos can be appeared in the dynamics on the quasi-minimal set which is the closure of a Poisson stable trajectory. Therefore, the Poincaré chaos is referred to as the dynamics on the quasi-minimal set of trajectory initiated from unpredictable point.

The definition of an unpredictable function is as follows.

Definition 10.1.2 ([7]) A uniformly continuous and bounded function $\varphi : \mathbb{R} \to \mathbb{R}^m$ is unpredictable if there exist positive numbers ϵ, δ and sequences $\{t_n\}$, $\{\tau_n\}$ both of which diverge to infinity such that $\|\varphi(t + t_n) - \varphi(t)\| \to 0$ as $n \to \infty$ uniformly on compact subsets of $\mathbb{R}$, and $\|\varphi(t + t_n) - \varphi(t)\| \geq \epsilon$ for each $t \in [\tau_n - \delta, \tau_n + \delta]$ and $n \in \mathbb{N}$.

To determine unpredictable functions, we apply the uniform convergence topology on compact subsets of the real axis. This topology allows us to construct Bebutov dynamical system on the set of the bounded functions [5, 54]. Consequently, the unpredictable functions imply presence of the Poincaré chaos.

10.1.6 The Role of Chaos in Global Weather and Climate

The topic of weather and climate is one of the most profoundly important issues concerning the international community. It becomes very actual since the catastrophic phenomena such as global warming, hurricanes, droughts, and floods. This is why weather and climate are agenda of researches in physics, geography, meteorology, oceanography, hydrodynamics, aerodynamics, and other fields. The problem is global, that is a comprehensive model would include the interactions of all major climate system components, howsoever, for a specific aspect of the problem, an appropriate model combination can be considered [61]. In the second half of the last century, it was learned [34] that the weather dynamics is irregular and sensitive to initial conditions. Thus the chaos was considered as a characteristic of weather which cannot be ignored. Moreover, chaos can be controlled [40, 43]. These all make us optimistic that the researches of weather and climate considering chaos effect may be useful not only for the deep comprehension of their processes but also for control of them. In the research [1], it has shown how a local control of chaos can be expanded globally.

It is not wrong to say that in meteorological studies, chaos is considered as a severe limiting factor in the ability to predict weather events accurately [51]. Beside this one can say that chaos is also a responsible factor for climate change if it is considered as a weather consequence. This is true, firstly, because of the weather unpredictability, since predictability can be considered as a useful feature of climate with respect to living conditions, and secondly, as the small weather change may cause a global climate change in time. Accordingly, it is possible to say that the control of weather, even a limited artificial one, bring us to a change of climate.

The chaotic behavior has also been observed in several models of ENSO [39]. Presence of chaos in the dynamic of this climate event provides other evidence of the unpredictable nature of the global weather. Besides the Lorenz chaos of atmosphere, “Vallis chaos” takes place in the hydrosphere. Without exaggerating, we can say that chaos seems to be inherent at the essence of any deterministic climate model. Therefore, unpredictability can be globally widespread phenomenon through constructive interactions between the components of the climate system.

To give a sketch how chaos is related globally to weather and climate, we will use information on dynamics of ENSO which will mainly utilize the Vallis model as well as the SST advection equation and the Lorenz equations. They will be properly coupled to have the global effect. It is apparent that, in the next research, the models will possibly be replaced by more developed ones, but our main idea is to demonstrate a feasible approach to the problem by constructing a special net of differential equations system. Obviously, one can consider the net as an instrument which can be subdued to an improvement by arranging new perturbation connections.

Proceeding from aforementioned remarks and as a part of the scientific work, we focus on one possible aspect of global weather and climate dynamics based on El Niño phenomenon. To address this aim, we first review the Vallis model research for El Niño in Sect. 10.1.3, then, in Sect. 10.2 we analyze the presence of chaos in isolated models for the SST advection equation. In Sect. 10.3, the extension of chaos in hydrosphere is discussed through coupling of advection equation, the Vallis model, and also mixing advection equation with the Vallis model. In paper [2] chaos as a global phenomenon in atmosphere was considered, but it is clear that, to say about the weather of the globe one should take into account the hydrosphere as well as the interaction processes between the atmosphere and seas. For this reason

Sect. 10.4 is written where chaos extension from ocean to air and vice versa is discussed on the base of the Lorenz and Vallis models. So, finalizing the introduction we can conclude that the present chapter is considered as an attempt to give a sketch of the global effects of chaos on weather and climate. These results are supposed to be useful for geographers, oceanographers, climate researchers, and those mathematician who are looking for chaotic models and theoretical aspects of chaos researching.

10.2 Unpredictable Solution of the Advection Equation

In this section we study the presence of Poincaré chaos in the dynamics of Eq. (10.1.3). We expect that the behavior of the solutions of (10.1.3) depends on the function f and the current velocity components u, v, w, which are used in the equation. From Eq. (10.1.5), we see that the forcing term f depends mainly on the heat fluxes between the sea surface and atmosphere which is governed by SST, air temperature, and wind speed, as well as between layers of sea which is caused by sea temperature gradient and vertical (entrainment) velocity. Therefore, this forcing term can be a natural source of noise and irregularity. Ocean currents are mainly driven by wind forces, as well as temperature and salinity differences [19]. Thence again we can deduce that the irregular fluctuations of wind may be reflected in the behavior of SST.

To demonstrate the role of the function f in the dynamics of Eq. (10.1.3), let us take into account the equation

$$\frac{\partial T}{\partial t} + u\frac{\partial T}{\partial x} + v\frac{\partial T}{\partial y} + w\frac{\partial T}{\partial z} = -0.7\,T + 0.3\,w_1\,T + 5\,\sin(xt), \quad (10.2.6)$$

where the current velocity components are defined by $u = \sin(\frac{x}{2}) + \sin(t) + 3$, $v = -0.02$, and $w = -\frac{1}{2}\cos(\frac{x}{2})z$.

Figure 10.3 represents the solution of (10.2.6) corresponding to the initial data $T(0, 0, 0, 0) = 0.5$. It is seen in Fig. 10.3 that the solution of Eq. (10.2.6) has an irregular oscillating behavior, whereas in the absence of the term $5\sin(xt)$ in the

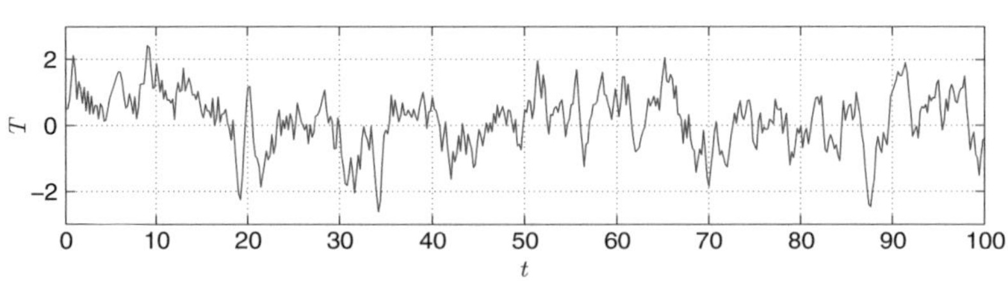

Fig. 10.3 The solution of Eq. (10.2.6) with the initial condition $T(0, 0, 0, 0) = 0.5$. The figure shows that the forcing term f has a significant role in the dynamics of (10.1.3)

function f, the solution approaches the steady state. Even though the behavior of this numerical solution depends on the step size of the numerical scheme used, this situation leads us to consider that the forcing term f has a dominant role in the behavior of SST.

To investigate the existence of an unpredictable solution in the dynamics of Eq. (10.1.3) theoretically, let us apply the method of characteristics. If we parameterize the characteristics by the variable t and suppose that the initial condition is given by $T(t_0, x, y, z) = \Phi(x, y, z)$, where t_0 is the initial time, then we obtain the system

$$\begin{aligned} \frac{dx}{dt} &= u(t, x, y, z, T), \\ \frac{dy}{dt} &= v(t, x, y, z, T), \\ \frac{dz}{dt} &= w(t, x, y, z, T), \\ \frac{dT}{dt} &= f(t, x, y, z, T), \end{aligned} \tag{10.2.7}$$

with the initial conditions

$$x(t_0) = x_0, \quad y(t_0) = y_0, \quad z(t_0) = z_0, \quad T(t_0, x_0, y_0, z_0) = \Phi(x_0, y_0, z_0).$$

In system (10.2.7), we assume that u, v, w, and f are functions of x, y, z, t, and T, and they have the forms

$$\begin{aligned} u &= a_1 x + a_2 y + a_3 z + a_4 T + U(x, y, z, T), \\ v &= b_1 x + b_2 y + b_3 z + b_4 T + V(x, y, z, T), \\ w &= c_1 x + c_2 y + c_3 z + c_4 T + W(x, y, z, T), \\ f &= d_1 x + d_2 y + d_3 z + d_4 T + F(x, y, z, T), \end{aligned} \tag{10.2.8}$$

where a_i, b_i, c_i, d_i, $i = 1, 2, 3, 4$, are real constants and the functions U, V, W, F are continuous in all their arguments. System (10.2.7) can be expressed in the form

$$X'(t) = AX(t) + Q(t), \tag{10.2.9}$$

in which

$$X(t) = \begin{bmatrix} x \\ y \\ z \\ T \end{bmatrix}, \quad A = \begin{bmatrix} a_1 & a_2 & a_3 & a_4 \\ b_1 & b_2 & b_3 & b_4 \\ c_1 & c_2 & c_3 & c_4 \\ d_1 & d_2 & d_3 & d_4 \end{bmatrix}, \quad Q = \begin{bmatrix} U \\ V \\ W \\ F \end{bmatrix}. \tag{10.2.10}$$

The following theorem is needed to verify the existence of Poincaré chaos in the dynamics of Eq. (10.1.3).

Theorem 10.1 ([6]) *Consider the system of ordinary differential equations*

$$X'(t) = A\,X(t) + G(X(t)) + H(t), \tag{10.2.11}$$

where the $n \times n$ constant matrix A has eigenvalues all with negative real parts, the function $G : \mathbb{R}^n \to \mathbb{R}^n$ is Lipschitzian with a sufficiently small Lipschitz constant, and $H : \mathbb{R} \to \mathbb{R}^n$ is a uniformly continuous and bounded function. If the function $H(t)$ is unpredictable, then system (10.2.11) possesses a unique uniformly exponentially stable unpredictable solution, which is uniformly continuous and bounded on the real axis.

In the remaining parts of the present section, we will discuss the unpredictability when SST is chaotified by external irregularity. For that purpose let us consider the logistic map

$$\eta_{j+1} = 3.91\,\eta_j\,(1-\eta_j),\ \ j \in \mathbb{Z}. \tag{10.2.12}$$

According to Theorem 4.1 in [6], the map (10.2.12) is Poincaré chaotic such that it possesses an unpredictable trajectory.

Next, we define a function $\phi(t)$ by

$$\phi(t) = \int_{-\infty}^{t} e^{-2(t-s)}\gamma^*(s)\,ds, \tag{10.2.13}$$

where

$$\gamma^*(t) = \begin{cases} 1.5, & \zeta^*_{2j} < t \le \zeta^*_{2j+1}, \quad j \in \mathbb{Z}, \\ 0.3, & \zeta^*_{2j-1} < t \le \zeta^*_{2j}, \quad j \in \mathbb{Z}, \end{cases} \tag{10.2.14}$$

is a relay function. In (10.2.14), the sequence $\{\zeta^*_j\}$ of switching moments is generated through the equation $\zeta^*_j = j + \eta^*_j,\ j \in \mathbb{Z}$, where $\{\eta^*_j\}$ is an unpredictable trajectory of the logistic map (10.2.12).

One can confirm that $\phi(t)$ is bounded such that $\sup\limits_{t\in\mathbb{R}} |\phi(t)| \le 3/4$. It was shown in paper [6] that the function $\phi(t)$ is the unique uniformly exponentially stable unpredictable solution of the differential equation

$$\vartheta'(t) = -2\vartheta(t) + \gamma^*(t). \tag{10.2.15}$$

It is not an easy task to visualize the unpredictable function $\phi(t)$. Therefore, in order to illustrate the chaotic dynamics, we take into account the differential equation

$$\vartheta'(t) = -2\vartheta(t) + \gamma(t), \tag{10.2.16}$$

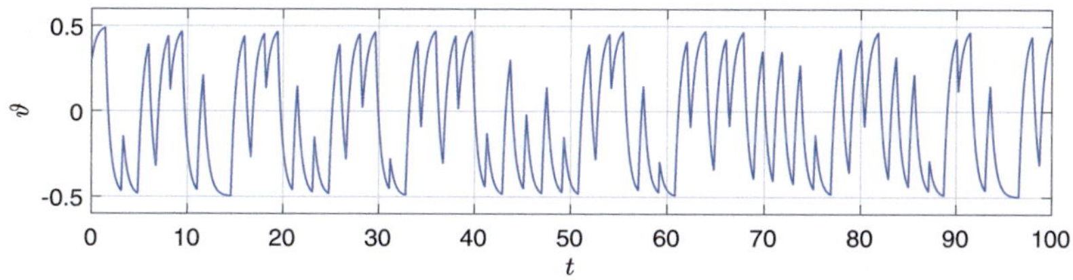

Fig. 10.4 The solution of Eq. (10.2.16) with $\vartheta(0) = 0.3$. The figure supports that the function $\phi(t)$ is unpredictable

where

$$\gamma(t) = \begin{cases} 1.5, & \zeta_{2j} < t \leq \zeta_{2j+1}, \quad j \in \mathbb{Z}, \\ 0.3, & \zeta_{2j-1} < t \leq \zeta_{2j}, \quad j \in \mathbb{Z}, \end{cases} \tag{10.2.17}$$

and the sequence $\{\zeta_j\}$ satisfies the equation $\zeta_j = j + \eta_j, \ j \in \mathbb{Z}$, in which $\{\eta_j\}$ is a solution of (10.2.12) with $\eta_0 = 0.4$. The coefficient 3.91 used in the logistic map (10.2.12) and the initial data $\eta_0 = 0.4$ were considered for shadowing analysis in the paper [27].

We depict in Fig. 10.4 the solution of Eq. (10.2.16) with $\vartheta(0) = 0.3$. It is seen in Fig. 10.4 that the behavior of the solution is irregular, and this support that the function $\phi(t)$ is unpredictable.

10.2.1 Unpredictability Due to the Forcing Source Term

Let us perturb Eq. (10.1.3) with the unpredictable function $\phi(t)$ defined by (10.2.13) and set up the equation

$$\frac{\partial T}{\partial t} + u\frac{\partial T}{\partial x} + v\frac{\partial T}{\partial y} + w\frac{\partial T}{\partial z} = f(t, x, y, z, T) + \psi(\phi(t)), \tag{10.2.18}$$

where u, v, w, and f are in the form of (10.2.8) and $\psi : [-3/4, 3/4] \to \mathbb{R}$ is a continuous function.

Using the method of characteristics, one can reduce Eq. (10.2.18) to system (10.2.7) that can be expressed in the form of (10.2.11) with

$$G(X(t)) = \begin{bmatrix} U \\ V \\ W \\ F \end{bmatrix}, \qquad H(t) = \begin{bmatrix} 0 \\ 0 \\ 0 \\ \psi(\phi(t)) \end{bmatrix}.$$

According to the result of Theorem 5.2 in [6], if there exist positive constants L_1 and L_2 such that

$$L_1 |s_1 - s_2| \leq |\psi(s_1) - \psi(s_2)| \leq L_2 |s_1 - s_2| \tag{10.2.19}$$

for all $s_1, s_2 \in [-3/4, 3/4]$, then the function $H(t)$ is also unpredictable.

Now, in Eq. (10.2.18), let us take $u = -0.03x + 0.1\sin(\frac{x}{80}) + 0.4$, $v = -0.01y - 0.05\sin(y)$, $w = -0.02z + (0.05\cos(y) - 0.00125\cos(\frac{x}{80}))z$, $\psi(s) = 6s$, and $f(t, x, y, z, T) = -1.5T + 0.1w_2T$. Since the conditions of Theorem 10.1 are valid and inequality (10.2.19) holds for these choices of ψ, f, u, v, and w, Eq. (10.2.18) exhibits Poincaré chaos.

In order to simulate the chaotic behavior, we consider the equation

$$\frac{\partial T}{\partial t} + u\frac{\partial T}{\partial x} + v\frac{\partial T}{\partial y} + w\frac{\partial T}{\partial z} = f(t, x, y, z, T) + \psi(\vartheta(t)), \tag{10.2.20}$$

where $\vartheta(t)$ is the function depicted in Fig. 10.4, and u, v, w, f, ψ are the same as above. Figure 10.5 shows the solution $T(t, x, y, z)$ of (10.2.20) corresponding to the initial condition $T(0, 0, 0, 0) = 0.5$. It is seen in Fig. 10.5 that the behavior of the solution is chaotic, and this supports the result of Theorem 10.1 such that Eq. (10.2.18) admits an unpredictable solution.

Next, we will visualize the chaotic dynamics in the integral surface of SST. For that purpose, we omit the term of the meridional advection $v\frac{\partial T}{\partial y}$ in (10.2.18), which has less effect on SST compared with the zonal and vertical advections [14], and set up the equation

$$\frac{\partial T}{\partial t} + u\frac{\partial T}{\partial x} + w\frac{\partial T}{\partial z} = -1.5\,T + w\,T + 6\,\vartheta(t), \tag{10.2.21}$$

where $u = 1.2 + 0.1\sin(\frac{x}{80}) + 0.05\sin(3t)$ and $w = 0.1 - 0.00125\cos(\frac{x}{80})z$. In (10.2.21), $\vartheta(t)$ is again the function shown in Fig. 10.4.

We apply a finite difference scheme to solve Eq. (10.2.21) directly. In such a scheme, we need to specify boundary conditions along with an initial condition. Using the initial condition $T(0, x, z) = 5$ and the boundary conditions $T(t, 0, z) =$

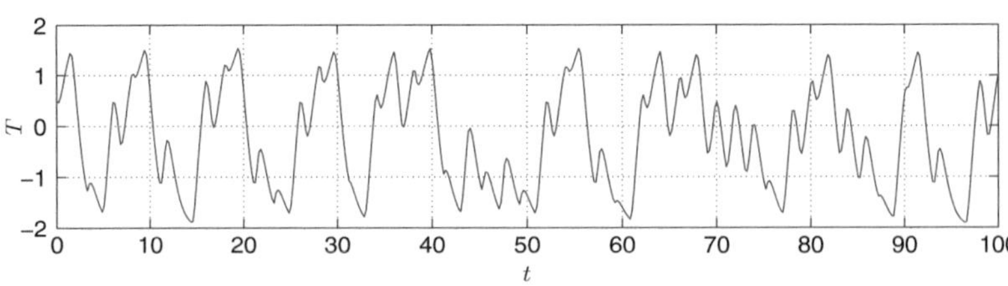

Fig. 10.5 The solution of Eq. (10.2.20) with the initial condition $T(0, 0, 0, 0) = 0.5$. The figure reveals the presence of an unpredictable solution in the dynamics of (10.2.18)

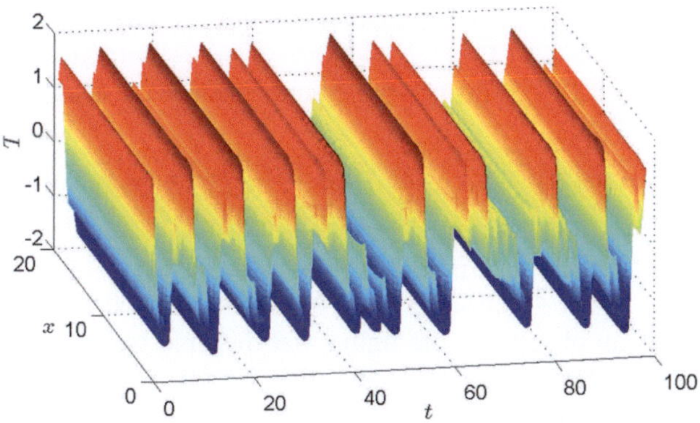

Fig. 10.6 The integral surface of (10.2.21). The chaotic behavior in the SST is observable in the figure

$T(t, x, 0) = 0.5$, we represent in Fig. 10.6 the integral surface of (10.2.21) with respect to t, x, and the fixed value $z = 0$ for $5 \le x \le 20$ and $0 \le t \le 100$. Figure 10.6 supports the result of Theorem 10.1 one more time such that Poincaré chaos is present in the dynamics.

10.2.2 Unpredictability Due to the Current Velocity

This subsection is devoted to the investigation of SST when the current velocity behaves chaotically. Here, we will make use of the unpredictable function $\phi(t)$ defined by (10.2.13) to apply perturbations to the zonal and vertical components of current velocity in Eq. (10.1.3).

We begin with considering the equation

$$\frac{\partial T}{\partial t} + [u + \psi(\phi(t))]\frac{\partial T}{\partial x} + v\frac{\partial T}{\partial y} + w\frac{\partial T}{\partial z} = f(t, x, y, z, T), \qquad (10.2.22)$$

where, in a similar way to (10.2.18), u, v, w, and f are in the form of (10.2.8), and $\psi : [-3/4, 3/4] \to \mathbb{R}$ is a continuous function.

One can confirm that Theorem 10.1 can be used to verify the existence of Poincaré chaos in the dynamics of (10.2.22) since it can be reduced by means of the method of characteristics to a system of the form (10.2.11) with

$$H(t) = \begin{bmatrix} \psi(\phi(t)) \\ 0 \\ 0 \\ 0 \end{bmatrix},$$

which is an unpredictable function provided that ψ satisfies the condition (10.2.19).

In order to demonstrate the chaotic dynamics of (10.2.22), we take $u = -0.003\,x + 0.2\sin(\frac{x}{3}) + 0.4$, $v = -0.001\,y$, $w = -0.002\,z - \frac{0.2}{3}\cos(\frac{x}{3})\,z$, $\psi(s) = 3s$, $f = -1.5\,T - 3\sin(3x) + 0.2$, and consider the equation

$$\frac{\partial T}{\partial t} + [u + \psi(\vartheta(t))]\frac{\partial T}{\partial x} + v\frac{\partial T}{\partial y} + w\frac{\partial T}{\partial z} = f(t, x, y, z, T), \qquad (10.2.23)$$

where $\vartheta(t)$ is the function shown in Fig. 10.4.

The time series of the solution of (10.2.23) with $T(0, 0, 0, 0) = 0.5$ is depicted in Fig. 10.7. One can observe in the figure that the time series is chaotic, and this confirms the result of Theorem 10.1 such that Eq. (10.2.22) possesses an unpredictable solution. More precisely, the perturbation of the zonal velocity component in Eq. (10.1.3) by the unpredictable function $\psi(\phi(t))$ affects the dynamics in such a way that the perturbed equation (10.2.22) is Poincaré chaotic.

Next, we will examine the case when the vertical velocity component in Eq. (10.1.3) is perturbed by the unpredictable function $\phi(t)$. For this aim we set up the equation

$$\frac{\partial T}{\partial t} + u\frac{\partial T}{\partial x} + v\frac{\partial T}{\partial y} + [w + \psi(\phi(t))]\frac{\partial T}{\partial z} = f(t, x, y, z, T), \qquad (10.2.24)$$

where the function $\psi : [-3/4, 3/4] \to \mathbb{R}$ is continuous. If we take $u = -0.001\,x + 0.2\sin(\frac{x}{3}) + 0.4$, $v = -0.001\,y$, $w = -0.03z - \frac{0.2}{3}\cos(\frac{x}{3})z$, $\psi(s) = 3s$, and $f = -1.7\,T + 0.5\,z + 1.6$, then Eq. (10.2.24) admits an unpredictable solution in accordance with Theorem 10.1.

We represent in Fig. 10.8 the solution of the equation

$$\frac{\partial T}{\partial t} + u\frac{\partial T}{\partial x} + v\frac{\partial T}{\partial y} + [w + \psi(\vartheta(t))]\frac{\partial T}{\partial z} = f(t, x, y, z, T), \qquad (10.2.25)$$

corresponding to the initial data $T(0, 0, 0, 0) = 0.5$. Here, we use the same u, v, w, ψ, and f as in (10.2.24), and $\vartheta(t)$ is again the function whose time series is depicted in Fig. 10.4. The irregular fluctuations seen in the figure uphold the result of Theorem 10.1.

We end up this subsection by illustrating the influence of the chaotic current velocity on the integral surface of SST. Figure 10.9a shows the integral surface of

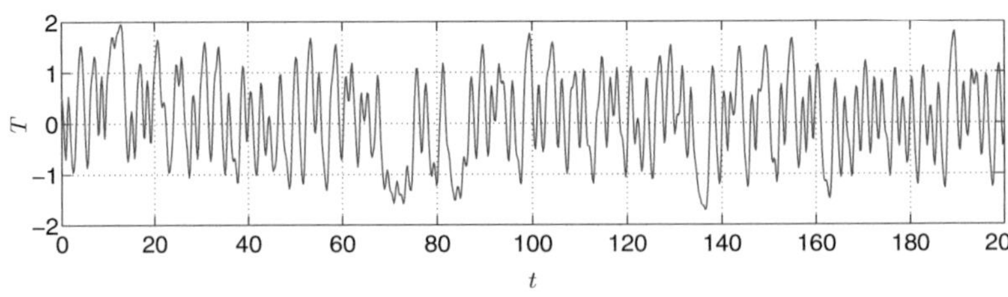

Fig. 10.7 The solution of (10.2.22) with $T(0, 0, 0, 0) = 0.5$. The chaotic behavior of the solution is apparent in the figure

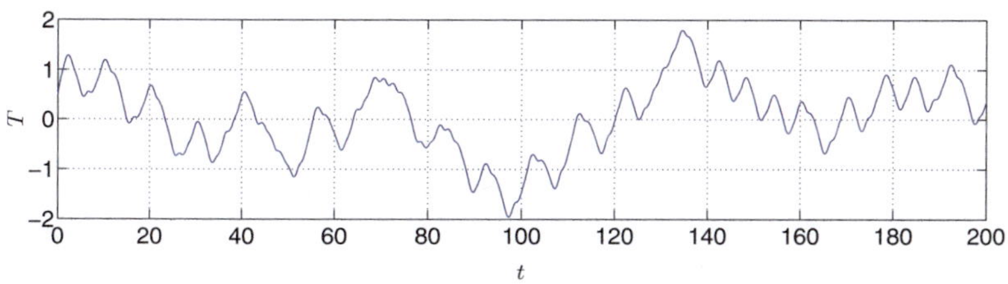

Fig. 10.8 Chaotic behavior of SST due to the perturbation of the vertical component of current velocity. The figure shows the solution of (10.2.25) with $T(0, 0, 0, 0) = 0.5$

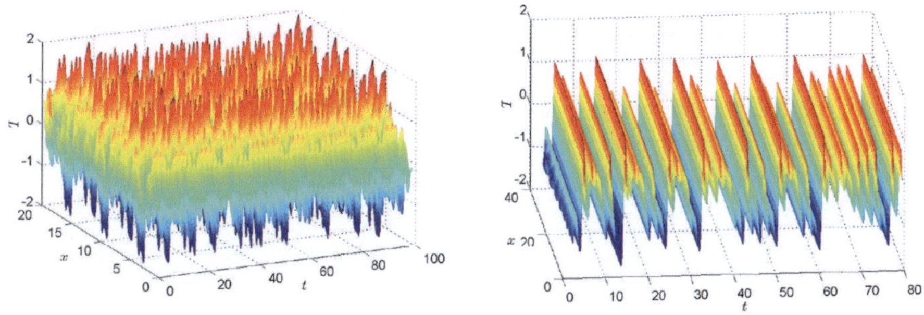

(a) (b)

Fig. 10.9 Chaotic behavior of SST due to the current velocity with initial condition $T(0, x, y, z) = \sin(xz) + 1$, and boundary conditions $T(t, 0, y, z) = T(t, x, y, 0) = 0.5$. Both pictures in (**a**) and (**b**) reveal that chaotic behavior in the current velocity leads to the presence of chaos in SST. (**a**) The integral surface of (10.2.22) at $z = 0$. (**b**) The integral surface of (10.2.24) at $z = 1.5$

(10.2.22) with $u = 1.5 + 0.5 \sin x$, $v = 0$, $w = 1 - 0.5 \cos x$, $\psi(s) = 2s$, and $f = -1.2T - 3\sin(3x)$ at $z = 0$. The initial condition $T(0, x, y, z) = \sin(xz) + 1$ and the boundary conditions $T(t, 0, y, z) = T(t, x, y, 0) = 0.5$ are utilized in the simulation. One can see in Fig. 10.9a that the SST has chaotic behavior in keeping with the result of Theorem 10.1. On the other hand, using the same initial and boundary conditions, we represent in Fig. 10.9b the integral surface of (10.2.24) with $u = 1$, $v = 0$, $w = 1$, $\psi(s) = 2s$, and $f = -1.2T + 3\sin(3z)$ at $z = 1.5$. Figure 10.9b also manifests that the applied perturbation on the vertical component of current velocity make the Eq. (10.2.24) behave chaotically even if it is initially non-chaotic in the absence of the perturbations.

10.3 Chaotic Dynamics of the Global Ocean Parameters

Chaotic behavior may transmit from one model to another [1]. This transmission interprets, for instance, why the unpredictability in one stock market or in the weather of one area is affected by another. Chaos in SST may be gained from another endogenous chaotic system like air temperature or wind speed. We can deal with the global ocean as a finite union of subregions. Each of these subregions may be controlled by different models depending on the position and circumstances. An assumption of the existence of chaotic and non-chaotic subregions for SST behavior is very probable. However, it seems quite unreasonable to imagine a predictable SST for one region, whereas its neighbor region is characterized by an unpredictable SST. The mutual effect in SST between neighbor regions can be seen by coupling their controlling models.

10.3.1 Extension of Chaos in Coupled Advection Equations

In this part of the chapter we deal with the extension of chaos in coupled advection equations. For that purpose, we consider a Poincaré chaotic advection equation of the form (10.2.18) as the drive, and we take into account the equation

$$\frac{\partial \tilde{T}}{\partial t} + \tilde{u}\frac{\partial \tilde{T}}{\partial x} + \tilde{v}\frac{\partial \tilde{T}}{\partial y} + \tilde{w}\frac{\partial \tilde{T}}{\partial z} = \tilde{f}(t, x, y, z, \tilde{T}) + g(T) \tag{10.3.26}$$

as the response, in which g is a continuous function and T is a solution of the drive equation (10.2.18). We assume that the response does not possess chaos in the absence of the perturbation, i.e., we suppose that the advection equation

$$\frac{\partial \tilde{T}}{\partial t} + \tilde{u}\frac{\partial \tilde{T}}{\partial x} + \tilde{v}\frac{\partial \tilde{T}}{\partial y} + \tilde{w}\frac{\partial \tilde{T}}{\partial z} = \tilde{f}(t, x, y, z, \tilde{T}) \tag{10.3.27}$$

is non-chaotic.

Figure 10.10 shows the extension of unpredictability between neighbor regions schematically. We assume that the dynamics of the chaotic region is governed by the drive equation (10.2.18), which has an unpredictable solution, and the dynamics of the non-chaotic region is governed by Eq. (10.3.27). The coupling between these two equations leads to the transmission of unpredictability such that the response system (10.3.26) possesses chaos.

To demonstrate the extension of chaos numerically, let us consider the response equation (10.3.26) with $u = 1.2$, $v = 0$, $w = 0.3$, $f = -1.5\tilde{T} + 0.2$, and $g(T) = T$. Using the solution T of Eq. (10.2.21) satisfying $T(0, 0, 0, 0) = 0.5$ as the perturbation in Eq. (10.3.26), we depict in Fig. 10.11 the solution $\tilde{T}$ of

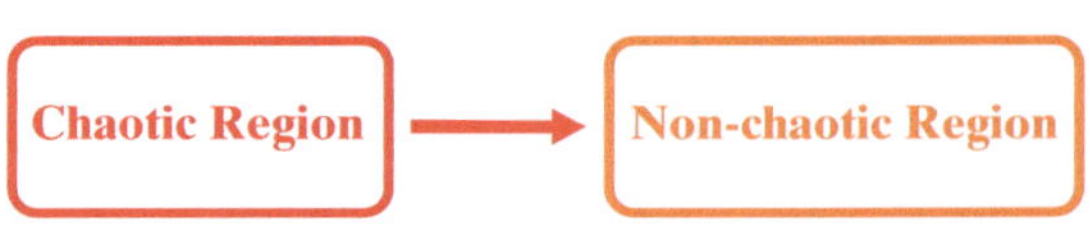

Fig. 10.10 Chaos extension between neighbor regions

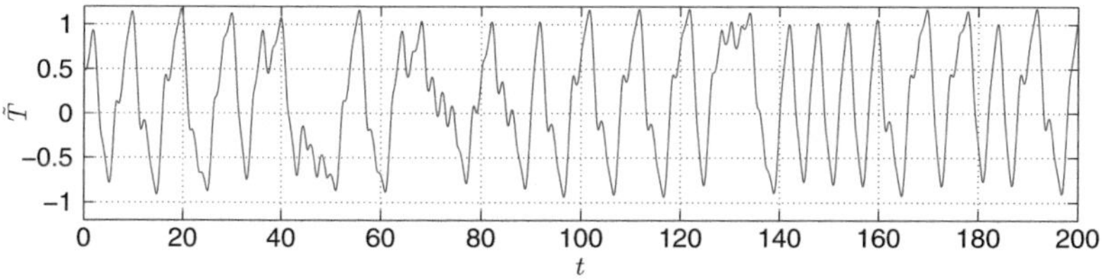

Fig. 10.11 The solution of the response Eq. (10.3.26) with initial condition $T(0, 0, 0, 0) = 0.5$. The figure manifests the extension of chaos in the coupled system (10.2.18)–(10.3.26)

(10.3.26) corresponding to the initial data $\tilde{T}(0, 0, 0, 0) = 0.5$. Figure 10.11 reveals the extension of chaos in the coupled system (10.2.18)–(10.3.26).

10.3.2 Coupling of the Advection Equation with Vallis Model

The Lorenz-like form of the Vallis model is given by [66]

$$\begin{aligned} \frac{du}{dt} &= B\,T_d - C\,u, \\ \frac{dT_d}{dt} &= u\,T_s - T_d, \\ \frac{dT_s}{dt} &= -u\,T_d - T_s + 1, \end{aligned} \tag{10.3.28}$$

where u represents the zonal velocity, $T_d = (T_e - T_w)/2$, $T_s = (T_e + T_w)/2$, T_e and T_w are the SST in the eastern and western ocean, respectively, and B and C are constants. System (10.3.28) is comparable to the Lorenz system and it was shown by Vallis [65, 66] that (10.3.28) with the parameters $B = 102$ and $C = 3$ is chaotic. In paper [25] the authors studied the system (10.3.28) with the same parameters and by using the computer-assisted proofs that follow the standard Mischaikow–Mrozek–Zgliczynski approach they located, in the dynamics of the system, topological horseshoes in iterates of Poincaré return maps such that chaos was detected. They considered the chaos with the standard Li–Yorke conditions and said that the dynamics is complicated at least as the dynamics of the full shift on the space of two symbols. The existence of chaos in the dynamics of the Vallis system was also investigated in other studies such as [9, 15].

Next, we take into account the equations

$$\frac{\partial T_1}{\partial t} + 1.2\frac{\partial T_1}{\partial x} + 0.3\frac{\partial T_1}{\partial z} = -1.2\,T_1 - 1 + 2\sin x, \tag{10.3.29}$$

$$\frac{\partial T_2}{\partial t} + 1.2\frac{\partial T_2}{\partial x} + 0.3\frac{\partial T_2}{\partial z} = -2\,T_2 + 4\sin x, \tag{10.3.30}$$

$$\frac{\partial T_3}{\partial t} + 0.6\frac{\partial T_3}{\partial x} + 0.5\frac{\partial T_3}{\partial z} = -2\,T_3 - 1 + 3\sin x, \tag{10.3.31}$$

and

$$\frac{\partial T_4}{\partial t} + 1.2\frac{\partial T_4}{\partial x} + 0.3\frac{\partial T_4}{\partial z} = -1.5\,T_4. \tag{10.3.32}$$

One can verify that the Eqs. (10.3.29), (10.3.30), (10.3.31), and (10.3.32) are all non-chaotic such that they admit asymptotically stable regular solutions. By applying perturbations to these equations, we set up the following ones:

$$\frac{\partial T_1}{\partial t} + 1.2\frac{\partial T_1}{\partial x} + 0.3\frac{\partial T_1}{\partial z} = -1.2\,T_1 - 1 + 2\sin x + 4.6\,T_s, \tag{10.3.33}$$

$$\frac{\partial T_2}{\partial t} + (1.2 + 0.8\,u)\frac{\partial T_2}{\partial x} + 0.3\frac{\partial T_2}{\partial z} = -2\,T_2 + 4\sin x, \tag{10.3.34}$$

$$\frac{\partial T_3}{\partial t} + (0.6 + u)\frac{\partial T_3}{\partial x} + 0.5\frac{\partial T_3}{\partial z} = -2\,T_3 - 1 + 3\sin x + 4\,T_s, \tag{10.3.35}$$

$$\frac{\partial T_4}{\partial t} + 1.2\frac{\partial T_4}{\partial x} + 0.3\frac{\partial T_4}{\partial z} = -1.5\,T_4 + 2.7\,T_2, \tag{10.3.36}$$

where (u, T_d, T_s) is the solution of the chaotic Vallis model (10.3.28) with $B = 102$, $C = 3$ and the initial conditions $u(0) = 2$, $T_d(0) = 0.2$, and $T_s(0) = 0.4$.

In Eq. (10.3.33) the forcing term is perturbed by the SST average, T_s, whereas in Eq. (10.3.34) the zonal velocity of Vallis model, u, is used as perturbation. On the other hand, in Eq. (10.3.35) both the forcing term and the zonal velocity components are perturbed with the solution of (10.3.28). Moreover, the solution T_2 of (10.3.34) is used as a perturbation in the forcing term of Eq. (10.3.36). The appearance of the zonal velocity u of the model (10.3.28) in the coefficients of Eqs. (10.3.34) and (10.3.35) looks reasonable if one remembers that the parts of the ocean surface under consideration are adjoining to each other, and consequently, the zonal velocity u perturbs its counterpart in the neighbor region from 1.2 to $1.2 + 0.8u$ in Eq. (10.3.34) and from 0.6 to $0.6 + u$ in Eq. (10.3.35). Furthermore, the perturbations with T_s in Eqs. (10.3.33) and (10.3.35) can be attributed to the heat transfer between the neighbor regions because of the structure of the original Eq. (10.1.3). A schematic representation of these coupled systems is given in Fig. 10.12.

Figure 10.13a and b, respectively, shows the solutions T_2, T_3 of Eqs. (10.3.34) and (10.3.35), respectively. The initial data $T_2(0, 0, 0, 0) = 0.5$ and $T_3(0, 0, 0, 0) =$

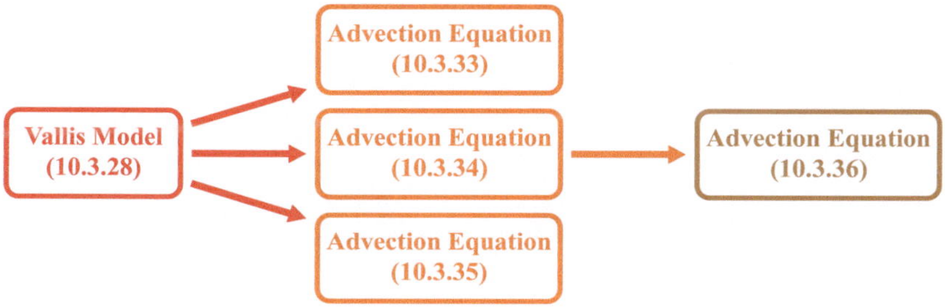

Fig. 10.12 Chaos extension through coupled systems

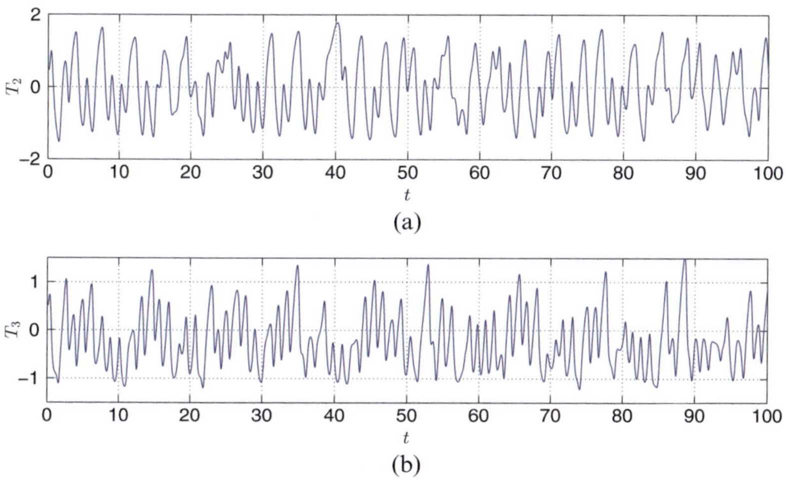

Fig. 10.13 The extension of the chaotic behavior by Eqs. 10.3.34 and 10.3.35. (**a**) The time series of the solution of Eq. 10.3.34. (**b**) The time series of the solution of Eq. 10.3.35. The initial data $T_2(0, 0, 0, 0) = 0.5$ and $T_3(0, 0, 0, 0) = 0.5$ are used

0.5 are used in the simulation. Figure 10.13 reveals that the chaos of the model (10.3.28) is extended by Eqs. (10.3.34) and (10.3.35).

On the other hand, we depict in Fig. 10.14a and b the 3-dimensional integral surfaces corresponding to Equations (10.3.33) and (10.3.36), respectively. Here, we make use of the boundary conditions $T_1(0, x, z) = T_1(t, 0, z) = T_1(t, x, 0) = 0.5$ and $T_4(0, x, z) = T_4(t, 0, z) = T_4(t, x, 0) = 0.5$. The figure confirms one more time that the chaos of system (10.3.28) is extended.

10.3.3 Coupling of Vallis Models

Our purpose in this subsection is to demonstrate numerically our suggestion that chaos can be extended between the regions of some global climate variabilities. We

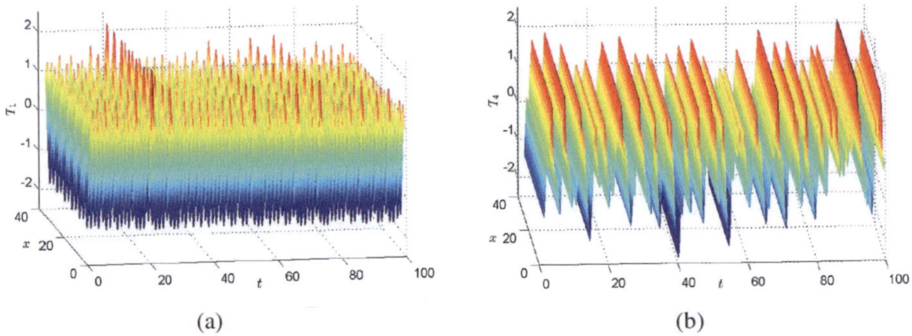

Fig. 10.14 Extension of chaos by Eqs. 10.3.33 and 10.3.36. (**a**) The integral surface of Eq. 10.3.33. (**b**) The integral surface of Eq. 10.3.36

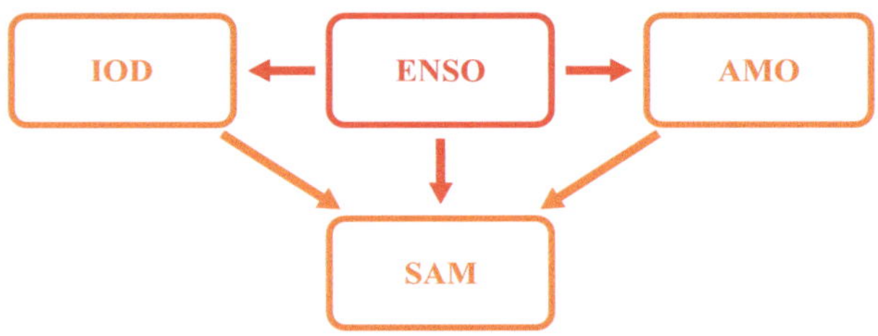

Fig. 10.15 Diagram of possible chaos extensions through global climate patterns regions

assume that there are intermediate subregions located between these main regions and chaos can transmit from one region to another in a sequential way.

We also suggest that the IOD can be described by a Vallis model in the form of (10.3.28) with parameters appropriate to the Indian Ocean. Evaluation of these parameters is rather difficult. However, for simplicity we can choose these values such that system (10.3.28) does not exhibit chaotic behavior. Similar arguments can also be supposed for the AMO and SAM.

A diagram of possible chaos extensions between the regions of IOD, AMO, and SAM is shown in Fig. 10.15. In this diagram, the Vallis model representing ENSO is assumed to be the main source of the chaotic behavior, while the Vallis models representing the behaviors of the regions of IOD, AMO, and SAM are all assumed to be initially non-chaotic when interactions do not occur between the regions.

To demonstrate the extension of chaos, let us consider the perturbed Vallis system

$$\begin{aligned}
\frac{d\tilde{u}}{dt} &= \tilde{B}\,\tilde{T}_d - \tilde{C}\,\tilde{u} + 1.5\,u,\\
\frac{d\tilde{T}_d}{dt} &= \tilde{u}\,\tilde{T}_s - \tilde{T}_d + 0.3\,T_d,\\
\frac{d\tilde{T}_s}{dt} &= -\tilde{u}\,\tilde{T}_d - \tilde{T}_s + 1 + 0.2\,T_s,
\end{aligned} \tag{10.3.37}$$

where (u, T_d, T_s) is the solution of the chaotic Vallis system (10.3.28) with $B = 102$ and $C = 3$ corresponding to the initial conditions $u(0) = 2$, $T_d(0) = 0.2$, and $T_s(0) = 0.4$. We use the parameters $\tilde{B} = 20$ and $\tilde{C} = 7$ in (10.3.37) and assume that the unperturbed Vallis model

$$
\begin{aligned}
\frac{d\tilde{u}}{dt} &= \tilde{B}\,\tilde{T}_d - \tilde{C}\,\tilde{u},\\
\frac{d\tilde{T}_d}{dt} &= \tilde{u}\,\tilde{T}_s - \tilde{T}_d,\\
\frac{d\tilde{T}_s}{dt} &= -\tilde{u}\,\tilde{T}_d - \tilde{T}_s + 1,
\end{aligned}
\tag{10.3.38}
$$

represents the SST and zonal velocity variabilities associated with the dynamics of IOD. With these parameter values one can verify that the system (10.3.38) has an asymptotically stable equilibrium point at $(1.363, 0.477, 0.350)$, and therefore, it is non-chaotic. Figure 10.16 shows the trajectory of (10.3.38) corresponding to the initial conditions $\tilde{u}(0) = 2$, $\tilde{T}_d(0) = 0.2$, $\tilde{T}_s(0) = 0.4$, and it confirms the presence of the asymptotically stable equilibrium point. It is rigorously proven in paper [1] that the Li–Yorke chaos can be transmitted from a chaotic generator to a non-chaotic replicator with asymptotically stable equilibrium, and since the chaos exhibited by (10.3.28) satisfies the Li–Yorke conditions, system (10.3.37) will inherit the same chaotic behavior as (10.3.28). In Fig. 10.17, we represent the time series of $\tilde{u}$, $\tilde{T}_d$, and $\tilde{T}_s$ coordinates of the solution of system (10.3.37). One can see in Fig. 10.17 that system (10.3.37) possesses chaotic behavior.

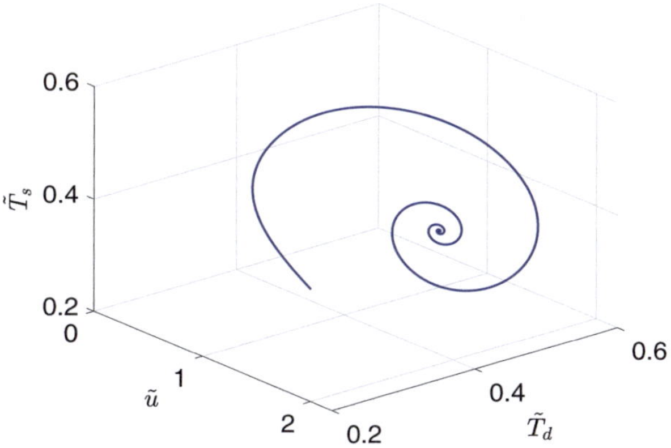

Fig. 10.16 The asymptotically stable equilibrium of system (10.3.38)

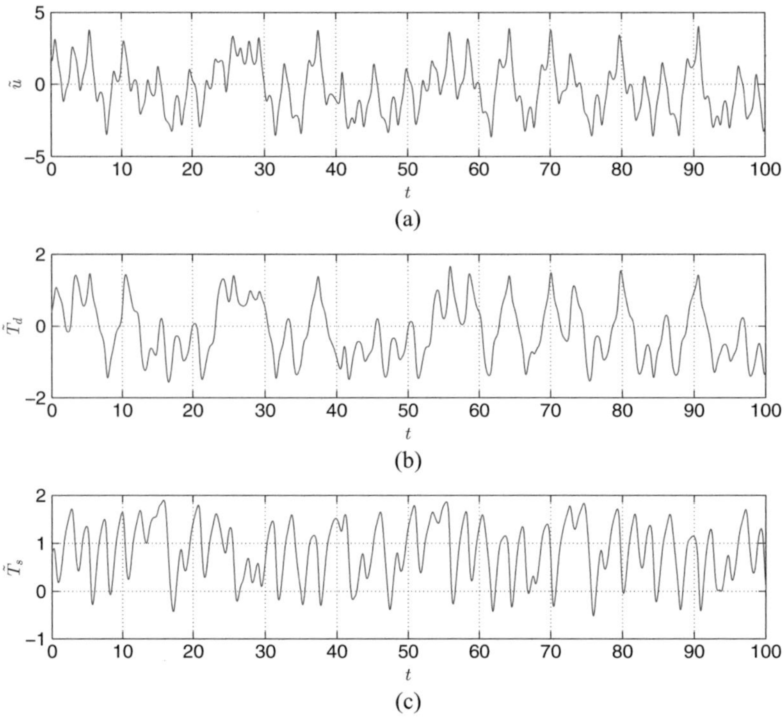

Fig. 10.17 The solution of system (10.3.37). The figure reveals chaos extension between the pair of Vallis systems (10.3.28) and (10.3.37)

10.4 Ocean-Atmosphere Unpredictability Interaction

In this section, we discuss the possibility of the "vertical" extension of unpredictability, i.e., the transmission of chaotic dynamics from ocean to atmosphere and vice versa. To demonstrate this interaction we apply the Lorenz system for the atmosphere and the Vallis model for the ocean. Vallis model is constructed for the domain length of 7500 km, however, depending on the method of construction, the model can be applied for more localized region to be compatible with the Lorenz model. There are two interacted regions shown in Fig. 10.18, the atmosphere box $\mathscr{L}$ and the ocean box $\mathscr{V}$, whose dynamics are governed the Lorenz system (10.1.1) and the Vallis system (10.3.28), respectively.

Heat and momentum exchanges are two important ways of interaction between ocean and atmosphere. The heat exchange is mainly controlled by the air-sea temperature gradient, and, on the other hand, the momentum transfer is determined by the sea-surface stress caused by wind and currents [24]. These characteristics are represented in both Lorenz system (10.1.1) and Vallis model (10.3.28). Two

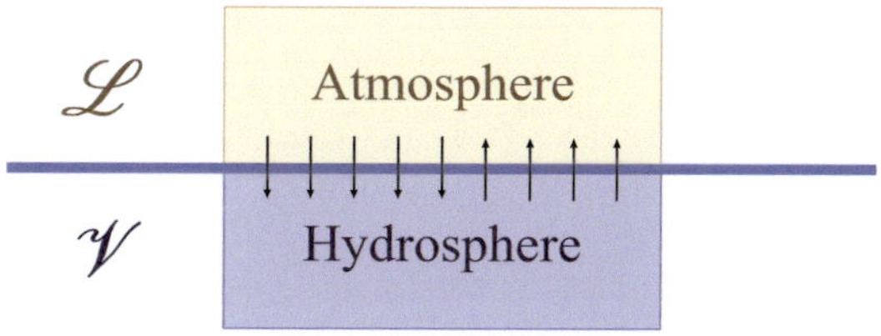

Fig. 10.18 Schematic representation of ocean–atmosphere interactions

coordinates in the Lorenz system represent temperature, whereas the third one is related to velocity, and the same could be said for the Vallis system. Therefore, the interaction between ocean and atmosphere can be modeled by coupling the Lorenz and Vallis models.

Let us consider the coupled Lorenz–Vallis systems

$$\begin{aligned}\frac{dx}{dt} &= \sigma(y - x) + f_1(u, T_d, T_s),\\ \frac{dy}{dt} &= x(r - z) - y + f_2(u, T_d, T_s),\\ \frac{dz}{dt} &= xy - bz + f_3(u, T_d, T_s),\end{aligned} \tag{10.4.39}$$

and

$$\begin{aligned}\frac{du}{dt} &= B\,T_d - C\,u + g_1(x, y, z),\\ \frac{dT_d}{dt} &= u\,T_s - T_d + g_2(x, y, z),\\ \frac{dT_s}{dt} &= -u\,T_d - T_s + 1 + g_3(x, y, z),\end{aligned} \tag{10.4.40}$$

where $f_i, g_i, i = 1, 2, 3$, are continuous functions. The coupled model (10.4.39)–(10.4.40) is in a sufficiently general form of interaction between the $\mathscr{L}$ and $\mathscr{V}$ regions shown in Fig. 10.18, where the functions $f_i, g_i, i = 1, 2, 3$ are given in most general form.

To demonstrate the transmission of chaos between the atmosphere and ocean, we consider specific forms of the coupled model (10.4.39)–(10.4.40). This technique relies on the theoretical investigations of replication of chaos introduced in [1].

In the case of upward transmission of chaos from the ocean to the atmosphere, we consider (10.4.39) with specific choices of the perturbation functions f_1, f_2, and f_3 to set up the following system:

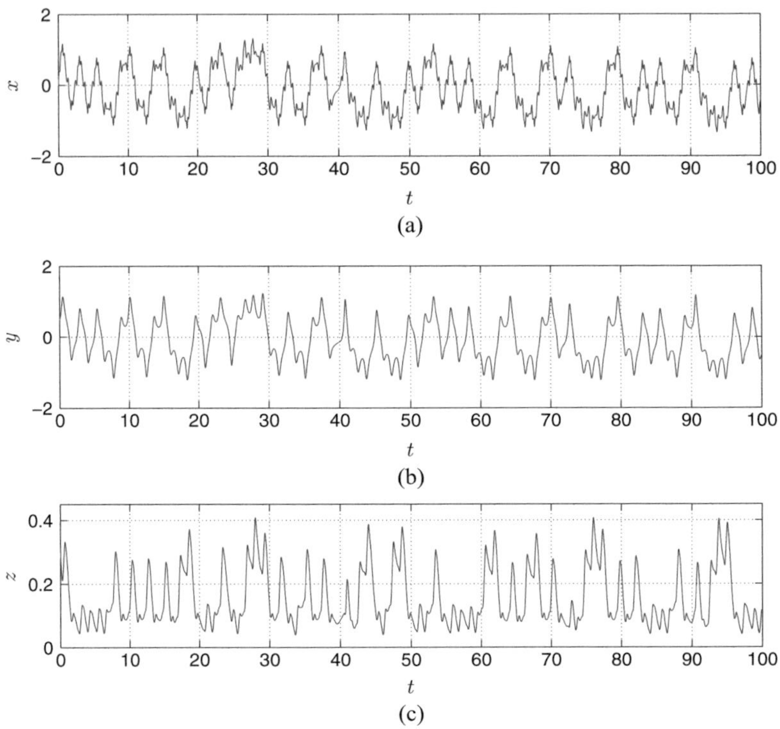

Fig. 10.19 The chaotic solution of the perturbed Lorenz system (10.4.41)

$$\begin{aligned}\frac{dx}{dt} &= \sigma(y-x) + 3\sin u,\\ \frac{dy}{dt} &= x(r-z) - y + 6\,T_d,\\ \frac{dz}{dt} &= xy - b\,z + 0.5\,T_s^2,\end{aligned} \tag{10.4.41}$$

where (u, T_d, T_s) is the solution of the chaotic Vallis system (10.3.28) with $B = 102$, $C = 3$, and the initial data $u(0) = 2$, $T_d(0) = 0.2$, $T_s(0) = 0.4$. We use the parameter values $\sigma = 10$, $r = 0.35$, and $b = 8/3$ in (10.4.41) such that the corresponding unperturbed Lorenz system (10.1.1) does not possess chaos [59].

Figure 10.19 shows the time series of the x, y, and z components of the solution of system (10.4.41). The initial data $x(0) = 0$, $y(0) = 0.5$, $z(0) = 0.3$ are used in the figure. The irregular behavior in each component reveals that the chaotic behavior of the atmosphere can be gained from the chaoticity of the hydrosphere characteristics.

For the downward chaos transmission from the atmosphere to the ocean, we consider the perturbed Vallis system

$$
\begin{aligned}
\frac{du}{dt} &= B\,T_d - C\,u + 0.7x, \\
\frac{dT_d}{dt} &= u\,T_s - T_d + 0.3\cos y + 0.4y, \\
\frac{dT_s}{dt} &= -u\,T_d - T_s + 1 + 0.5z,
\end{aligned} \tag{10.4.42}
$$

where (x, y, z) is the solution of the Lorenz system (10.1.1) with the parameters $\sigma = 10$, $r = 28$, and $b = 8/3$ and the initial data $x(0) = 0$, $y(0) = 1$, $z(0) = 0$. System (10.1.1) possesses a chaotic attractor with these choices of the parameter values [34, 59].

Let us take $B = 20$ and $C = 7$ in system (10.4.42). One can verify in this case that the corresponding unperturbed system (10.3.28) is non-chaotic such that it possesses an asymptotically stable equilibrium. Figure 10.20 depicts the solution of (10.4.42) with $u(0) = 2$, $T_d(0) = 0.2$, and $T_s(0) = 0.4$. It is seen in Fig. 10.20 that the chaotic behavior of the Lorenz system (10.1.1) is transmitted to (10.4.42). In other words, system (10.4.42) admits chaos even if it is initially non-chaotic in the absence of the perturbation.

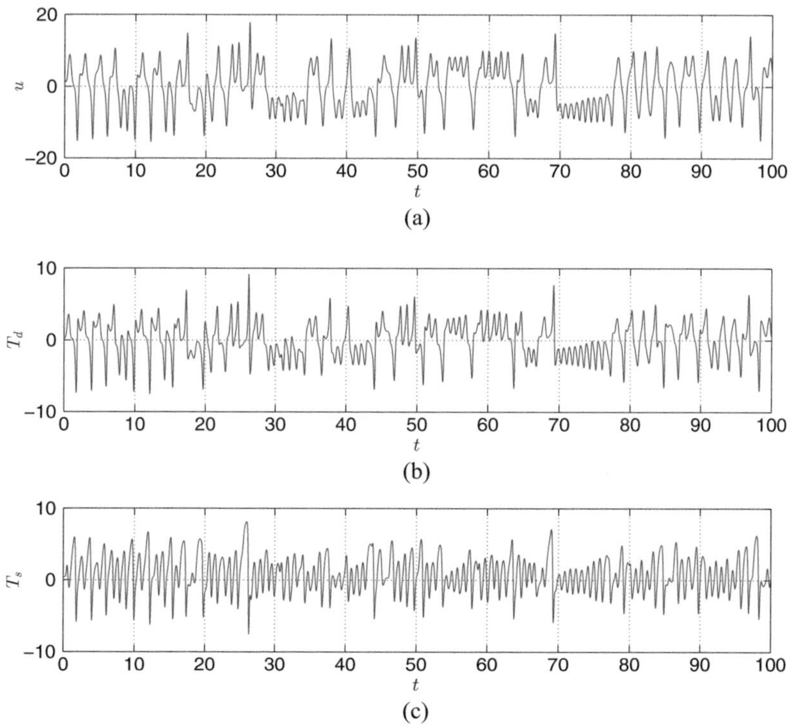

Fig. 10.20 Chaotic behavior of system (10.4.42)

10.5 Notes

In this chapter we discuss the possible unpredictable behavior of climate variables on a global scale. Some ENSO-like climate variabilities have a significant influence on global weather and climate. ENSO variability is suggested to be chaotic by many studies. The well-known Vallis ENSO chaotic model is one among several ENSO models that exhibit irregular behavior. The presence of chaos in ENSO can be indicated by the behavior of SST as well as ocean current velocity. We describe the dynamics of SST by the advection equation. The forcing term, based on ocean–atmosphere interaction, and the current velocity in this equation can be a source of unpredictability in SST. We prove the presence of chaos in SST dynamics by utilizing the concept of unpredictable function. The relationship and interaction between the climate variabilities, like the ones between ENSO and IOD, have attracted attention in recent literature. Constructing and understanding the dynamic models driving these phenomena are the main steps to investigate the mutual influences between these global events. The SST anomalies are closely linked to some climate variabilities teleconnections in different parts of the global ocean. We suggest that the hydrosphere characteristics can behave chaotically through the possibility of transmission of chaos between ocean neighbor subregions. We verified this transmission by different "toy" couples of advection equations and Vallis models. The simulations of these couples show that unpredictability can be transmitted from a local region controlled by a chaotic model into its neighbor which is described by a non-chaotic model.

The mechanism of unpredictability extension can be interpreted in terms of physical operations. The simplicity of the models under consideration, namely the Vallis model and the advection equation, allows to make the physics much clearer. The onset of ENSO is accompanied with zonal SST gradient over the equatorial Pacific Ocean. The same situation applies to IOD in Indian Ocean and other similar climate patterns. This distribution of SST leads to heat transport by convection in the mixed layer and ocean circulation through its effect on surface wind and ocean atmospheric circulation. These physical processes, which include heat, mass, and momentum transfer, can be accompanied with "*chaos transfer*." We believe that this thought still needs a consistent theoretical framework to understand all features of such operation, principally the intrinsicness of chaos for these physical quantities. Nevertheless, the presented mechanism of unpredictability extension could be seen as a step towards this goal. Further steps can be performed by including different models for more climate components.

We proposed to apply the same technique for the "vertical" unpredictability exchange between atmosphere and hydrosphere. In this case, the Lorenz system and the Vallis model are assigned for the atmosphere and ocean, respectively. Physically, this exchange may be done in the midst of interaction between ocean and atmosphere associated with, for example, heat exchange. By this procedure, the global unpredictability of oceanic oscillation can be viewed as accompaniment to weather unpredictability.

Our approach provides a basic frame for mathematical interpretation to the irregular behavior of some global climate characteristics. It gives a way to link the local unpredictability in a component of climate system to more global scope. Further investigation can be done by including different models for more climate components. Another important and interesting problem is *controlling weather*. Even though the weather is too complicated to modify, a vital step can be taken toward this goal by modifying the ENSO oscillation through control of chaos in its models and study the "extension of the control" between ENSO-like models and weather models. Chaos control in Lorenz system is still not effectively developed in the literature, where the most proposed methods are mainly depend on forcing the system into a single stable periodic behavior [18, 70], and this is not adequate for real life applications. It is known that the chaos control can be achieved by using small perturbation to some parameters or variables of the system. This idea may be practically applied by making a small local artificial effect in atmosphere or hydrosphere. If we consider the positive tenor of the Lorenz's famous question, "Does the flap of a butterfly's wing in Brazil set off a tornado in Texas?" we can say that the small artificial climate change may prevent the occurrence or at least decrease the intensity of some extreme weather events such as cyclones, hurricanes, droughts, and floods.

References

1. M.U. Akhmet, M.O. Fen, Replication of chaos. Commun. Nonlinear Sci. Numer. Simul. **18**, 2626–2666 (2013)
2. M. Akhmet, M.O. Fen, Extension of Lorenz unpredictability. Int. J. Bifurcat. Chaos **25**, 1550126 (2015)
3. M. Akhmet, M.O. Fen, *Replication of Chaos in Neural Networks, Economics and Physics* (Higher Education Press, Beijing; Springer, Heidelberg, 2016)
4. M. Akhmet, M.O. Fen, Unpredictable points and chaos. Commun. Nonlinear Sci. Numer. Simul. **40**, 1–5 (2016)
5. M. Akhmet, M.O. Fen, Existence of unpredictable solutions and chaos. Turk. J. Math. **41**, 254–266 (2017)
6. M. Akhmet, M.O. Fen, Poincaré chaos and unpredictable functions. Commun. Nonlinear Sci. Numer. Simul. **48**, 85–94 (2017)
7. M. Akhmet, M.O. Fen, Non-autonomous equations with unpredictable solutions. Commun. Nonlinear Sci. Numer. Simul. **59**, 657–670 (2018)
8. M. Akhmet, M.O. Fen, E.M. Alejaily, Extension of sea surface temperature unpredictability. Ocean Dynamics **69**, 145–156 (2019)
9. B.S.T. Alkahtani, A. Atangana, Chaos on the Vallis model for El Niño with fractional operators. Entropy **18**, 100 (2016)
10. V. Barale, M. Gade, *Remote Sensing of the European Seas* (Springer Science Business Media, B.V., 2008)
11. D.S. Battisti, Dynamics and thermodynamics of a warming event in a coupled tropical atmosphere-ocean model. J. Atmos. Sci. **45**, 2889–2919 (1988)
12. S.K. Behera, J.J. Luo, S. Masson, S.A. Rao, H. Sakuma, T. Yamagata, A CGCM study on the interaction between IOD and ENSO. J. Climate **19**, 1688–1705 (2006)

13. J. Bjerknes, Atmospheric teleconnections from the equatorial. Pacific. Mon. Weather Rev. **97**, 163–172 (1969)
14. F. Bonjean, Influence of sea currents on the sea surface temperature in the tropical Pacific ocean. J. Phys. Oceanogr. **31**, 943–961 (2001)
15. M. Borghezan, P.C. Rech, Chaos and periodicity in Vallis model for El Niño. Chaos Solitons Fractals **97**, 15–18 (2017)
16. L. Bunge, A.J. Clarke, A verified estimation of the El Niño index Niño-3.4 since 1877. J. Climate **22**, 3979–3992 (2009)
17. D.P. Chambers, B.D. Tapley, R.H. Stewart, Anomalous warming in the Indian Ocean coincident with El Niño. J. Geophys. Res. **104**, 3035–3047 (1999)
18. M. Chen, D. Zhou, Y. Shang, Nonlinear feedback control of Lorenz system. Chaos Solitons Fractals **21**, 295–304 (2004)
19. D. Coley, *Energy and Climate Change: Creating a Sustainable Future* (Wiley, Chichester, UK, 2008)
20. R.L. Devaney, *An Introduction to Chaotic Dynamical Systems* (Addison-Wesley, USA, 1989)
21. D. Eamus, A. Huete, Q. Yu, *Vegetation Dynamics: A Synthesis of Plant Ecophysiology, Remote Sensing and Modelling* (Cambridge University Press, New York, 2016)
22. L. Eppelbaum, I. Kutasov, A. Pilchin, *Applied Geothermics, Lecture Notes in Earth System Sciences* (Springer, Berlin, 2014)
23. M.J. Feigenbaum, Universal behavior in nonlinear systems. Los Alamos Sci./Summer **1**, 4–27 (1980)
24. B. Gallego, P. Cessi, Exchange of heat and momentum between the atmosphere and the ocean: A minimal model of decadal oscillations. Clim. Dyn. **16**, 479–489 (2000)
25. B.M. Garay, B. Indig, Chaos in Vallis' asymmetric Lorenz model for El Niño. Chaos Solitons Fractals **75**, 253–262 (2015)
26. P.R. Gent, M.A. Cane, A reduced gravity, primitive equation model of the upper equatorial ocean. J. Comput. Phys. **81**, 444–480 (1989)
27. S.M. Hammel, J.A. Yorke, C. Grebogi, Do numerical orbits of chaotic dynamical processes represent true orbits? J. Complexity **3**, 136–145 (1987)
28. M. Jochum, R. Murtugudde, Temperature advection by tropical instability waves. J. Phys. Oceanogr. **36**, 592–605 (2006)
29. S.H. Kellert, *In the Wake of Chaos: Unpredictable Order in Dynamical Systems* (University of Chicago Press, Chicago, 1994)
30. W.S. Kessler, L.M. Rothstein, D. Chen, The annual cycle of SST in the eastern tropical Pacific as diagnosed in an OGCM. J. Climate **11**, 777–799 (1998)
31. B.P. Kirtman, N. Perlin, L. Siqueira, Ocean eddies and climate predictability. Chaos **27**, 126902 (2017)
32. C. Lehr, P.J. Ward, M. Kummu, Impact of large-scale climatic oscillations on snowfall-related climate parameters in the world's major downhill ski areas: a review. Mt. Res. Dev. **32**, 431–445 (2012)
33. T.Y. Li, J.A. Yorke, Period three implies chaos. Am. Math. Monthly **82**, 985–992 (1975)
34. E.N. Lorenz, Deterministic nonperiodic flow. J. Atmos. Sci. **20**, 130–141 (1963)
35. L.E. Lucas, D.E. Waliser, R. Murtugudde, Mechanisms governing sea surface temperature anomalies in the eastern tropical Pacific Ocean associated with the boreal winter Madden-Julian Oscillation. JGR-Oceans **115**, C05012 (2010)
36. R. Lukas, E. Lindstrom, The mixed layer of the western equatorial Pacific ocean. J. Geophys. Res. **96**, 3343–3357 (1991)
37. J.J. Luo, R.C. Zhang, S.K. Behera, Y. Masumoto, F.F. Jin, R. Lukas, T. Yamagata, Interaction between El Niño and extreme Indian Ocean Dipole. J. Climate **23**, 726–742 (2010)
38. M. Münnich, M.A. Cane, S.E. Zebiak, A study of self-excited oscillations in a tropical ocean–atmosphere system, Part II: Nonlinear cases. J. Atmos. Sci. **48**, 1238–1248 (1991)
39. J.D. Neelin, M. Latif, El Niño dynamics. Phys. Today **51**, 32–36 (1998)
40. E. Ott, C. Grebogi, J.A. Yorke, Controlling chaos. Phys. Rev. Lett. **64**, 1196–1199 (1990)

41. C. Penland, L. Matrasova, A balance condition for stochastic numerical models with application to the El Niño-Southern Oscillation. J. Climate **7**, 1352–1372 (1994)
42. J.F. Petersen, D. Sack, R.E. Garbler, *Physical Geography* (Cengage Learning, Boston, 2017)
43. K. Pyragas, Continuous control of chaos by self-controlling feedback. Phys. Lett. A **170**, 421–428 (1992)
44. N.A. Rayner, D.E. Parker, E.B. Horton, C.K. Folland, L.V. Alexander, D.P. Rowell, E.C. Kent, A. Kaplan, Global analyses of sea surface temperature, sea ice, and night marine air temperature since the late nineteenth century. J. Geophys. Res. **108**, 4407 (2003)
45. R.C. Robinson, *An Introduction to Dynamical Systems: Continuous and Discrete* (American Mathematical Society, Providence, 2012)
46. C. Rosenzweig, D. Hillel, *Climate Variability and the Global Harvest: Impact of El-Nino and Other Oscillations on Agro-ecosystems* (Oxford University Press, New York, 2008)
47. M. Roxy, S. Guladi, H.K.L. Drobhlar, A. Navarra, Seasonality in the relationship between El Niño and Indian Ocean dipole. Clim. Dyn. **37**, 221–236 (2011)
48. S. Sen Roy, R.B. Singh, *Climate Variability, Extreme Events and Agriculture Productivity in Mountain Regions* (Oxford and 1BH Publication, New Delhi, 2002)
49. N.H. Saji, B.N. Goswami, P.N. Vinayachandran, T. Yamagata, A dipole mode in the tropical Indian Ocean. Nature **401**, 360–362 (1999)
50. E. Sander, J.A. Yorke, Period-doubling cascades galore. Ergod. Theory Dyn. Syst. **31**, 1249–1267 (2011)
51. R. Saravanan, Seasonal-to-decadal prediction using climate models: successes and challenges, in *Large-Scale Disasters: Prediction, Control, and Mitigation*, ed. by M. Gad-el-Hak (Cambridge University Press, New York, 2008), pp. 318–328
52. J. Schmandt, J. Clarkson, G.R. North, *The Impact of Global Warming on Texas*, 2nd edn. (University of Texas Press, Austin, Texas, 2011)
53. E. Schöll, H.G. Schuster (eds.), *Handbook of Chaos Control* (Wiley-VCH, Weinheim, Germany, 2008)
54. G.R. Sell, *Topological Dynamics and Ordinary Differential Equations* (Van Nostrand Reinhold Company, London, 1971)
55. F.M. Selten, F.J. Schevenhoven, G.S. Duane, Simulating climate with a synchronization-based supermodel. Chaos **27**, 126903 (2017)
56. J. Sheinbaum, Current theories on El Niño-Southern oscillation: A review. Geofisica Internacional-Mexico **42**, 291–306 (2003)
57. M.L. Shen, N. Keenlyside, B.C. Bhatt, G.S. Duane, Role of atmosphere-ocean interactions in supermodeling the tropical Pacific climate. Chaos **27**, 126704 (2017)
58. G. Siedler, J. Church, J. Gould (eds.), *Ocean Circulation and Climate: Observing and Modelling the Global Ocean* (Academic Press, CA, USA, 2001)
59. C. Sparrow, *The Lorenz Equations: Bifurcations, Chaos and Strange Attractors* (Springer, New York, 1982)
60. J.W. Stevenson, P.P. Niiler, Upper ocean heat budget during the Hawaii-to-Tahiti shuttle experiment. J. Phys. Oceanogr. **13**, 1894–1907 (1983)
61. T. Stocker, *Introduction to Climate Modelling* (Springer, Berlin, 2011)
62. S.H. Strogatz, *Nonlinear Dynamics and Chaos: With Applications to Physics, Biology, Chemistry, and Engineering* (Addison-Wesley, Reading, MA, 1994)
63. M.F. Stuecker, F.F. Jin, A. Timmermann, El Niño-Southern Oscillation frequency cascade. Proc. Natl. Acad. Sci. USA **112**, 13490–13495 (2015)
64. E. Tziperman, L. Stone, M.A. Cane, H. Jarosh, El Niño chaos: Overlapping of resonance between the seasonal cycle and the Pacific ocean-atmosphere oscillator. Science **264**, 72–74 (1994)
65. G.K. Vallis, El Niño: A chaotic dynamical system? Science **232**, 243–245 (1986)
66. G.K. Vallis, Conceptual models of El Niño and the southern oscillation. J. Geophys. **93**, 13979–13991 (1988)

67. M. Vuille, R.D. Garreaud, Ocean-atmosphere interactions on interannual to decadal timescales, in *The SAGE Handbook of Environmental Change*, ed. by J.A. Matthews, P.J. Bartlein, K.R. Briffa, A.G. Dawson, A. De Vernal, T. Denham, S.C. Fritz, O. Frank (Sage, London, 2011), pp. 471–496
68. J. Willebrand, D.L.T. Anderson (eds.), *Modelling Oceanic Climate Interactions* (Springer, Berlin, Heidelberg, 1993)
69. T. Yamagata, S.K. Behera, J.J. Luo, S. Masson, M.R. Jury, S.A. Rao, Coupled ocean-atmosphere variability in the tropical Indian Ocean, Earth's climate: The ocean-atmosphere interaction. Geophys. Monogr. **147**, 189–212 (2004)
70. H.T. Yau, C.L. Chen, Chaos control of Lorenz systems using adaptive controller with input saturation. Chaos Solitons Fractals **34**, 1567–1574 (2007)
71. S.E. Zebiak, M.A. Cane, A model El Niño-southern oscillation. Mon. Wea. Rev. **115**, 2262–2278 (1987)

Chapter 11
Fractals: Dynamics in the Geometry

In this chapter, dynamics are constructed for fractals utilizing the motion associated with differential equations. Firstly, we introduce an algorithm to map fractals by developing a mapping iteration on the basis of Fatou–Julia iteration. Because of the close link between mappings, differential equations, and dynamical systems, one can introduce dynamics for a fractal through differential equations such that it becomes points of the solution trajectory. In the present chapter, Julia set, Mandelbrot set, and Sierpinski fractals are considered as initial points for the trajectories of the dynamics. The characterization of fractals as trajectory points of the dynamics can help to enhance and widen the scope of their applications in physics and engineering. The results of this chapter are published in [2–4].

11.1 Introduction

French mathematicians Pierre Fatou and Gaston Julia in 1917–1918 invented a special iteration in the complex plane [17, 24] such that new geometrical objects with unusual properties can be built. The iteration is called Fatou–Julia iteration (FJI) [32] or sometimes "Escape Time Algorithm (ETA)" [8]. One of the famous fractals constructed by the iteration is the Julia set. Besides the iteration of rational maps, there are various ways to construct fractal shapes. The well-known self-similar fractals like Sierpinski gasket, Sierpinski carpet, and Koch curve are constructed by means of a simple recursive process which consists in iteratively removing shrinking symmetrical parts from an initial shape. These types of geometrical fractals can also be produced by an Iterated Function System (IFS) [9, 22], which is defined as collections of affine transformations.

There are two sides of the fractal research related to the present chapter. The first one is FJI and the second one is the proposal by Mandelbrot to consider dimension as a criterion for fractals. In our approach, both factors are crucial as we apply the

M. Akhmet et al., *Dynamics with Chaos and Fractals*, Nonlinear Systems and Complexity 29, https://doi.org/10.1007/978-3-030-35854-9_11

FJI for the construction of the sets and the dimension factor to confirm that the built sets are fractals. In previous studies the iteration and the dimension factors were somehow separated, since self-similarity provided by the iteration has been self-sufficient to recognize fractals, but in our research the similarity is not true in general. We have to emphasize that there is a third player on the scene, the modern state of computers' power. Their roles are important for the realization of our idea exceptionally for continuous dynamics. One can say that the instrument is at least of the same importance for application of our idea to fractals as for realization of FJI in Mandelbrot and Julia sets. Nevertheless, we expect that the present chapter will significantly increase the usage of computers for fractal analysis. Moreover, beside differential equations, our suggestions will affect the software development for fractals investigation and applications [15, 42].

Studying the problem, we have found that fractal-like appearances can be observed in ancient natural philosophy. Let us consider the Achilles and tortoise in the Zeno's Paradox [45], (see Fig. 11.1a).

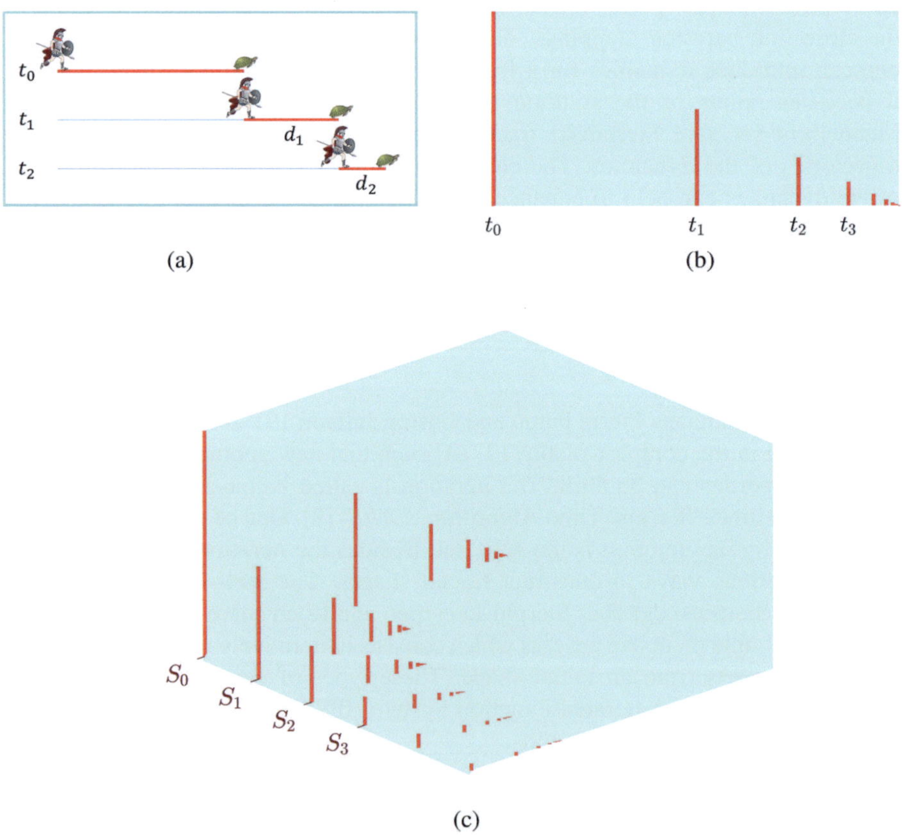

Fig. 11.1 The dynamics of the Zeno's Paradox. (**a**) Achilles and the tortoise dynamics. (**b**) The state, S_0, of the dynamics. (**c**) Pseudo-fractal trajectory

In the paradox, Achilles is observed at the initial moment $t_0 = 0$ with the distance d_0 from the tortoise. Suppose that Achilles runs at a constant speed, two times faster than the tortoise, then he would reach the previous position of the tortoise at moments t_1, $t_2 = 3t_1/2$, $t_3 = 7t_1/4$, ... with distances $d_1 = d_0/2$, $d_2 = d_0/4$, $d_3 = d_0/8$, ... from the tortoise, respectively. Now contemplate Fig. 11.1b, where the heights of the red lines are proportional to the distance of Achilles from the tortoise at the fixed moments, and denote the diagram by S_0. The set S_0 demonstrates the entire dynamics for $t \geq t_0$. Fix $i \geq 0$, and let S_i be the similar diagram which consists of all the lines for the moments which are not smaller than t_i.

Let us consider the collection of the states $\{S_i\}$, $i \geq 0$. One can assume that there exists a map $\mathscr{B}$ such that the equations

$$S_{i+1} = \mathscr{B}\, S_i, \;\; i = 0, 1, 2, \ldots,$$

which symbolize a dynamics, are valid. It is easily seen that S_0 is self-similar to each of its parts S_i, $i > 0$. Nevertheless, the Hausdorff dimension of the set S_0 is equal to one. For this reason, we call S_i, $i \geq 0$, *pseudo-fractals*, due to the similarity. The trajectory $\{S_i\}$, $i \geq 0$, is also a pseudo-fractal. The sketch of the trajectory is seen in Fig. 11.1c.

The present chapter is an extension of the ancient paradigm, since we will investigate dynamics having all points of a trajectory as well as the trajectory itself fractals.

11.2 Fatou–Julia Iteration

Involvement of the dynamics of iterative maps in fractal construction was a critical step made by Fatou and Julia [17, 24]. They described what we call today FJI. The iteration is defined over a domain $\mathscr{D} \subseteq \mathbb{C}$ by

$$z_{n+1} = F(z_n), \tag{11.2.1}$$

where $F : \mathscr{D} \to \mathscr{D}$ is a given function for the construction of the fractal set $\mathscr{F}$. The points $z_0 \in \mathscr{D}$ are included in the set $\mathscr{F}$ depending on the boundedness of the sequence $\{z_n\}$, $n = 0, 1, 2, \ldots$, and we say that the set $\mathscr{F}$ is constructed by FJI.

In practice one cannot verify the boundedness for infinitely long iterations. This is why in simulation we fix an integer k and a bounded subset $M \subset \mathbb{C}$, and denote by $\mathscr{F}_k$ the collection of all points $z_0 \in \mathscr{D}$ such that the points z_n where the index n is between 1 and k, $n = 1, 2, \ldots, k$, belong to M. In what follows we call the set $\mathscr{F}_k$ the *kth approximation* of the set $\mathscr{F}$.

The most popular fractals, Julia and Mandelbrot sets, are generated using the iteration of the quadratic map $F(z_n) = z_n^2 + c$, where c is a complex number. The so-called filled-in Julia set, $\mathscr{K}_c$, is constructed by including only the points $z_0 \in \mathbb{C}$ such that the sequence z_n is bounded [1]. Moreover, in the simulation, those points

$z_0 \in \mathbb{C}$ where $\{z_n\}$ is divergent are colored in a different way, correspondingly to the rate of divergence. The term Julia set $\mathscr{J}_c$, usually denotes the boundary of the filled Julia set, i.e., $\mathscr{J}_c = \partial \mathscr{K}_c$.

In the case of the Mandelbrot set, $\mathscr{M}$, we include in $\mathscr{M}$ the points $c \in \mathbb{C}$ such that $\{z_n(c)\}$, $z_0(c) = 0$, is bounded. Here again, the points $c \in \mathbb{C}$ corresponding to divergent sequences z_n are plotted in various colors depending on the rate of the divergence.

11.3 How to Map Fractals

To describe our way for mapping of fractals, let us consider a fractal set $\mathscr{F} \subseteq A \subset \mathbb{C}$, constructed by the following FJI,

$$z_{n+1} = F(z_n), \tag{11.3.2}$$

where $F : A \to A$ is not necessarily a rational map. We suggest that the original fractal $\mathscr{F}$ can be transformed "recursively" into a new fractal set. For that purpose, we modify the FJI, and consider iterations to be of the form

$$f^{-1}(z_{n+1}) = F\big(f^{-1}(z_n)\big), \tag{11.3.3}$$

or explicitly,

$$z_{n+1} = f\Big(F\big(f^{-1}(z_n)\big)\Big), \tag{11.3.4}$$

where f is a one-to-one map on A. Next, we examine the convergence of the sequence $\{z_n\}$ for each $z_0 \in f(A)$. Denote by $\mathscr{F}_f$ the set which contains only the points z_0 corresponding to the bounded sequences. Moreover, other points can be plotted in different colors depending on the rate of the divergence of $\{z_n\}$. To distinct the iterations (11.3.4) from the Fatou–Julia iterations let us call the first ones *Fractals Mapping Iterations* (FMI) [2–4]. It is clear that FJI is a particular FMI, when the function is the identity map. The mapping of fractals is a difficult problem which depends on infinitely long iteration processes, and has to be accompanied with sufficient conditions to ensure that the image is again fractal.

The next theorem is the main instrument for the detection of fractal mappings. Accordingly, we call it *Fractal Mapping Theorem* (FMT).

Theorem 11.1 ([2–4]) *If f is a bi-Lipschitz function, i.e., there exist numbers $l_1, l_2 > 0$ such that*

$$l_1|u - v| \le |f(u) - f(v)| \le l_2|u - v| \tag{11.3.5}$$

for all $u, v \in A$, then $\mathscr{F}_f = f(\mathscr{F})$.

Proof Fix an arbitrary $w \in \mathscr{F}_f$. There exists a bounded sequence $\{w_k\}$ such that $w_0 = w$ and $f^{-1}(w_{k+1}) = F(f^{-1}(w_k)$. Let us denote $z_k = f^{-1}(w_k)$. Our purpose is to show that $\{z_k\}$ is a bounded sequence. Indeed

$$|z_k - z_0| = |f^{-1}(w_k) - f^{-1}(w_0)| \leq \frac{1}{l_1}|w_k - w_0|.$$

Hence, the boundedness of $\{w_k\}$ implies the same property for $\{z_k\}$, and therefore, we have $z_0 = f^{-1}(w) \in \mathscr{F}$.

Now, assume that $w \in f(\mathscr{F})$. There is $z \in \mathscr{F}$ such that $f(z) = w$ and a bounded sequence $\{z_k\}$ such that $z_0 = z$ and $z_{k+1} = F(z_k)$. Consider, $w_0 = w$ and $w_k = f(z_k),\ k \geq 0$. It is clear that the sequence $\{w_k\}$ satisfies the iteration (11.3.3) and moreover

$$|w_k - w_0| = |f(z_k) - f(z_0)| \leq l_2|z_k - z_0|.$$

Consequently, $\{w_k\}$ is bounded, and $w \in \mathscr{F}_f$. □

The following two simple propositions are required.

Lemma 11.3.1 ([16]) *If f is a bi-Lipschitz function, then*

$$\dim_H f(A) = \dim_H A,$$

where $\dim_H$ *denotes the Hausdorff dimension.*

Lemma 11.3.2 ([3]) *If $f : A \to \mathbb{C}$ is a homeomorphism, then it maps the boundary of A onto the boundary of $f(A)$.*

It is clear that a bi-Lipschitz function is a homeomorphism.

Shishikura [44] proved that the Hausdorff dimension of the boundary of the Mandelbrot set is 2. Moreover, he showed that the Hausdorff dimension of the Julia set corresponding to $c \in \partial\mathscr{M}$ is also 2.

It implies from the above discussions that if f is a bi-Lipschitz function and $\mathscr{F}$ is either a Julia set or the boundary of the Mandelbrot set, then their images $\mathscr{F}_f$ are fractals. In what follows, we will mainly use functions, which are bi-Lipschitzian except possibly in neighborhoods of single points.

To emphasize the novelty concerning the FMI algorithm (change of coordinates), we suggest the Table 11.1 below, which illustrates the differences between a routine map and the newly constructed fractal mapping.

Let us give comments on the content of Table 11.1 and explain why the mapping of fractals is different than those which can be accepted as routine maps. Consider the two-dimensional case for a set $\mathscr{F}$. In the routine map, the preimage $\mathscr{F}$ is simply given by equation (i) in the table, while for the fractal-preimage description we need the procedure (a) which theoretically consists of an infinite number of steps. The most important difference is manifested in the image description. In the routine mapping, the image can be defined by formula (iii). However, to describe the fractal-

Table 11.1 The differences between a routine mapping and the fractal mapping

Routine mapping	Fractal mapping
(i) The preimage description $\mathscr{F} = \{(x, y) : g(x, y) = 0\}$	**(a)** The fractal-preimage description $\mathscr{F} = \{(x, y) : (x_n, y_n),$ where $(x_0, y_0) = (x, y)$ and $(x_{n+1}, y_{n+1}) = g(x_n, y_n)$, is bounded$\}$
(ii) The map definition $f : (u, v) = f(x, y)$	**(b)** The map definition $f : (u, v) = f(x, y)$
(iii) The image description $\mathscr{F}_f = \{(u, v) : g(f^{-1}(u, v)) = 0\}$	**(c)** The fractal-image description $\mathscr{F}_f = \{(u, v) : f^{-1}(u_n, v_n),$ where $(u_0, v_0) = (u, v)$ and $f^{-1}(u_{n+1}, v_{n+1}) = g(f^{-1}(u_n, v_n))$, is bounded$\}$

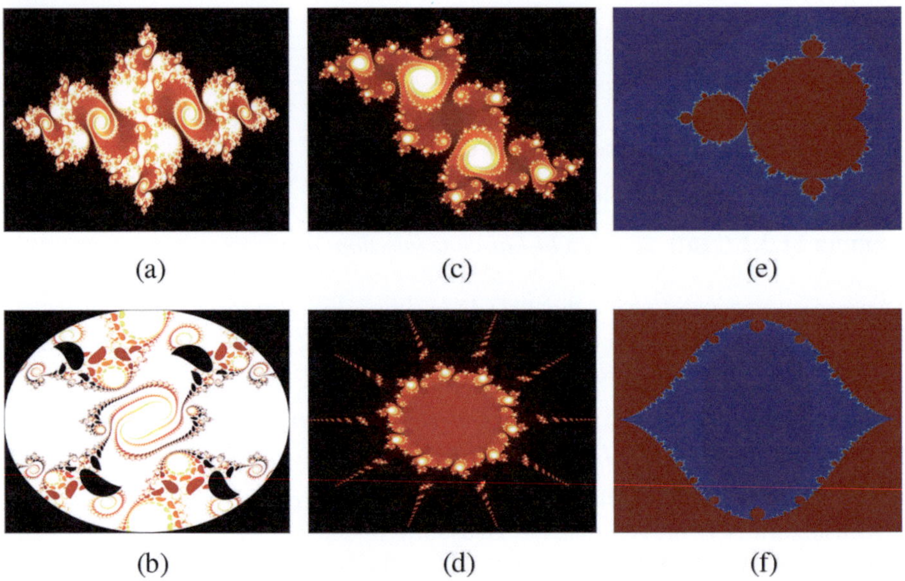

Fig. 11.2 Julia and Mandelbrot sets with their images

image, one should involve the infinite process of the preimage construction (a). The novelty of our approach is exactly in finding how the infinite process can be involved to define the fractal mapping algorithm (c).

Now, we apply FMI to a Julia set $\mathscr{J}$, and the iteration will be in the form

$$f^{-1}(z_{n+1}) = \left[f^{-1}(z_n)\right]^2 + c, \tag{11.3.6}$$

with various functions f and values of c. The resulting fractals $\mathscr{J}_f = f(\mathscr{J})$ are depicted in Fig. 11.2b and d. They are mapped by $f(z) = \cos^{-1}\left(\frac{1}{z} - 1\right)$, $c = -0.7589 + 0.0735i$, and $f(z) = \left(\sin^{-1} z\right)^{\frac{1}{5}}$, $c = -0.175 - 0.655i$ from the Julia sets in Fig. 11.2a and c, respectively.

For mapping of the Mandelbrot set, we propose the FMI

$$z_{n+1} = z_n^2 + f^{-1}(c). \tag{11.3.7}$$

Along the lines of the proof of Theorem 11.1, one can show that if the map f is bi-Lipschitzian, then the iteration (11.3.7) defines the relation $f(\mathscr{M}) = \mathscr{M}_f$, where $\mathscr{M}_f$ is a new fractal. Figure 11.2f shows a fractal mapped by $f(c) = \left(\frac{1}{c} - 1\right)^{\frac{1}{2}}$ from the Mandelbrot set in Fig. 11.2e.

11.4 Dynamics for Julia Sets

11.4.1 Discrete Dynamics

Discrete fractal dynamics means simply iterations of mappings introduced in the last section. Let us consider a discrete dynamics with a bi-Lipschitz iteration function f and a Julia set $\mathscr{J}_0 = \mathscr{J}$ as an initial fractal for the dynamics. The trajectory

$$\mathscr{J}_0, \mathscr{J}_1, \mathscr{J}_2, \mathscr{J}_3, \ldots,$$

is obtained by the FMI

$$z_{n+1} = f^k\Big([f^{-k}(z_n)]^2 + c\Big) \tag{11.4.8}$$

such that $\mathscr{J}_{k+1} = f(\mathscr{J}_k)$, $k = 0, 1, 2, 3, \ldots$. The last equation is a fractal propagation algorithm.

The computational procedure of the numerical simulation for the discrete fractal trajectory of Eq. (11.4.8) is summarized in Algorithm 1. Figure 11.3, which is obtained by using Algorithm 1, shows the trajectory and its points at $k = 1$ and $k = 5$ for the function $f(z) = z^2 + ac + b$ with $a = 0.6$, $b = 0.02 - 0.02i$ and $c = -0.175 - 0.655i$.

11.4.2 Continuous Dynamics

To demonstrate a continuous dynamics A_t with real parameter t and fractals, we use the differential equation

$$\frac{dz}{dt} = g(z), \tag{11.4.9}$$

Algorithm 1: Discrete dynamics simulation

Define the map $f(t)$ and its inverse $f^{-1}(t)$
Specify the initial Julia set $\mathcal{J}_0$ by setting the parameter c
Set the upper bound b
Set the number of maximum iterations j_m
Create a mesh with $N_p = N \times M$ elements on the domain of $\mathcal{J}_0$
Set the number of the map iterations K
for $k = 1$ *to* K **do**
    Initiate the image matrix $Z = \text{zeros}(M, N)$
    for $n = 1$ *to* N_p **do**
        Pick a point (x, y) from the domain
        Let $z = x + iy$
        Set $j = 0$
        while $(j < j_m)$ *and* $(x^2 + y^2 < b^2)$ **do**
            Let $j = j + 1$
            Find the image $z = f^k\Big(\big(f^{-k}(z)\big)^2 + c\Big)$
            Compute $x = \text{Re}(z)$ and $y = \text{Im}(z)$
        end
        Assign the image matrix elements, $Z(n) = j$
    end
    Display the image Z on the x, y, k coordinates
end

such that $A_t z = \phi(t, z)$, where $\phi(t, z)$ denotes the solution of (11.4.9) with $\phi(0, z) = z$. Thus, we will construct dynamics of sets $A_t\mathcal{F}$, where a fractal $\mathcal{F}$ is the initial value. To be in the course of the previous sections, we define a map $f(z) = A_t z$ and the equation

$$A_{-t}(z_{n+1}) = [A_{-t}(z_n)]^2 + c.$$

Thus the FMI (11.3.4) in this case will be in the form

$$z_{n+1} = A_t\Big([A_{-t}(z_n)]^2 + c\Big). \tag{11.4.10}$$

In what follows, we assume that the map A_t is bi-Lipschitzian. This is true, for instance, if the function g in (11.4.9) is Lipschitzian. Then the set $A_t\mathcal{F}$ for each fixed t is a fractal determined by the FMI, and we can say about continuous fractal dynamics. Algorithm 2, which is provided in the Appendix, gives the general outline of the computational steps of the numerical simulation for trajectories generated by the FMI (11.4.10). The Algorithm is used to produce the trajectories shown in Figs. 11.4 and 11.5.

As an example we consider the differential equation $dz/dt = -z,\ 0 \le t \le 1$, with the flow $A_t z = ze^{-t}$. It represents a contraction mapping when it is applied to the iteration (11.4.10), whereas the unstable dynamical system $A_t z = ze^t$

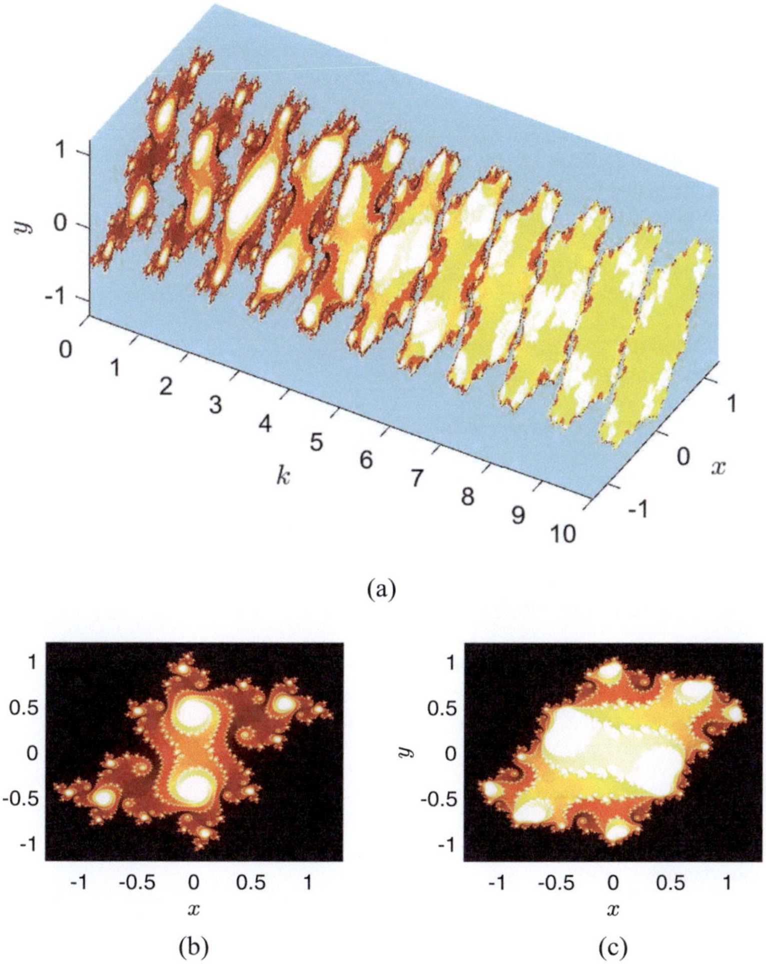

Fig. 11.3 The discrete trajectory. (**a**) Discrete fractal trajectory. (**b**) $k = 1$. (**c**) $k = 5$

corresponding to the differential equation $dz/dt = z$ represents an expansion mapping.

Figures 11.4 and 11.5a contain fractal trajectories of the dynamics with the initial Julia set $\mathscr{J}$, corresponding to $c = -0.175 - 0.655i$. The initial fractal and the point of the expansion at $t = 1$ are seen in parts (b) and (c) of the Fig. 11.5, respectively.

Now, we will focus on the autonomous system of differential equations

$$\begin{aligned} \frac{dx}{dt} &= -y + x(4 - x^2 - y^2), \\ \frac{dy}{dt} &= x + y(4 - x^2 - y^2). \end{aligned} \tag{11.4.11}$$

Algorithm 2: Continuous dynamics simulation

Find the solution $\phi(t, z)$ of the used differential equation
Define the map as $f(z) = A_t z$, where $A_t z = \phi(t, z)$
Define the inverse of the map as $f^{-1}(z) = A_{-t} z$
Specify the initial Julia set $\mathscr{J}_0$ by setting the parameter c
Set the upper bound b
Set the number of maximum iterations j_m
Create a mesh with $N_p = N \times M$ elements on the domain of $\mathscr{J}_0$
Set the temporal domain $t_0 \le t \le T$, with a step size Δt
Compute the number of image iteration $N_t = T/\Delta t + 1$
Set $t = t_0$
for $k = 1$ *to* N_t **do**
    Initiate the image matrix $Z = \text{zeros}(M, N)$
    for $n = 1$ *to* N_p **do**
        Pick a point (x, y) from the domain
        Let $z = x + iy$
        Set $j = 0$
        while $(j < j_m)$ *and* $(x^2 + y^2 < b^2)$ **do**
            Let $j = j + 1$
            Find the image $z = A_t\Big(\big(A_{-k}(z)\big)^2 + c\Big)$
            Compute $x = \text{Re}(z)$ and $y = \text{Im}(z)$
        end
        Assign the image matrix elements, $Z(n) = j$
    end
    Display the image Z on the x, y, t coordinates
    Let $t = t + \Delta t$
end

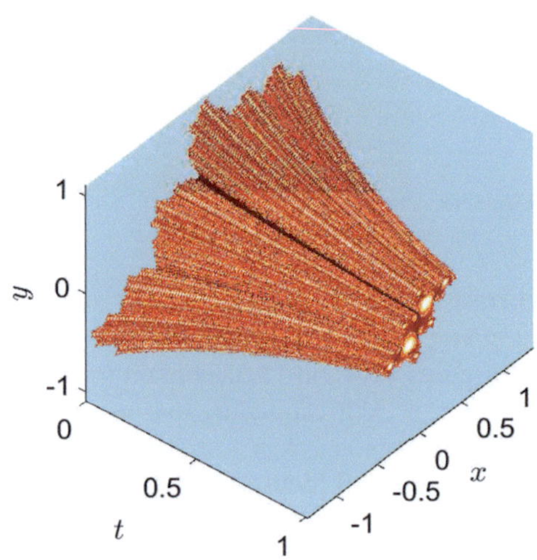

Fig. 11.4 Fractals of the continuous dynamics $\mathscr{J} e^{-t}$

The solution of the last system in polar coordinates with initial conditions $\rho(0) = \rho_0$ and $\varphi(0) = \varphi_0$ is given by

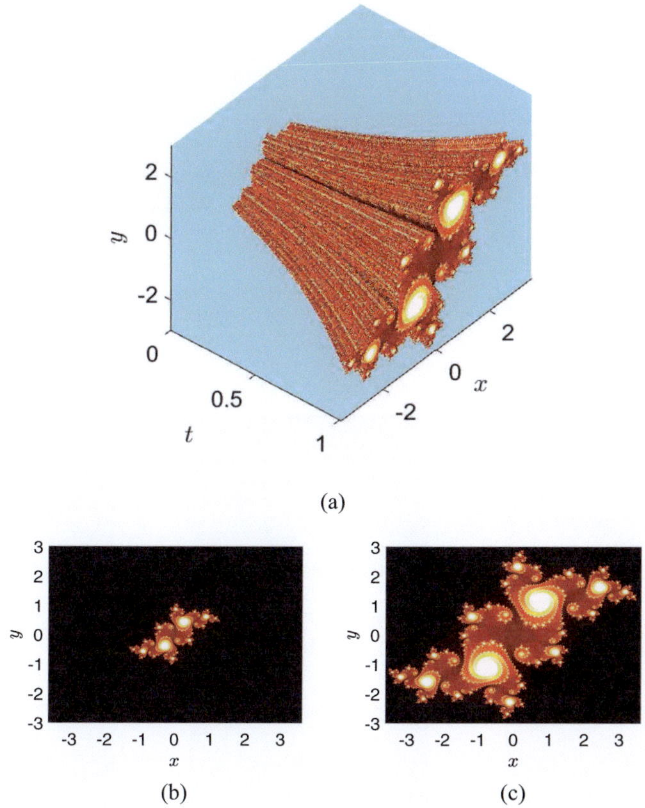

Fig. 11.5 Fractals of the continuous dynamics $\mathscr{J}e^t$. (**b**) $t = 0$. (**c**) $t = 1$

$$\rho(t) = 2e^{4t}\left(\frac{4}{\rho_0^2} + e^{8t} - 1\right)^{-\frac{1}{2}},$$

$$\varphi(t) = t + \varphi_0.$$

Thus, the map can be constructed by

$$A_t z = x(t) + iy(t), \tag{11.4.12}$$

where

$$x(t) = \rho(t)\cos(\varphi(t)),$$

$$y(t) = \rho(t)\sin(\varphi(t)),$$

$$z = \rho_0\cos(\varphi_0) + i\rho_0\sin(\varphi_0).$$

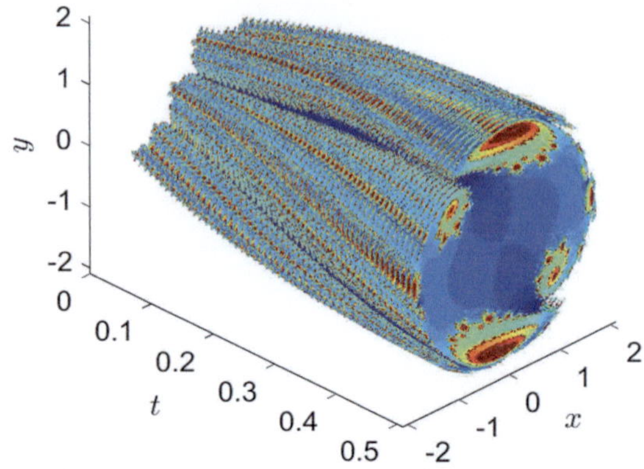

Fig. 11.6 The fractal trajectory for (11.4.11)

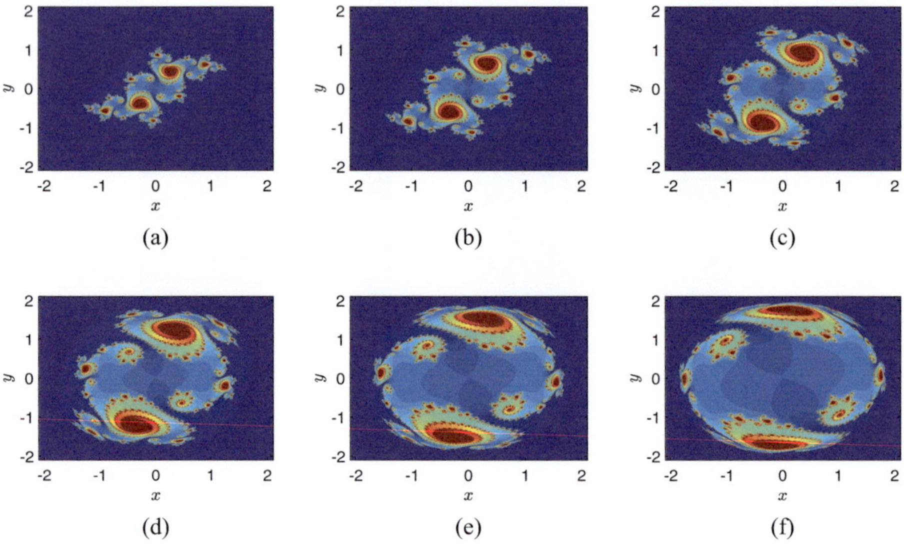

Fig. 11.7 The points of the fractal trajectory for (11.4.11). (**a**) $t = 0$. (**b**) $t = 0.1$. (**c**) $t = 0.2$. (**d**) $t = 0.3$. (**e**) $t = 0.4$. (**f**) $t = 0.5$

In Fig. 11.6, the fractal trajectory of system (11.4.11) is seen with the Julia set as the initial fractal. The parts (a)–(f) of Fig. 11.7 represent various points of the trajectory.

Next, let us consider the non-autonomous differential equation

$$\frac{dz}{dt} = az + (\cos t + i \sin t), \tag{11.4.13}$$

and the map

$$A_t z = (z + \frac{a+i}{1+a^2})e^{at} - \frac{a+i}{1+a^2}(\cos t + i \sin t), \tag{11.4.14}$$

which is determined by the solutions of Eq. (11.4.13).

The map is not of a dynamical system since there is no group property for non-autonomous equations, in general. This is why, Eq. (11.4.10) cannot be used for fractal mapping along the solutions of the differential equation (11.4.13). However, for the moments of time $2\pi n$, $n = 1, 2, \ldots$, which are multiples of the period, the group property is valid, and therefore iterations by Eq. (11.4.10) determine a fractal dynamics at the discrete moments. In the future, finding conditions to construct fractals by non-autonomous systems might be an interesting theoretical and application problem. We have applied the map with $a = 0.01$ and the Julia set corresponding to $c = -0.175 - 0.655i$ as the initial fractal. The results of the simulation are seen in Fig. 11.8. Since the moment $t = \pi/2$ is not a multiple of the period, the section in part (a) of the Fig. 11.9 does not seem to be a fractal, but in part (b) of the figure, the section is a Julia set.

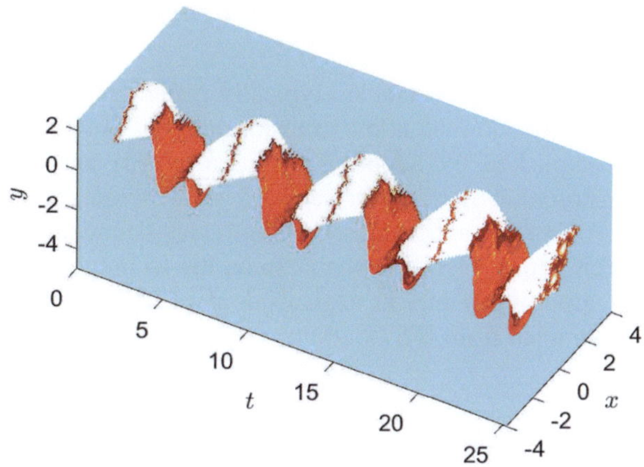

Fig. 11.8 The parametric set $A_t \mathscr{J}$ for (11.4.13)

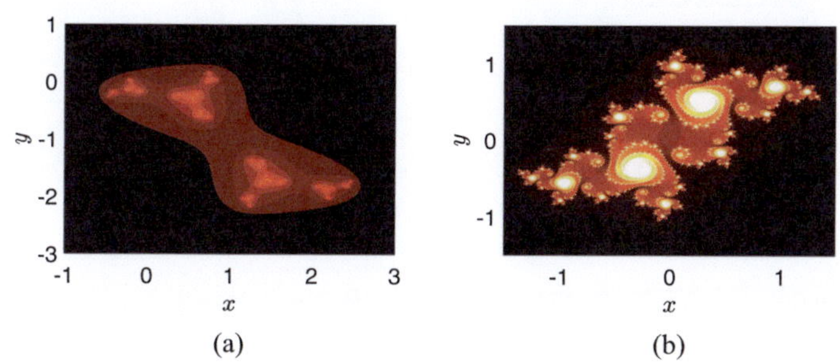

Fig. 11.9 The sections of the parametric set $A_t \mathscr{J}$ for (11.4.13). **(a)** $t = \frac{\pi}{2}$. **(b)** $t = 2\pi$

11.5 Dynamics Motivated by Sierpinski Fractals

Fatou–Julia iteration is an effective instrument to construct fractals. Famous Julia and Mandelbrot sets are strong confirmations of this. In this section, we use the paradigm of FJI to construct and map Sierpinski fractals. The process of the construction of the Sierpinski gasket starts with dividing an initial solid triangle into four identical triangles and removing the central one. In the next iterations, the same procedure is repeated to each of the remaining triangles from the preceding iteration. In an analogous way to the gasket, the construction of the Sierpinski carpet starts with dividing an initial solid square into nine identical squares and removing the central one. Similarly, each next iteration is a repetition of the same procedure to each of the remaining squares from the preceding iteration. Figure 11.10 shows the two Sierpinski fractals for finite iterations, which are constructed by the methods introduced and discussed in this section.

11.5.1 Construction of the Fractals

The Sierpinski fractals are typically generated using IFS [9, 22], which is defined as a collection of affine transformations. Another way of construction can be accomplished by adopting the idea of FJI and developing some schemes for constructing Sierpinski fractals. The technique of the FJI is based on detecting the points of a fractal set through the boundedness of their iterations under a specific map. Here, we shall extend the technique to include any possible criterion for grouping points in a given domain. It is worthy to mention that FJI can be constructed from IFS [8, 9].

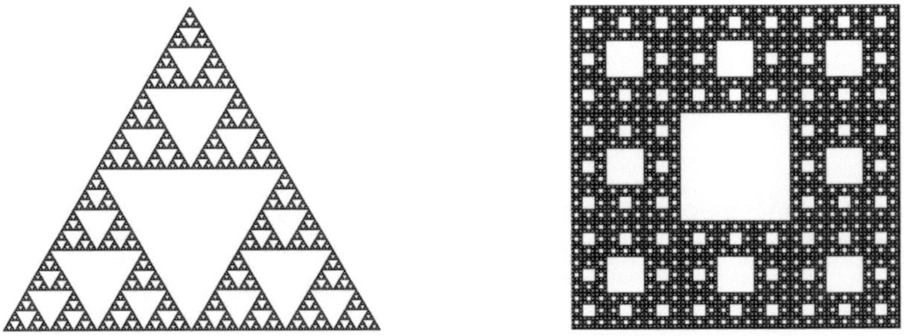

(a) (b)

Fig. 11.10 Sierpinski Fractals. (**a**) Sierpinski gasket. (**b**) Sierpinski carpet

11.5.1.1 Sierpinski Carpets

At first glance, the Sierpinski carpet seems to be a two-dimensional version of the middle third Cantor set. To discuss this thought, let us consider the tent map T defined on the interval $\mathscr{I} = [0, 1]$ such that

$$T(x) = \begin{cases} 3x, & \text{if } x \leq \frac{1}{2}, \\ 3(1-x), & \text{if } x > \frac{1}{2}. \end{cases} \tag{11.5.15}$$

The Julia set corresponding to the map (11.5.15) is the middle third Cantor set [32]. For planar fractals we consider the FJI defined by the two-dimensional tent map

$$(x_{n+1}, y_{n+1}) = \left(\frac{3}{2} - 3\left|x_n - \frac{1}{2}\right|, \frac{3}{2} - 3\left|y_n - \frac{1}{2}\right|\right), \tag{11.5.16}$$

with the initial square $\mathscr{D} = [0, 1] \times [0, 1]$. If we exclude each point (x_0, y_0) whose iterated values (x_n, y_n) escape from $\mathscr{D}$, i.e., at least one coordinate escapes, $x_n > 1$ or $y_n > 1$ for some $n \in \mathbb{N}$, we shall get a Cantor dust. This set is simply the Cartesian product of the Cantor set with itself, and it is a fractal possessing both self-similarity and fractional dimension. Figure 11.11a shows the 3rd approximation of the Cantor dust generated by (11.5.16).

Let us now modify this procedure such that a point (x_0, y_0) is excluded from $\mathscr{D}$ if both of its coordinates' iterations (x_n, y_n) escape from the initial set, that is, if $x_n > 1$ and $y_n > 1$ for some $n \in \mathbb{N}$. This procedure for iteration (11.5.16) will give a kind of two-dimensional Cantor set shown in Fig. 11.11b with the 3rd approximation. More similar object to the Sierpinski carpet can be obtained by considering simultaneous escape of both coordinates, viz., (x_0, y_0) is excluded only if $x_n > 1$ and $y_n > 1$ at the same iteration n. Figure 11.11c shows the 5th approximations of the resulting set. This set is clearly not a fractal from the dimension point of view. The self-similarity is also not satisfied over the whole set. However, a special type of self-similarity can be observed where the corners replicate the whole shape.

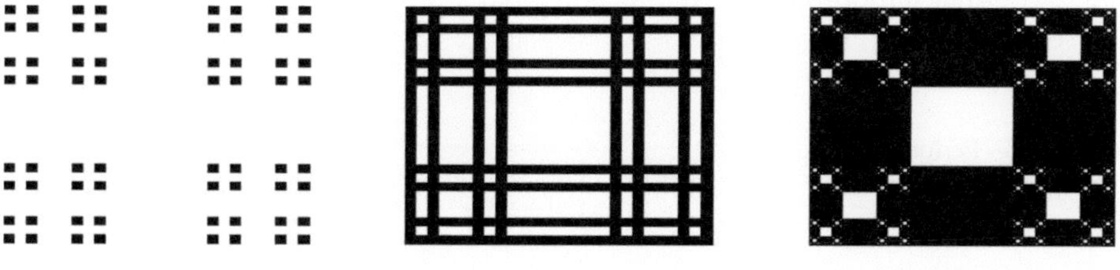

(a) (b) (c)

Fig. 11.11 Approximations of planar sets generated by (11.5.16) with different conditions of grouping the points

The construction of the Sierpinski carpet cannot possibly be performed through any arrangement of two-dimensional Cantor set and, therefore, a different strategy should be considered. To this end, we shall use maps that construct sets which are similar to Cantor sets in the generation way but different in structure. A suitable set for generating the Sierpinski carpet started with the initial set $\mathscr{I} = [0, 1]$. The first iteration involves subdividing $\mathscr{I}$ into three equal intervals and removing the middle open interval $(\frac{1}{3}, \frac{2}{3})$. In the second iteration the middle interval is restored and each of the three intervals are again subdivided into three equal subintervals then we remove the middle open intervals $(\frac{1}{9}, \frac{2}{9})$, $(\frac{4}{9}, \frac{5}{9})$, and $(\frac{7}{9}, \frac{8}{9})$. We continue in the same manner for the succeeding iterations. Figure 11.12 illustrates the first three stages of construction of the set. The purpose of such sets is to cut out successively smaller parts (holes) in the Sierpinski fractals kind. This is why we call these types of sets "*perforation sets*."

To construct perforation sets, we use the modified tent map

$$F(x) = \begin{cases} 3\,[x(\text{mod}\,1)], & \text{if } x \le \frac{1}{2} \text{ or } x > 1, \\ 3(1-x), & \text{if } \frac{1}{2} < x \le 1. \end{cases}$$

The point $x \in \mathscr{I}$ is excluded from the kth approximation of the set if its kth iteration $F^k(x)$ does not belong to $\mathscr{I}$. For the Sierpinski carpet we use a two-dimensional version of the modified tent map defined on the domain $\mathscr{D} = [0, 1] \times [0, 1]$. We consider the iteration

$$\begin{aligned} x_{n+1} &= \begin{cases} 3\,[x_n(\text{mod}\,1)], & \text{if } x_n \le \frac{1}{2} \text{ or } x_n > 1, \\ 3(1-x_n), & \text{if } \frac{1}{2} < x_n \le 1, \end{cases} \\ y_{n+1} &= \begin{cases} 3\,[y_n(\text{mod}\,1)], & \text{if } y_n \le \frac{1}{2} \text{ or } y_n > 1, \\ 3(1-y_n), & \text{if } \frac{1}{2} < y_n \le 1. \end{cases} \end{aligned} \tag{11.5.17}$$

To generate the Sierpinski carpet we exclude any point $(x_0, y_0) \in \mathscr{D}$ if its iteration (x_n, y_n) under (11.5.17) escapes from $\mathscr{D}$ such that $x_n > 1$, $y_n > 1$ for some natural number n. Figure 11.10b shows the 6th approximation of the Sierpinski carpet generated by iteration (11.5.17). The iteration looks very similar to the FJI but with different criterion for grouping the points. However, it can be classified under FJI type.

Fig. 11.12 Perforation set

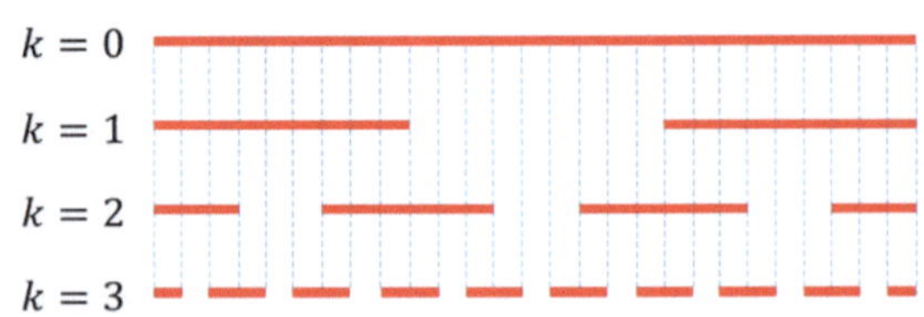

Another scheme can be developed by using a map to generate a sequence for each point in a given domain and then applying a suitable criterion to group the points. For that purpose, let us introduce the map

$$\psi_n(x) = B\sin(A_n x), \tag{11.5.18}$$

where $A_n = \pi a^{n-1}$, $B = \csc\frac{\pi}{b}$, and a, b are parameters. The recursive formula is defined as follows:

$$\begin{aligned}\psi_0(x_0) &:= x_0,\\ x_n &= \psi_n(x_0),\ \ n = 1, 2, \ldots .\end{aligned}$$

To construct the perforation set, we start with the interval $\mathscr{I} = [0, 1]$, and include in the kth approximation of the set each point $x_0 \in \mathscr{I}$ that satisfies $|x_k| \le 1$. Thus, for Sierpinski carpet, we use a two-dimensional version of the map (11.5.18) which can be defined in the form

$$\psi_n(x, y) = \big(B\sin(A_n x),\ \ B\sin(A_n y)\big). \tag{11.5.19}$$

The procedure here is to determine the image sequence (x_n, y_n) of each point $(x_0, y_0) \in \mathscr{D}$, i.e.,

$$(x_n, y_n) = \psi_n(x_0, y_0). \tag{11.5.20}$$

If we choose $\mathscr{D} = [0, 1] \times [0, 1]$, the point (x_0, y_0) is excluded from the set if the condition

$$|x_n| > 1,\ \ |y_n| > 1 \tag{11.5.21}$$

is satisfied for some $n \in \mathbb{N}$.

For the values of the parameters $a = 3$ and $b = 3$, the scheme gives the classical Sierpinski carpet (the simulation result for the 6th approximation is identical to Fig. 11.10b). Figure 11.13 shows other carpets generated by (11.5.19) with different values of the parameters a and b, and for limited stages. The colors that appear in the parts (d) and (f) of the figure are related to the sequences generated by (11.5.20) such that the color of each point in the carpets depends on the smallest number of stages n that satisfies condition (11.5.21).

This scheme is quite different than the usual procedure of FJI since iterations are not utilized and a different criterion is applied for grouping the points. However, we shall see later that the idea of the FMI can be applied for this type which allows to map and introduce dynamics for the constructed carpets.

A more similar iteration to that of Fatou and Julia can be constructed by finding the inverse $x = \psi_n^{-1}(x_n)$ in (11.5.18) and then substituting in $x_{n+1} = \psi_{n+1}(x)$ to get the autonomous iteration

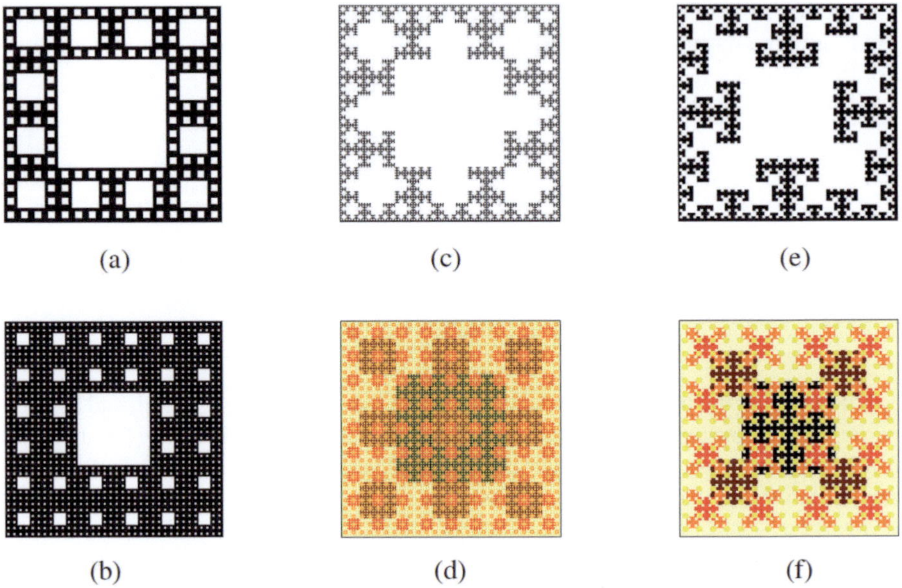

Fig. 11.13 Sierpinski carpets. (**a**) $a = b = 4$. (**b**) $a = 6,\ b = 3$. (**c**) $a = 3,\ b = 4$. (**d**) $a = 3,\ b = 4$. (**e**) $a = 2,\ b = 3.5$. (**f**) $a = 2,\ b = 3.5$

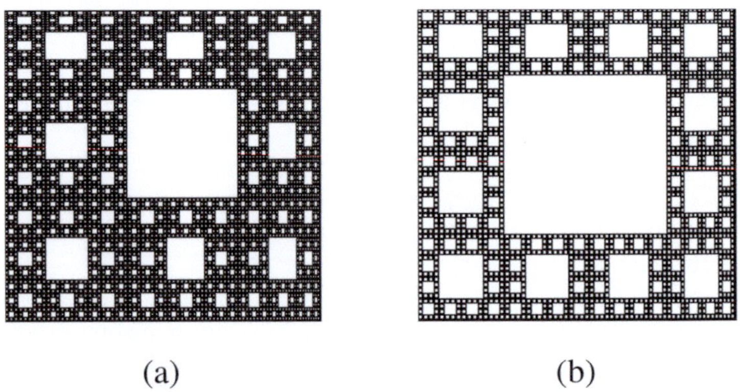

Fig. 11.14 Sierpinski carpets by the FJI (11.5.22). (**a**) $a = 3,\ b = 3$. (**b**) $a = 4,\ b = 4$

$$x_{n+1} = B \sin\left(a \sin^{-1} \frac{x_n}{B}\right). \tag{11.5.22}$$

The carpets shown in Fig. 11.14 are simulated by using a two-dimensional form of iteration (11.5.22), and they are irregular types of fractals. These fractals have asymmetric similarities and they are not categorized under random fractals [18].

11.5.1.2 Sierpinski Gasket

For constructing the Sierpinski gasket we again use the perforation sets, and in this case we introduce a special coordinate system shown in Fig. 11.15a. The system consists of three non-rectangular plane axes denoted by x', x'', and y, and the thick red lines in the figure represent the perforation set constructed in Fig. 11.15b.

We start with the initial set $\mathscr{D}$ which is a unit equilateral triangle defined by $\mathscr{D} = \{(x, y) \in \mathbb{R}^2 : \frac{-1}{\sqrt{3}}y \leq x \leq \frac{1}{\sqrt{3}}y,\ 0 \leq y \leq \frac{\sqrt{3}}{2}\}$. For each point $(x, y) \in \mathscr{D}$, we detect the triple (x', x'', y), where x' and x'' are the projections of (x, y) on the x' and x'' axes, respectively. To examine whether the point (x, y) belongs to the gasket, we set up the recursive formula

$$(x_n', x_n'', y_n) = \Big(\alpha\big(A_n x'\big),\ \beta\big(A_n x''\big),\ \gamma\big(A_n y\big)\Big), \tag{11.5.23}$$

where α, β, and γ are functions, $A_n = \frac{2}{\sqrt{3}}\pi a^n$, and a is a parameter. The point (x, y) is excluded from the resulting gasket if $x_n' > 0,\ x_n'' > 0$ and $y_n < 0$ for some $n \in \mathbb{N}$.

For $\alpha(x) = \beta(x) = \gamma(x) = \sin x$ and $a = 2$, the resulting gasket is the classical Sierpinski gasket. Figure 11.10a shows the 8th approximation of the Sierpinski gasket generated by iteration (11.5.23). Examples of other gaskets with different choices of the functions α, β, γ and the parameter a are shown in Fig. 11.16.

11.5.2 Mappings

In this part of the research we give procedures for mapping the Sierpinski fractals through the schemes introduced in the preceding section.

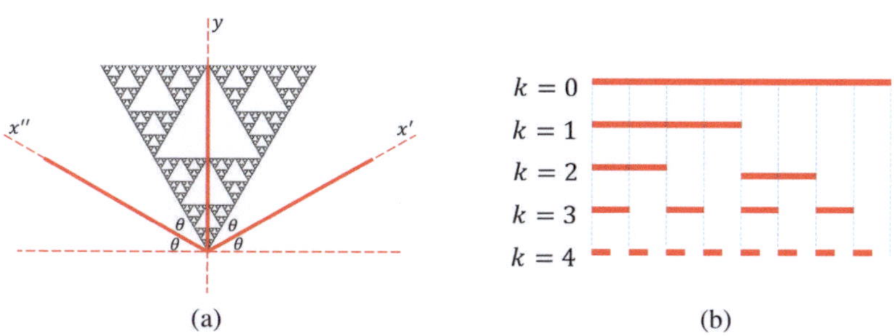

Fig. 11.15 Sierpinski gasket construction. (**a**) Coordinate system for Sierpinski gasket construction. (**b**) Perforation set

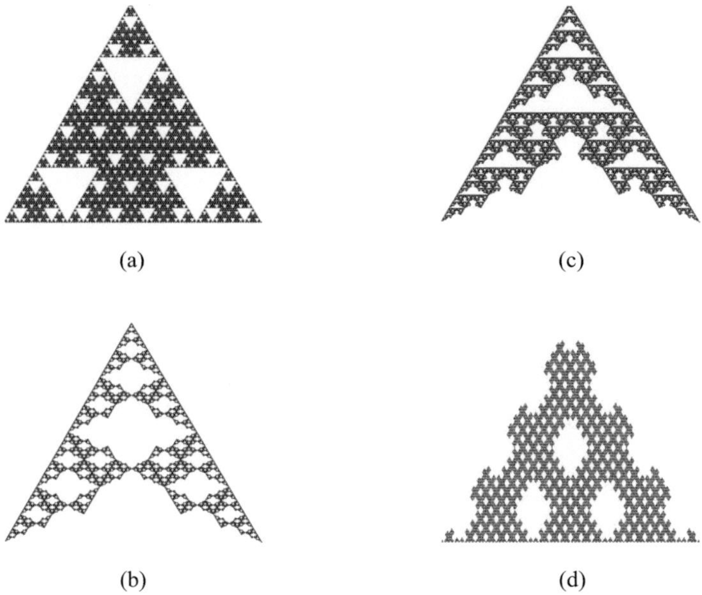

Fig. 11.16 Sierpinski gaskets. (a) $\alpha(x) = \beta(x) = \gamma(x) = \sin(x),\ a = 4$. (b) $\alpha(x) = \beta(x) = \gamma(x) = \cos(x),\ a = 2$. (c) $\alpha(x) = \tan(x),\ \beta(x) = \gamma(x) = \cos(x),\ a = 2$. (d) $\alpha(x) = \tan^{-1}(x),\ \beta(x) = \gamma(x) = \cos(x),\ a = 7$

11.5.2.1 Mapping of Carpets

To map the carpets generated by the scheme (11.5.19), we use the idea of FMI. Let $\Phi : \mathscr{D} \to \mathscr{D}'$ be an invertible function defined by

$$\Phi(x, y) = (\phi_1, \phi_2)(x, y), \tag{11.5.24}$$

with the inverse

$$\Phi^{-1}(\xi, \eta) = (\phi_3, \phi_4)(\xi, \eta). \tag{11.5.25}$$

Then the fractal mapping scheme can be defined as

$$\Phi^{-1}(\xi_n, \eta_n) = \psi_n\big(\Phi^{-1}(\xi_0, \eta_0)\big). \tag{11.5.26}$$

This scheme transforms a carpet $\mathscr{F}$ into a new carpet $\mathscr{F}_\Phi$, and the following theorem shows that the set $\mathscr{F}_\Phi$ is merely the image of $\mathscr{F}$ under the map Φ.

Theorem 11.2 $\mathscr{F}_\Phi = \Phi(\mathscr{F})$.

Proof First, we show that $\mathscr{F}_\Phi \subseteq \Phi(\mathscr{F})$. Let $(\xi, \eta) \in \mathscr{F}_\Phi$, which means that for formula (11.5.26), if $(\xi_0, \eta_0) = (\xi, \eta)$, then at least one of $|u_n|$ and $|v_n|$ is less than or equal to 1 for all $n \in \mathbb{N}$, where $(u_n, v_n) = \Phi^{-1}(\xi_n, \eta_n)$. This implies that $(u, v) = (u_0, v_0) \in \mathscr{F}$. Thus $(\xi, \eta) \in \Phi(\mathscr{F})$.

For the reverse inclusion, suppose that $(\xi, \eta) \in \Phi(\mathscr{F})$, i.e., there exists $(x, y) \in \mathscr{F}$ such that $\Phi(x, y) = (\xi, \eta)$ and $(x_0, y_0) = (x, y)$ with formula (11.5.20) in which at least one of $|x_n|$ and $|y_n|$ is less than or equal to 1 for all $n \in \mathbb{N}$. This directly implies that the sequence $\Phi^{-1}(\xi_n, \eta_n) = (x_n, y_n)$ satisfies (11.5.26) and $(\xi, \eta) \in \mathscr{F}_\Phi$. □

The following question arises here: Is the mapped carpet a fractal? The answer is "*yes*" if the map Φ satisfies a bi-Lipschitz condition. This result is stated in Lemma 11.3.1.

For our next examples, we shall use bi-Lipschitz functions to ensure that the mapped carpets are fractals. In order to obtain the mapped Sierpinski carpet $\mathscr{F}_\Phi$, we restrict the domain of (11.5.26) only to the points (ξ, η) that belong to the mapped domain $\mathscr{D}'$, thus Eq. (11.5.26) becomes

$$(\xi_n, \eta_n) = \Phi\big(\psi_n(x_0, y_0)\big), \quad (x_0, y_0) \in \mathscr{D}.$$

More precisely, by using Eqs. (11.5.19) and (11.5.25) in (11.5.26), we have

$$\Phi^{-1}(\xi_n, \eta_n) = \left(\frac{\sin(a^{n-1}\pi\phi_3(\xi_0, \eta_0))}{\sin(\frac{\pi}{b})}, \frac{\sin(a^{n-1}\pi\phi_4(\xi_0, \eta_0))}{\sin(\frac{\pi}{b})}\right).$$

Now letting $\frac{\sin(a^{n-1}\pi\phi_3(\xi_0,\eta_0))}{\sin(\frac{\pi}{b})} = X_n$ and $\frac{\sin(a^{n-1}\pi\phi_4(\xi_0,\eta_0))}{\sin(\frac{\pi}{b})} = Y_n$ and using Eq. (11.5.24), we get

$$(\xi_n, \eta_n) = \Big(\phi_1(X_n, Y_n),\ \phi_2(X_n, Y_n)\Big). \tag{11.5.27}$$

The semi-iteration (11.5.27) is applied for each point $(\xi_0, \eta_0) \in \mathscr{D}'$, and the point is excluded from the image set $\mathscr{F}_\Phi$ if $|\phi_3(\xi_n, \eta_n)| > 1,\ |\phi_4(\xi_n, \eta_n)| > 1$ for some $n \in \mathbb{N}$. Figures 11.17 and 11.18 show different examples for mappings of carpets by $\Phi(x, y) = (x^2 + y^2,\ x - y)$ and $\Phi(x, y) = (\sin x + y,\ \cos x)$, respectively. The colors displayed in these figures are produced in a similar way to those in Fig. 11.13.

11.5.2.2 Mapping of Gaskets

Let us give examples of mappings of gaskets using formula (11.5.23). Figure 11.19a shows the mapped Sierpinski gasket by the map $\Phi(x, y) = (x^2 - y, x + y^2)$, whereas Fig. 11.19b represents the mapped gasket depicted in Fig. 11.16b by the map $\Phi(x, y) = (x + y^2, x - 2y^{\frac{2}{3}})$.

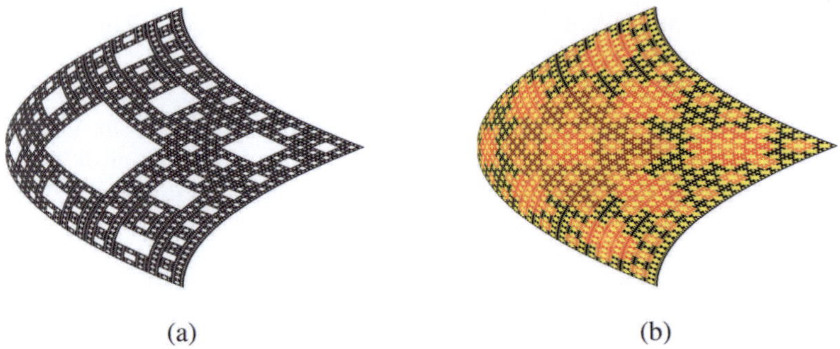

Fig. 11.17 The images of carpets with (**a**): $a = b = 3$, (**b**): $a = 3,\ b = 4$

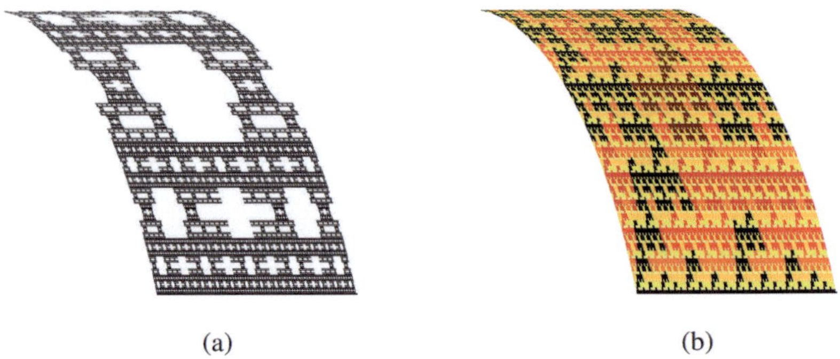

Fig. 11.18 The images of carpet with (**a**): $a = 2,\ b = 3.5$, (**b**): $a = 2,\ b = 1.5$

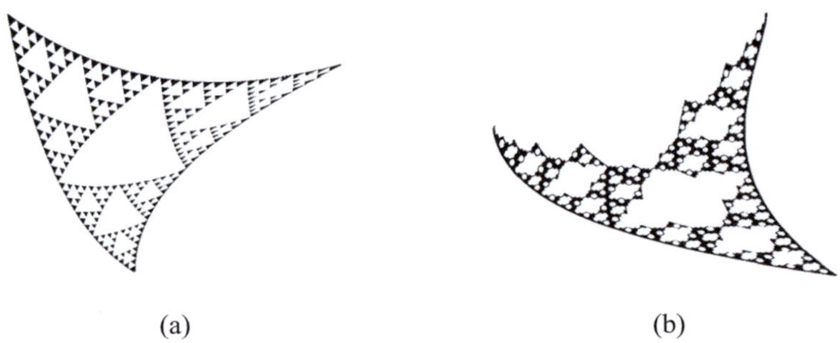

Fig. 11.19 Mappings of gaskets. (**a**) Image of Sierpinski gasket. (**b**) Image of gasket with $\alpha(x) = \beta(x) = \gamma(x) = \cos x,\ a = 2$

11.5.3 Dynamics

Based on the fractal mapping iteration, we introduce dynamics in fractals. As in Sect. 11.4.1, discrete dynamics can be constructed for Sierpinski fractals by iterating the mappings introduced in the previous section. In other words, if we start, for example, with the Sierpinski carpet as an initial set, $\mathscr{S}_0$, and iterate the map Φ in FMI, we shall have the image sets

$$\mathscr{S}_m = \Phi^m(\mathscr{S}_0), \ m = 1, 2, 3, \dots .$$

Thus the discrete dynamics will consist of these fractal sets, $\mathscr{S}_m$, as points of a trajectory.

For continuous dynamics, the idea is to use the motion of a dynamical system with a fractal as an initial set. The motion of dynamical system is defined by $\mathscr{A}_t \mathbf{x}_0 = \varphi(t, \mathbf{x}_0)$, where φ is the solution of a two-dimensional system of ordinary differential equations

$$\mathbf{x}' = g(t, \mathbf{x}), \tag{11.5.28}$$

with $\varphi(0, \mathbf{x}_0) = \mathbf{x}_0$.

In the case of the Sierpinski carpet, we iteratively apply a motion $\mathscr{A}_t$ to the scheme (11.5.19) in the way

$$\mathscr{A}_{-t}(\xi_n, \eta_n) = \psi_n\big(\mathscr{A}_{-t}(\xi, \eta)\big),$$

where $\mathscr{A}_t(x, y) = (A_t x, B_t y)$. Through this procedure, we construct dynamics of sets $A_t\mathscr{F}$, where the Sierpinski carpet $\mathscr{F}$ is the initial value. Thus, the differential equations are involved in fractals such that the latter become points of the solution trajectory. If the map A_t is bi-Lipschitzian (this is true, for instance, if the function g in (11.5.28) is Lipschitzian) then the set $A_t\mathscr{F}$ for each fixed t is a fractal.

Let us now consider the Van der Pol equation

$$u'' + \mu(u^2 - 1)u' + u = 0, \tag{11.5.29}$$

where μ is a real constant known as the damping parameter. Using the variables $x = u$ and $y = u'$, one can show that Eq. (11.5.29) is equivalent to the autonomous system

$$\begin{aligned} x' &= y, \\ y' &= \mu(1 - x^2)y - x. \end{aligned} \tag{11.5.30}$$

Let us denote by $\big(x(t, x_0), y(t, y_0)\big)$ the solution of (11.5.30) with $x(0, x_0) = x_0, \ y(0, y_0) = y_0$. System (11.5.30) can be numerically solved to construct a dynamical system with the motion $\mathscr{A}_t(x_0, y_0) = (A_t x_0, B_t y_0)$ where $A_t x_0 = x(t, x_0)$ and $B_t y_0 = y(t, y_0)$. We apply this dynamics for an approximation of the Sierpinski carpet as an initial set. The trajectory of the Van der Pol dynamics with $\mu = 0.5$ and $0 \le t \le 8$ is shown in Fig. 11.20. Figure 11.21 exhibits the sections of the trajectory at the moments $t = 1$, $t = 3$, $t = 5$, and $t = 7$.

Fig. 11.20 Van der Pol dynamics of Sierpinski carpet

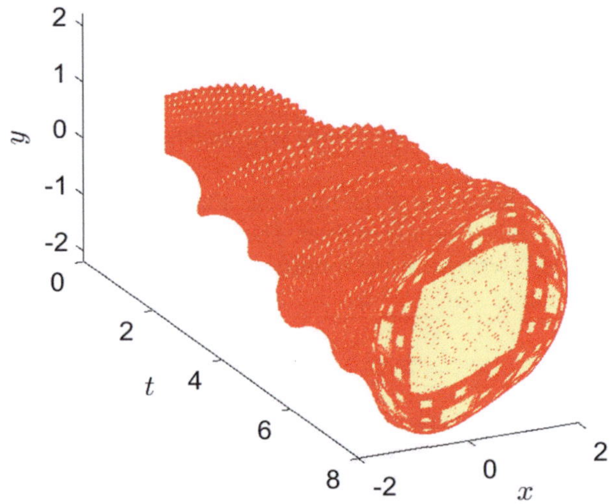

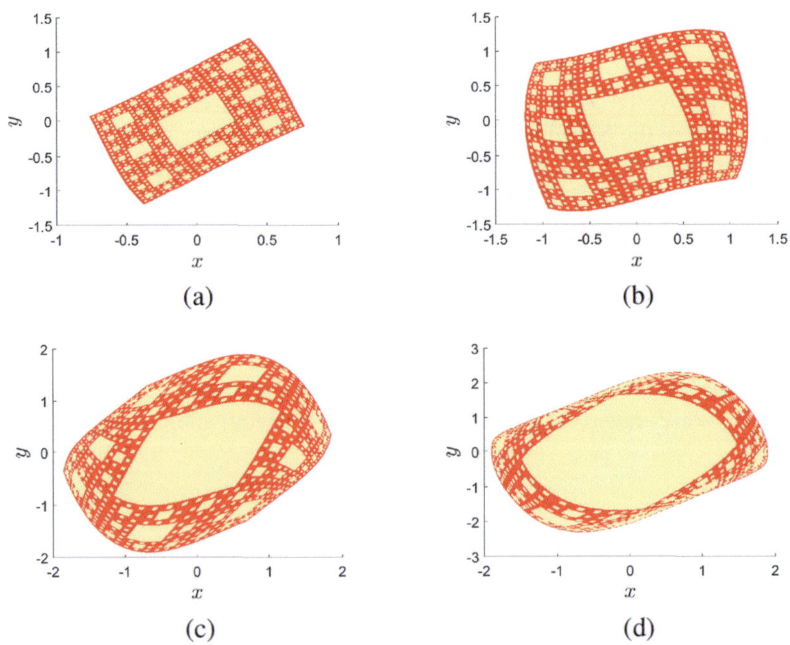

Fig. 11.21 Trajectory sections of the Van der Pol dynamics in Sierpinski carpet. (**a**) $t = 1$. (**b**) $t = 3$. (**c**) $t = 5$. (**d**) $t = 7$

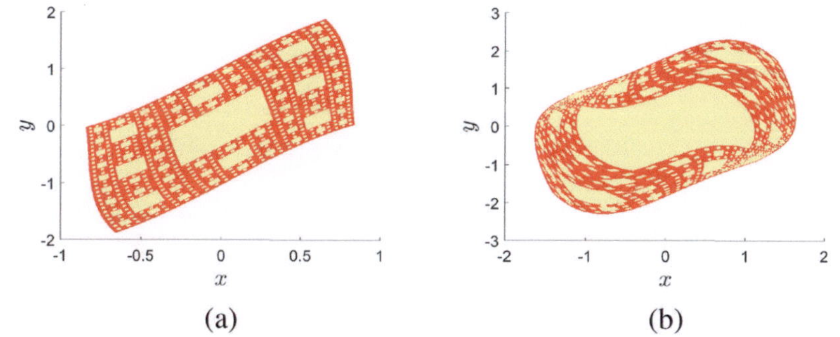

Fig. 11.22 Trajectory sections of the Van der Pol dynamics with $\mu = 1.3$. (**a**) $t = 1$. (**b**) $t = 3$

In Fig. 11.22 we again show two sections of the trajectory of the Van der Pol dynamics for the approximation of the Sierpinski carpet but with $\mu = 1.3$. Comparing these sections with their time counterpart in Fig. 11.21, we can observe the dissimilarity in the deformation rate in the structure of the Sierpinski carpet. This is attributed to that the value of the damping parameter reflects the degree of nonlinearity of the Van der Pol equation.

For dynamics in Sierpinski gasket, let us consider the Duffing equation

$$u'' + \delta u' + \beta u + \alpha u^3 = \gamma \cos \omega t,$$

where $\delta, \beta, \alpha, \gamma$, and ω are real parameters. The equation is equivalent to the non-autonomous system

$$\begin{aligned} x' &= y, \\ y' &= -\delta y - \beta x - \alpha x^3 + \gamma \cos \omega t. \end{aligned}$$

In a similar way to the mapping of gasket, we apply the dynamical system associated with the Duffing equation to an approximation of the Sierpinski gasket. The fractal trajectory for $0 \le t \le 3$ is shown in Fig. 11.23, whereas Fig. 11.24 displays the sections of the trajectory at the specific times $t = 0.8$, $t = 1.4$, $t = 2.0$, and $t = 2.6$. The values $\delta = 0.08$, $\beta = 0$, $\alpha = 1$, $\gamma = 0.2$, and $\omega = 1$ are used in the simulation.

11.6 Notes

Despite the intensive research of fractals lasts more than 35 years [31], there are still no results on mapping of the sets, and our research is the first one to consider the problem. To say about mathematical challenges connected to our suggestions, let us start with topological equivalence of fractals and consequently, normal forms.

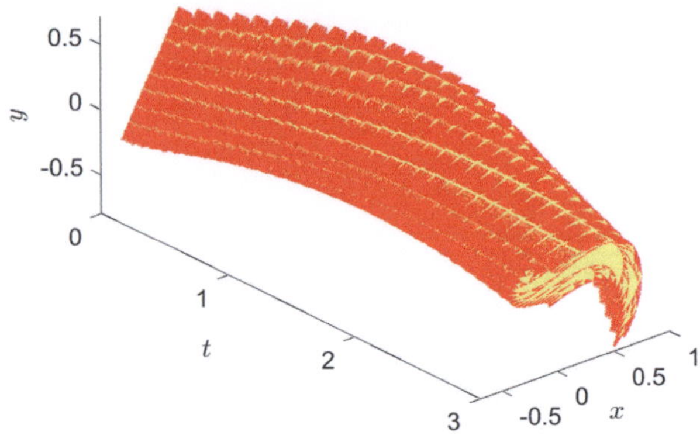

Fig. 11.23 Trajectory of the Duffing dynamics in Sierpinski gasket

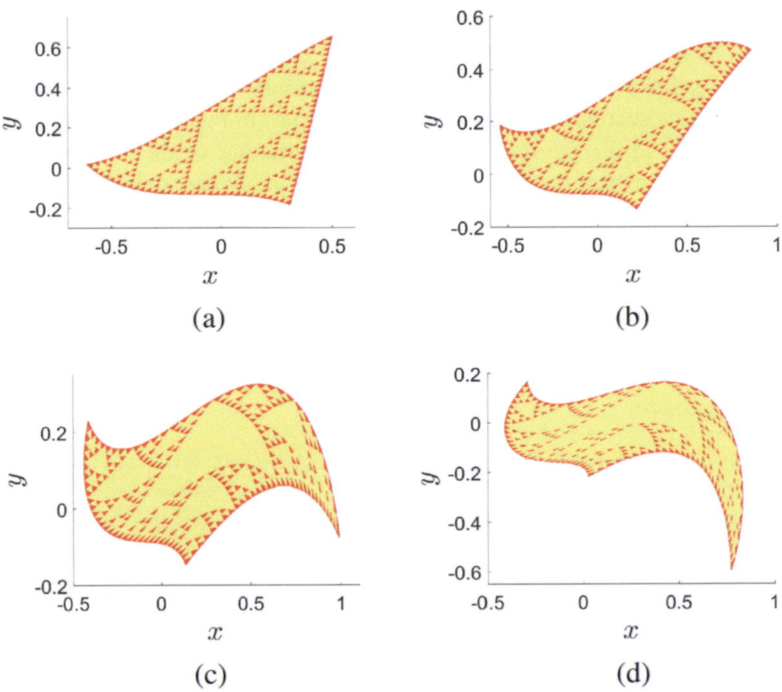

Fig. 11.24 Sections of the trajectory. (**a**) $t = 0.8$. (**b**) $t = 1.4$. (**c**) $t = 2.0$. (**d**) $t = 2.6$

Differential and discrete equations will be analyzed with new methods of fractal dynamics joined with dimension analysis. Next, the theory for dynamical systems which is defined as iterated maps can be developed. Therefore, mapping of fractals will be beneficial for new researches in hyperbolic dynamics, strange attractors, and ergodic theory [21, 48]. The developed approach will enrich the methods for the discovery and construction of fractals in the real world and industry such as nano-fiber engineering, 3D printing, biology, cosmology, biotechnologies, genetics, signal processing, civil engineering, etc. [12, 20, 29, 38, 47, 49].

Let us outline how our results can be useful for applications combined with already existing ones in the area. We start with the theory of scale relativity developed in [39, 40]. In this theory, fractals are considered as a geometric framework of atomic scale motions such that the quantum behavior can be viewed as particles moving on fractal trajectories. One can suppose that by composing the scale relativity theory with dynamics of fractals developed in our research, we will be able to understand better the fractal nature of the world. The expression of a scale-dependent physical quantity requires besides space-time variables, a scale variable. Thus, it is regarded as a fractal function but differentiable when the scale variable is nonzero, therefore, it can be a solution of differential equations involving its derivatives with respect to space-time and scale variables. For a fixed value of the scale variable the solution is a fractal. By varying the scale variable, one can set up an ordinary differential equation to construct a dynamical system. Then, by applying a mapping iteration whose initial set is a scale-dependent physical quantity with fixed a scale variable, one may show that the resulting surface is also a fractal. From another side, fractal dynamics determined by mapping iterations possibly can be a good instrument to study quantum mechanical properties. An example of such properties is the quantum interference of atoms and molecules. Fractal geometry has been used to study the interference patterns of waves such as in electromyography, diffraction grating, and color texture analysis [5, 11, 50]. In our case it would be interesting if one could perform simulations analogous to Young's experiment such that the interference occurs between two fractal trajectories. A possible connection between fractals mappings and quantum mechanics through the scale relativity theory can provide important applications for the former in various fields such as biology, cosmology, and fractal geodesics (see [40] and the relevant references listed therein).

Further applications can be done with another class of fractal functions which is defined as a family of real functions whose graphs are attractors for IFS [6–9, 35, 36]. This type is called fractal interpolation function where it is a continuous fractal function interpolating a given set of points [30]. The IFS is defined by a finite collection of affine mappings. The similarity between the IFS and FJI derives from the fact that both of them consist of infinite long iterations. Moreover, the graphs of some fractal functions constructed by IFS, likewise the Cantor set [36], can also be generated by FJI [32]. Additionally, the graphs of such functions can be mapped to graphs of new fractal functions by fractal mapping iterations and introduce dynamics on these sets. Furthermore, we suppose that the idea of mapping on the basis of FJI can be extended to IFS as they are iterative. Consequently, the results of the present research can be applied to various applications associated with fractal interpolation functions such as signal processing and modeling coastlines and shapes [13, 33, 34, 37, 51].

Owing to the important roles of Sierpinski fractals in several applications like weighted networks, trapping problems, antenna engineering, city planning, and urban growth [14, 25, 27, 46, 52], we expect that the results will be helpful in further fields of applications. One of the relevant applications involves optimization theory. Fractal geometry is used to solve some classes of optimization problems such as supply chain management and hierarchical design [26, 28]. In paper [26], for instance, the properties of a particular hierarchical structure are established. The authors constructed the relationship between the Hausdorff dimension of the optimal structure and loading for which the structure is optimized. The Hausdorff dimension is calculated through considering the self-similarity of the structure at different hierarchical levels. The self-similar fractals, like the Sierpinski gasket, are considered as effective tools for studying the hierarchical structures [27, 43]. Thus, finding a way to map this type of structures allows to create a new hierarchical structure with the same Hausdorff dimension but different mechanical properties if one considers bi-Lipschitz maps.

Finally, but not less important, another application field is partial differential equations with fractal boundaries [10, 19, 23, 41]. These equations are significantly useful for many fields such as electromagnetics, elasticity theory, and signal processing. For instance, the paper [23] deals with a relevant Brownian motion problem. The boundaries of the problem considered in the paper are of self-similar type such as Koch's snowflake curve. For partial differential equations, one can either apply a fractal mapping iteration extended to continuous dynamics in our research or develop the approach on the basis of IFS. Thus, in the future, one can not only develop numerical solutions of such problems but also confirm that the integral surfaces of the boundary value problems are fractals.

References

1. L.V. Ahlfors, *Lectures on Quasiconformal Mappings*, 2nd edn. (Amer. Math. Soc., Providence, RI, 2006)
2. M. Akhmet, M.O. Fen, E.M. Alejaily, Dynamics with fractals. Discontinuity Nonlinearity Complexity (in press)
3. M. Akhmet, M.O. Fen, E.M. Alejaily, Mapping Fatou-Julia Iterations. Proc. ICIME **2018**, 64–67 (2018)
4. M. Akhmet, M.O. Fen, E.M. Alejaily, Generation of fractals as Duffing equation orbits. Chaos **29**, 053113 (2019)
5. D. Bak, S.P. Kim, S.K. Kim, K-S. Soh, J.H. Yee, Fractal Diffraction Grating. ArXiv Physics e-prints, arXiv:physics/9802007
6. M.F. Barnsley, S. Demko, Iterated function systems and the global construction of fractals. Proc. R. Soc. Lond. Ser. A Math. Phys. Sci. **399**, 243–275 (1985)
7. M.F. Barnsley, Fractal functions and interpolation. Constr. Approx. **2**, 303–329 (1986)
8. M.F. Barnsley, *Fractals Everywhere* (Academic Press, London, 1988)
9. M.F. Barnsley, S. Demko, Iterated function systems and the global construction of fractals. Proc. R. Soc. Lond. Ser. A **399**, 243–275 (1985)
10. F.M. Borodich, A.Y. Volovikov, Surface integrals for domains with fractal boundaries and some applications to elasticity. Proc. R. Soc. Ser. A **456**, 1–24 (2000)
11. D. Casanova, J.B. Florindo, M. Falvo, O.M. Bruno, Texture analysis using fractal descriptors estimated by the mutual interference of color channels. Inf. Sci. 346–347, 58–72 (2016)
12. C. Cattani, Fractals and hidden symmetries in DNA. Math. Prob. Eng. **2010**, 1–31 (2010)
13. W.O. Cochran, J.C. Hart, P.J. Flynn, On approximating rough curves with fractal functions. Proc. Graphics Interface **1**, 65–72 (1998)

14. Y. Dong, M. Dai, D. Ye, Non-homogeneous fractal hierarchical weighted networks. Plos One. **10**, e0121946 (2015)
15. J. Encarnacao, H-O. Peitgen, G. Saka, G. Englert, *Fractal Geometry and Computer Graphics* (Springer, Berlin, 1992)
16. K. Falconer, *Fractal Geometry: Mathematical Foundations and Applications*, 2nd edn. (Wiley, Chichester, 2003)
17. P. Fatou, Sur les équations fonctionnelles, I, II, III. Bull. Soc. Math. France **47**, 161–271 (1919); **48**, 33–94 (1920); **48**, 208–314 (1920)
18. D.P. Feldman, *Chaos and Fractals: An Elementary Introduction* (Oxford University Press, UK, 2012)
19. G. Franceschetti, A. Iodice, D. Riccio, G. Ruello, Fractal surfaces and electromagnetic extended boundary conditions. IEEE Trans. Geoscience Remote Sensing **40**, 1018–1031 (2002)
20. P. Frankhauser, Fractals geometry of urban patterns and their morphogenesis. Discrete Dyn. Nat. Soc. **2**, 127–145 (1998)
21. J. Guckenheimer, J. Moser, S.E. Newhouse, *Dynamical Systems* (Birkhäuser, Boston, 1980)
22. J. Hutchinson, Fractals and self-similarity. Indiana Univ. J. Math. **30**, 713–747 (1981)
23. A. Jonsson, H. Wallin, Boundary value problems and Brownian motion on fractals. Chaos Solitons Fractals **8**, 191–205 (1997)
24. G. Julia, Mémoire suv l'itération des fonctions rationelles. J. Math. Pures Appl. **8**, 47–245 (1918)
25. A. Kansal, J. Kaur, Sierpinski gasket fractal array antenna. IJCSC **1**, 133–136 (2010)
26. D. Rayneau-Kirkhope, Y. Mao, R. Farr, Optimization of fractal space frames under gentle compressive load. Phys. Rev. E **87**, 063204 (2013)
27. J.J. Kozak, V. Balakrishnan, Analytic expression for the mean time to absorption for a random walker on the Sierpinski gasket. Phys. Rev. E **65**, 021105 (2002)
28. S. Liu, H. Dong, W. Zhao, Optimization model based on the fractal theory in supply chain management. Adv. Mat. Res. **694–697**, 3554–3557 (2013)
29. X.Y. Liu, P.D. Sawant, Determination of the fractal characteristic of nanofiber-network formation in supramolecular materials. Chem. Phys. Chem. **4**, 374–377 (2002)
30. D.C. Luor, Fractal interpolation functions with partial self similarity. J. Math. Anal. Appl. **464**, 911–923 (2018)
31. B.B. Mandelbrot, *Les Objets Fractals: Forme, Hasard, et Dimension* (Flammarion, Paris, 1975)
32. B.B. Mandelbrot, *Fractals and Chaos: The Mandelbrot Set and Beyond* (Springer, New York, 2004)
33. P. Manousopoulos, V. Drakopoulos, T. Theoharis, Curve fitting by fractal interpolation, in *Transactions on Computational Science I. Lecture Notes in Computer Science*, vol. 4750, ed. by M. L. Gavrilova, C. Tan (Springer, Berlin, Heidelberg, 2008), pp. 85–103
34. P. Manousopoulos, V. Drakopoulos, T. Theoharis, Parameter identification of 1d recurrent fractal interpolation functions with applications to imaging and signal processing. J. Math. Imaging Vision **40**, 162–170 (2011)
35. P. Massopust, *Fractal Functions, Fractal Surfaces, and Wavelets* (Academic Press, San Diego, 1994)
36. P. Massopust, Fractal functions and their applications. Chaos Solitons Fractals **8**, 171–190 (1997)
37. D.S. Mazel, Representation of discrete sequences with three-dimensional iterated function systems. IEEE Trans. Signal Process **42**, 3269–3271 (1994)
38. R. Noorani, *3D Printing: Technology, Applications, and Selection* (CRC Press, New York, 2017)
39. L. Nottale, *Fractal Space-Time and Microphysics Towards a Theory of Scale Relativity* (World Scientific, Singapore, 1993)
40. L. Nottale, *Scale Relativity and Fractal Space-Time: A New Approach 315 in Unifying Relativity and Quantum Mechanics* (Imperial College Press, London, 2011)

41. A.P. Pancheha, A.A. Puzenko, S.A. Puzenko, Boundary conditions for the electromagnetic field on a non-differentiable fractal surface. Phys. Lett. A **182**, 407–410 (1993)
42. H-O. Peitgen, D. Saupe (eds.), *The Science of Fractal Images* (Springer, New York, 1988)
43. J.A. Riera, Relaxation of hierarchical models defined on Sierpinski gasket fractals. J. Phys. A Math. Gen. **19**, L869–L873 (1986)
44. M. Shishikura, The Hausdorff dimension of the boundary of the Mandelbrot set and Julia sets. Ann. Math. **147**, 225–267 (1998)
45. K.J. Smith, *The Nature of Mathematics*, 13th edn. (Cengage Learning, Boston, 2017)
46. D. Triantakonstantis, Urban growth prediction modelling using fractals and theory of chaos. Open J. Civil Eng. **2**, 81–86 (2012)
47. J.L. Véhel, E. Lutton, C. Tricot (Eds.), *Fractals in Engineering: From Theory to Industrial Applications* (Springer, New York, 1997)
48. S. Wiggins, *Global Bifurcation and Chaos: Analytical Methods* (Springer, New York, Berlin, 1988)
49. G.W. Wornell, *Signal Processing with Fractals: A Wavelet-Based Approach* (Prentice-Hall, Upper Saddle River, NJ, 1996)
50. Z. Xu, S. Xiao, Fractal dimension of surface ENG and its determinants, in *The 19th Annual International Conference of the IEEE Engineering in Medicine and Biology Society*, vol. 4, pp. 1570–1573 (1997)
51. M.Y. Zhai, J.L. Fernández-Martínez, J.W. Rector, A new fractal interpolation algorithm and its applications to self-affine signal reconstruction. Fractals **19**, 355–365 (2011)
52. Z. Zhang, Y. Li, S. Gao, S. Zhou, J. Guan, M. Li, Trapping in scale-free networks with hierarchical organization of modularity. Phys. Rev. E **80**, 051120 (2009)

Chapter 12
Abstract Similarity, Fractals, and Chaos

To prove presence of chaos for fractals, a new mathematical concept of *abstract similarity* is introduced. As an example, the space of symbolic strings on a finite number of symbols is proved to possess the property. Moreover, Sierpinski fractals, Koch curve as well as Cantor set satisfy the definition. A natural similarity map is introduced and the problem of chaos presence for the sets is solved by considering the dynamics of the map. This is true for Poincaré, Li–Yorke, and Devaney chaos, even in multi-dimensional cases. Original numerical simulations which illustrate the results are delivered. The preliminary discussions in this chapter stem from [2].

12.1 Introduction

In this chapter, we concern with *self-similarity*. This concept is reflected in many problems that arise in various fields such as wavelets, fractals, and graph systems [16]. We define the abstract self-similar set as a collection of points in a metric space, where it can be considered as a union of infinite shrinking sets with notation that allows to introduce dynamics in the set and then prove chaos which is the most important result of the present chapter. For that purpose, a specific map over the invariant self-similar set is defined, and this map acts as the identity if the whole set is taken as its argument. This feature is equivalent to the property of self-similarity in the ordinary fractals, and this is the reason for calling this map the similarity map. We expect that the concept of abstract self-similarity will create new frontiers for chaos and fractals investigations, and we hope it will be a helpful tool in other fields such as harmonic analysis, discrete mathematics, probability, and operator algebras [16].

Our approach with respect to chaos is characterized by the priority of the domain over the map of chaos [3]. More precisely, the usual construction of chaos starts with description of a map with certain properties like unimodality, hyperbolicity,

M. Akhmet et al., *Dynamics with Chaos and Fractals*, Nonlinear Systems and Complexity 29, https://doi.org/10.1007/978-3-030-35854-9_12

period three, and topological conjugacy to a standard chaotic map. Next, chaotic behavior is discovered, and finally, the structure of the chaotic attractor is analyzed. For example, in Li–Yorke chaos, a map is firstly defined such that it has period three property, and then, the domain (scrambled set) for the chaos is characterized by specific properties. The same can be said for the chaos of unimodal maps, when the domain of chaos is a Cantor set. In the Smale horseshoe case, we can conclude that the map and the domain are simultaneously determined so that the construction of the domain started with an initial set and it is developed step-by-step using a particular map. Therefore, the structure of the domain and the nature of the map are mutually dependent on each other. For the chaos in symbolic dynamics, the domain is primarily described as infinite sequences of symbols then the map is introduced as a shift on the space. However, the map still has priority since the properties of the sequences are described with respect to the map. In the proposed approach, we first construct a domain in a metric space with specific conditions to be a suitable venue for manifestations of chaos. Thereafter, the similarity map is built on the basis of the invariance and self-similarity properties of the domain to define an abstract motion. This is why the map is appropriate for abstract self-similar set as well as for any fractal constructed through self-similarity, and thence proving chaos for these classes of fractals becomes possible. Moreover, we drop the continuity requirement for the motion since the chaotic map need not be continuous [1, 18]. In paper [18], for instance, the authors ignore the continuity of some chaotic maps during the discussion of chaos conditions on the product of semi-flows. We regard the continuity of a chaotic map as an important property only from the analytical side, that is to say, it is very useful for handling the map to prove presence of chaos [12, 23], however, it is not rigorously correct to consider it as an intrinsic property for chaos. Despite the discontinuity of the similarity map, opposite to our desire, one can recognize that presence of sensitivity and the irregular behavior of simulations make the discussion precious for theoretical investigations as well as for future applications. As implementations we consider different types of chaos, namely Devaney, Li–Yorke, and Poincaré, for self-similar fractals, Sierpinski fractals, Koch curve as well as Cantor set.

The concept of the abstract self-similarity in different forms has been realized for several problems in our papers [3, 4].

12.2 Abstract Self-Similarity

Let us consider the metric space $(\mathscr{F}, d)$, where $\mathscr{F}$ is a compact set and d is a metric. We assume that $\mathscr{F}$ is divided into m disjoint nonempty subsets, $\mathscr{F}_i,\ i = 1, 2, \ldots, m$, such that $\mathscr{F} = \cup_{i=1}^{m} \mathscr{F}_i$. In their own turn, the sets $\mathscr{F}_i$ $i = 1, 2, \ldots, m$, are divided into m disjoint nonempty subsets $\mathscr{F}_{ij},\ j = 1, 2, \ldots, m$, such that $\mathscr{F}_i = \cup_{j=1}^{m} \mathscr{F}_{ij}$. That is, in general we have $\mathscr{F}_{i_1 i_2 \ldots i_n} = \cup_{j=1}^{m} \mathscr{F}_{i_1 i_2 \ldots i_n j}$, for each natural number n, where all sets $\mathscr{F}_{i_1 i_2 \ldots i_n j},\ j = 1, 2, \ldots, m$, are nonempty and disjoint. We assume that for the sets $\mathscr{F}_{i_1 i_2 \ldots i_n}$, the *diameter condition* is valid. That is,

$$\max_{i_k=1,2,\ldots,m} \operatorname{diam}(\mathscr{F}_{i_1 i_2 \ldots i_n}) \to 0 \text{ as } n \to \infty, \tag{12.2.1}$$

where $\operatorname{diam}(A) = \sup\{d(\mathbf{x}, \mathbf{y}) : \mathbf{x}, \mathbf{y} \in A\}$, for a set A in $\mathscr{F}$.

Let us construct a sequence, p_n, of points in $\mathscr{F}$ such that $p_0 \in \mathscr{F}$, $p_1 \in \mathscr{F}_{i_1}$, $p_2 \in \mathscr{F}_{i_1 i_2}, \ldots, p_n \in \mathscr{F}_{i_1 i_2 \ldots i_n}$, $n = 1, 2, \ldots$. It is clear that,

$$\mathscr{F} \supset \mathscr{F}_{i_1} \supset \mathscr{F}_{i_1 i_2} \supset \ldots \supset \mathscr{F}_{i_1 i_2 \ldots i_n} \supset \mathscr{F}_{i_1 i_2 \ldots i_n i_{n+1}} \ldots, i_k = 1, 2, \ldots, m, k = 1, 2, \ldots.$$

That is, the sets form a nested sequence. Therefore, due to the compactness of $\mathscr{F}$ and the diameter condition, there exists a unique limit point for the sequence p_n. Denote the point as $\mathscr{F}_{i_1 i_2 \ldots i_n \ldots} \in \mathscr{F}$, accordingly to the indexes of the nested subsets. Conversely, it is easy to verify that each point $p \in \mathscr{F}$ admits a corresponding p_n and it can be written as $p = \mathscr{F}_{i_1 i_2 \ldots i_n \ldots}$, and this representation is a unique one due to the diameter condition. Finally, we have that

$$\mathscr{F} = \left\{\mathscr{F}_{i_1 i_2 \ldots i_n \ldots} : i_k = 1, 2, \ldots, m,\ k = 1, 2, \ldots\right\}, \tag{12.2.2}$$

and

$$\mathscr{F}_{i_1 i_2 \ldots i_n} = \bigcup_{j_k=1,2,\ldots,m} \mathscr{F}_{i_1 i_2 \ldots i_n j_1 j_2 \ldots}, \tag{12.2.3}$$

for fixed indexes $i_1, i_2, \ldots, i_n$.

The set $\mathscr{F}$ satisfies (12.2.2) and (12.2.3) is said to be the *abstract self-similar set* as well as the triple $(\mathscr{F}, d, \varphi)$ the *self-similar space*.

Let us introduce the map $\varphi : \mathscr{F} \to \mathscr{F}$ such that

$$\varphi(\mathscr{F}_{i_1 i_2 \ldots i_n \ldots}) = \mathscr{F}_{i_2 i_3 \ldots i_n \ldots}. \tag{12.2.4}$$

Considering iterations of the map, one can verify that

$$\varphi^n(\mathscr{F}_{i_1 i_2 \ldots i_n}) = \mathscr{F}, \tag{12.2.5}$$

for arbitrary natural number n and $i_k = 1, 2, \ldots, m,\ k = 1, 2, \ldots$. The relations (12.2.4) and (12.2.5) give us a reason to call φ a *similarity map* and the number n the *order of similarity*.

In the next example of our study and in the future studies, it is important to find the structure of abstract self-similar space for a given mathematical object.

Example 12.1 ([2]) Let us consider the space of symbolic strings of 0 and 1 [13, 23], which is defined by

$$\Sigma = \{s_1 s_2 s_3 \ldots : s_k = 0 \text{ or } 1\}.$$

The distance in Σ is defined by

$$d(s,t)=\sum_{k=1}^{\infty}\frac{|s_k-t_k|}{2^{k-1}}, \tag{12.2.6}$$

where $s = s_1s_2\ldots$ and $t = t_1t_2\ldots$ be two elements in Σ.

Considering the pattern of the self-similar set, we denote the elements of the set by $\Sigma_{s_1s_2\ldots} = s_1s_2\ldots$, and describe the nth order subsets of strings in Σ by

$$\Sigma_{s_1s_2\ldots s_n} = \{s_1s_2\ldots s_ns_{n+1}s_{n+2}\ldots : s_k = 0 \text{ or } 1\},$$

where $s_1, s_2, \ldots, s_n$ are fixed symbols. One can show that $d(s,t) \le \frac{1}{2^{n-1}}$ for any two elements $s, t \in \Sigma_{s_1s_2\ldots s_n}$. Moreover, $d(s_1s_2\ldots s_n000\ldots, s_1s_2\ldots s_n111\ldots) = \frac{1}{2^{n-1}}$. Therefore, $\text{diam}(\Sigma_{s_1s_2\ldots s_n}) = \frac{1}{2^{n-1}}$. Consequently,

$$\lim_{n\to\infty}\text{diam}(\Sigma_{s_1s_2\ldots s_n}) = \lim_{n\to\infty}\frac{1}{2^{n-1}} = 0,$$

and the diameter condition holds.

The similarity map for the space is the Bernoulli shift, $\sigma(s_1s_2s_3\ldots) = s_2s_3s_4\ldots$. That is,

$$\varphi(\Sigma_{s_1s_2s_3\ldots}) = \sigma(s_1s_2s_3\ldots).$$

On the basis of the above discussion, one can conclude that the triple (Σ, d, φ) is a self-similar space. This is a purely illustrative example since it makes us perceive how self-similarity can be defined for abstract objects which are not necessarily geometrical ones. The space of symbolic strings on two symbols has been considered, since it is the most basic example that frequently used to describe the dynamics on symbolic spaces. However, more generally, the space on m symbols can also be considered.

12.3 Similarity and Chaos

To prove chaos for the self-similar space, we assumed in this section the *separation condition*. Define the distance between two nonempty bounded sets A and B in $\mathscr{F}$ by $d(A,B) = \inf\{d(\mathbf{x},\mathbf{y}) : \mathbf{x} \in A, \mathbf{y} \in B\}$. The set $\mathscr{F}$ satisfies the separation condition of degree n if there exist a positive number ε_0 and a natural number n such that for arbitrary $i_1i_2\ldots i_n$ one can find $j_1j_2\ldots j_n$ so that

$$d\big(\mathscr{F}_{i_1i_2\ldots i_n}, \mathscr{F}_{j_1j_2\ldots j_n}\big) \ge \varepsilon_0. \tag{12.3.7}$$

We call ε_0 the *separation constant*.

In the following theorem, we prove that the similarity map φ possesses the three ingredients of Devaney chaos, namely density of periodic points, transitivity, and sensitivity. A point $\mathscr{F}_{i_1 i_2 i_3 \ldots} \in \mathscr{F}$ is periodic with period n if its index consists of endless repetitions of a block of n terms.

Theorem 12.1 ([2]) *If the separation condition holds, then the similarity map is chaotic in the sense of Devaney.*

Proof Fix a member $\mathscr{F}_{i_1 i_2 \ldots i_n \ldots}$ of $\mathscr{F}$ and a positive number ε. Find a natural number k such that $\mathrm{diam}(\mathscr{F}_{i_1 i_2 \ldots i_k}) < \varepsilon$ and choose a k-periodic element $\mathscr{F}_{i_1 i_2 \ldots i_k i_1 i_2 \ldots i_k \ldots}$ of $\mathscr{F}_{i_1 i_2 \ldots i_k}$. It is clear that the periodic point is an ε-approximation for the considered member. The density of periodic points is thus proved.

Next, utilizing the diameter condition, the transitivity will be proved if we show the existence of an element $\mathscr{F}_{i_1 i_2 \ldots i_n \ldots}$ of $\mathscr{F}$ such that for any subset $\mathscr{F}_{i_1 i_2 \ldots i_k}$ there exists a sufficiently large integer p so that $\varphi^p(\mathscr{F}_{i_1 i_2 \ldots i_n \ldots}) \in \mathscr{F}_{i_1 i_2 \ldots i_k}$. This is true since we can construct the sequence $i_1 i_2 \ldots i_n \ldots$ such that it contains all sequences of the type $i_1 i_2 \ldots i_k$ as blocks.

For sensitivity, fix a point $\mathscr{F}_{i_1 i_2 \ldots} \in \mathscr{F}$ and an arbitrary positive number ε. Due to the diameter condition, there exist an integer k and element $\mathscr{F}_{i_1 i_2 \ldots i_k j_{k+1} j_{k+2} \ldots} \neq \mathscr{F}_{i_1 i_2 \ldots i_k i_{k+1} i_{k+2} \ldots}$ such that $d(\mathscr{F}_{i_1 i_2 \ldots i_k i_{k+1} \ldots}, \mathscr{F}_{i_1 i_2 \ldots i_k j_{k+1} j_{k+2} \ldots}) < \varepsilon$. We choose $j_{k+1}, j_{k+2}, \ldots$ such that $d(\mathscr{F}_{i_{k+1} i_{k+2} \ldots i_{k+n}}, \mathscr{F}_{j_{k+1} j_{k+2} \ldots j_{k+n}}) > \varepsilon_0$, by the separation condition. This proves the sensitivity. □

For Poincarè chaos, Poisson stable motion is utilized to distinguish the chaotic behavior instead of the periodic motions in Devaney and Li–Yorke types. Existence of infinitely many unpredictable Poisson stable trajectories that lie in a compact set meet all requirements of chaos. Based on this, chaos can be appeared in the dynamics on the quasi-minimal set which is the closure of a Poisson stable trajectory. Therefore, the Poincarè chaos is referred to as the dynamics on the quasi-minimal set of trajectory initiated from unpredictable point. For more details we refer the reader to [7, 8].

Next theorem shows that the Poincarè chaos is valid for the similarity dynamics.

Theorem 12.2 ([2]) *If the separation condition is valid, then the similarity map possesses Poincarè chaos .*

The proof of the last theorem is based on the verification of Lemma 3.1 in [8] adopted to the similarity map.

In addition to the Devaney and Poincarè chaos, it can be shown that the Li–Yorke chaos also takes place in the dynamics of the map φ. The proof of the following theorem is similar to that of Theorem 6.35 in [12] for the shift map defined on the space of symbolic sequences.

Theorem 12.3 ([2]) *The similarity map is Li–Yorke chaotic if the separation condition holds.*

Example 12.2 ([2]) We have shown that the space of symbolic strings is a self-similar set in Example 12.1. One can see that $\Sigma = \Sigma_0 \cup \Sigma_1$, where $\Sigma_0 = \{0 s_2 s_3 \ldots\}$

and $\Sigma_1 = \{1s_2s_3\ldots\}$, hence,

$$\begin{aligned} d(\Sigma_0, \Sigma_1) &= \inf\{d(s,t) : s \in \Sigma_0,\ t \in \Sigma_1\} \\ &= d(000\ldots, 1000\ldots) \\ &= d(0111\ldots, 111\ldots) = 1. \end{aligned}$$

Therefore, the separation condition of degree 1 holds with the separation constant ε_0 equal to unity.

According to the results of this section, the Bernoulli shift is chaotic in the sense of Poincarè, Li–Yorke, and Devaney. That is, we confirm one more time the presence of chaos which have been proven for the dynamics in [6, 8, 13, 23].

12.4 Chaos in Fractals

As implementations of abstract self-similarity, we consider several examples of fractals, namely Sierpinski carpet, Sierpinski gasket, Koch curve, and Cantor set. This consists of two main tasks. The first one is to indicate abstract self-similarity for the fractals, and the second one is to ascertain chaos according to the results of the last section.

12.4.1 Chaos for Sierpinski Carpet

Let S be the Sierpinski carpet constructed in a unit square. In what follows, we are going to find the structure of the abstract self-similar space for the Sierpinski carpet. We shall denote the abstract set by the italic S. Let us start by dividing the carpet into eight subsets and denote them as $S_1, S_2, \ldots S_8$ (see Fig. 12.1a). The subsets will be determined such that any couple of adjacent subsets have common horizontal or vertical boundary line. For this reason, we shall use the *boundary agreement* such that: (i) The points of the common boundary of two horizontally adjacent subsets belong to the left one. (ii) The points of the common boundary of two vertically adjacent subsets belong to the lower one. Figure 12.1b illustrates the boundary agreement, (i) and (ii), for the subsets, S_5, S_7 and S_8. For clarification the boundaries are shown by black lines and we see that the common boundary points of S_5 and S_8 belong to S_5 not to S_8 and the common boundary points of S_7 and S_8 belong to S_7 not to S_8. In the second step, each subset S_i, $i = 1, 2, \ldots, 8$ is again subdivided into eight smaller subsets denoted as S_{ij}, $j = 1, 2, \ldots, 8$.

Continuing in the same manner, the subsets of higher order can inductively be determined such that at each nth step we have 8^n subsets notated as $S_{i_1 i_2 \ldots i_n}$, $i_k = 1, 2, \ldots, 8$. Figure 12.2a and b show, for example, the subsets of S_1 and subsets of S_{11}, respectively.

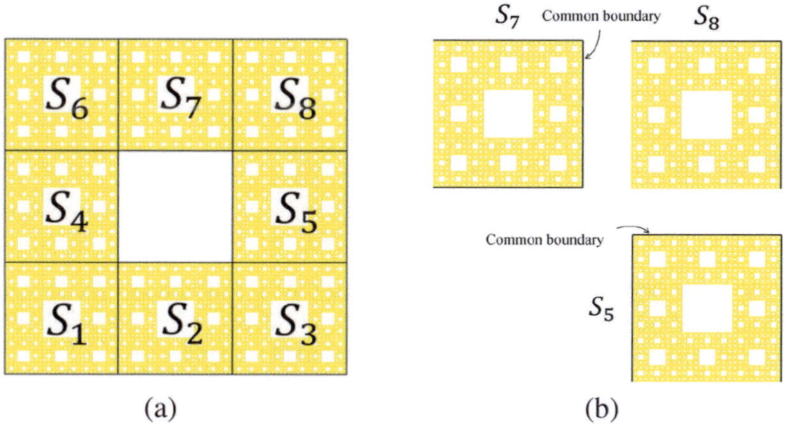

Fig. 12.1 (**a**) The first step of abstract self-similar set construction. (**b**) The illustration of the boundary agreement

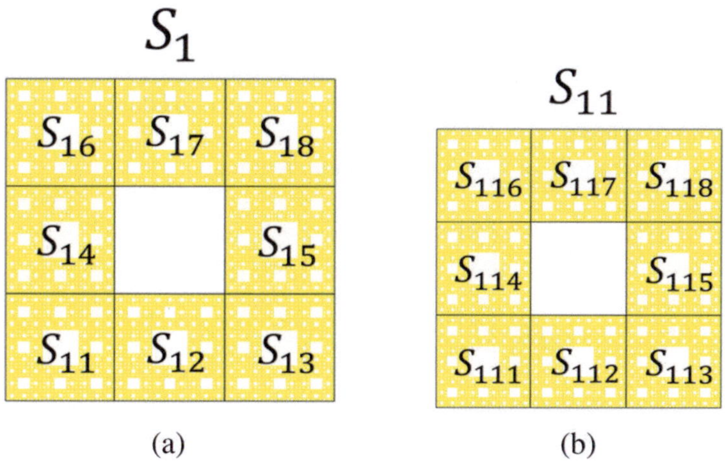

Fig. 12.2 Examples of the 2nd and the 3rd order subsets of the Sierpinski carpet

To determine the distance between the points of S, we will apply the corresponding Euclidean distance for the set S such that if $\mathbf{x} = (x_1, x_2)$ and $\mathbf{y} = (y_1, y_2)$ are two points in S, then, $d(\mathbf{x}, \mathbf{y}) = \sqrt{(x_1 - y_1)^2 + (x_2 - y_2)^2}$. The diameter of a subset at an nth step is equal to, $\text{diam}(S_{i_1 i_2 i_3 \ldots i_n}) = \sqrt{2}/3^n$, and therefore, it diminishes to zero as n tends to infinity, and the diameter condition holds. It is easy to check that each point in S has a unique presentation $S_{i_1 i_2 \ldots i_n \ldots}$. Hence, the set S can be written as

$$S = \left\{S_{i_1 i_2 \ldots i_n \ldots} : i_k = 1, 2, \ldots 8,\ k \in \mathbb{N}\right\}.$$

The separation condition of degree 1 is satisfied and the separation constant is

$$\varepsilon_0 = \left\{ \min\{d(S_i, S_j)\} : S_i, S_j \text{ are disjoint, } i, j = 1, 2, \ldots, 8 \right\} = \frac{1}{3}.$$

Let us now define the similarity map by

$$\varphi(S_{i_1 i_2 i_3 \ldots}) = S_{i_2 i_3 \ldots}.$$

Thus, we have shown that the triple (S, d, φ) is a self-similar space with the separation condition. In view of Theorems 12.1, 12.2, and 12.3, the similarity map S is chaotic in the sense of Poincaré, Li–Yorke, and Devaney.

12.4.2 A Chaotic Trajectory in the Sierpinski Carpet

In this section, we provide a geometric realization of the similarity map on the Sierpinski carpet and see how the map can be useful for visualizing the trajectories of the points of a self-similar set and indexing its subsets. A chaotic trajectory is seen as expected in the last section. In the Chap. 11, we adopt the idea of Fatou–Julia iteration and develop a scheme for constructing the Sierpinski carpet. The scheme is based on the iterations of the modified planar tent map

$$T(x) = \begin{cases} 3\,[x(\text{mod}1)] & x \le \frac{1}{2} \text{ or } x > 1, \\ 3(1-x) & \frac{1}{2} < x \le 1, \end{cases}$$

$$T(y) = \begin{cases} 3\,[y(\text{mod}1)] & y \le \frac{1}{2} \text{ or } y > 1, \\ 3(1-y) & \frac{1}{2} < y \le 1. \end{cases}$$

Depending on this, one can construct a map $\bar{T} = (\bar{T}_1, \bar{T}_2) : S \to S$ such that the set S is invariant,

$$\bar{T}_1(x) = \begin{cases} 3x & 0 \le x \le \frac{1}{3}, \\ 3x - 1 & \frac{1}{3} < x \le \frac{1}{2}, \\ 2 - 3(1-x) & \frac{1}{2} < x < \frac{2}{3}, \\ 3(1-x) & \frac{2}{3} \le x \le 1, \end{cases}$$
$$\bar{T}_2(y) = \begin{cases} 3y & 0 \le y \le \frac{1}{3}, \\ 3y - 1 & \frac{1}{3} < y \le \frac{1}{2}, \\ 2 - 3(1-y) & \frac{1}{2} < y < \frac{2}{3}, \\ 3(1-y) & \frac{2}{3} \le y \le 1. \end{cases} \tag{12.4.8}$$

Fig. 12.3 The trajectory of a point under the similarity map

This map is equivalent to the similarity map φ defined above, therefore, the trajectory of a point $\mathbf{x} \in S$ can be visualized using the map (12.4.8). Figure 12.3 shows an example of the trajectory for the center point $\mathbf{x}$ of the subset

$$S_{277311373132771824315158224617847648526563584625456271254233172162 44}.$$

The points of the trajectory are considered as the centers of the subsets which are determined by $\bar{T}^k(\mathbf{x}),\ k = 0, 1, 2, \ldots 68$. The idea of indexing of the subsets is illustrated in Example 12.3.

12.4.3 *Sierpinski Gasket as Chaos Domain*

To construct an abstract self-similar set on the basis of the Sierpinski gasket, let us consider the Sierpinski gasket generated in a unit equilateral triangle. We firstly divided the gasket into three smaller parts to be the first order subsets and denoted them by G_1, G_2, and G_3 as shown in Fig. 12.4a. By glancing at the figure, one can see that every two subsets share only a single point as a common boundary. For this reason we consider the following boundary agreement: The common boundary point of every couple of adjacent subsets belongs either to the left one or to the lower one. Applying the agreement, the subsets G_1, G_2, and G_3, become disjoint subsets of the desired abstract self-similar set G such that $G = \cup_{i=1}^{3} G_i$. Secondly, each subset, $G_i,\ i = 1, 2, 3$, is again subdivided into three subsets, and we notate them as $G_{ij},\ i, j = 1, 2, 3$, (see Fig. 12.4b). Taking into account the boundary agreement, we repeat the same procedure such that at each nth step, we denote the resultant subsets by $G_{i_1 i_2 \ldots i_n},\ i_k = 1, 2, 3$.

The subsets of the Sierpinski gasket described above have an inverse relationship with the construction-step variable, n, $\text{diam}(S_{i_1 i_2 i_3 \ldots i_n}) = 1/2^n$, from which one can verify the validity of the diameter condition.

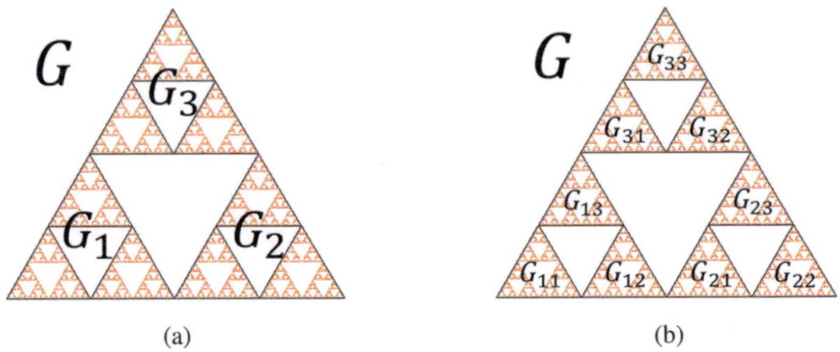

Fig. 12.4 The construction of abstract self-similar set corresponding to the Sierpinski gasket

Following the arguments of the abstract self-similarity, one can deduce that a point in G can be uniquely represented by $G_{i_1 i_2 \ldots i_n \ldots}$, and the abstract self-similar set G can be expressed as

$$G = \left\{G_{i_1 i_2 \ldots i_n \ldots} : i_k = 1, 2, 3,\ k \in \mathbb{N}\right\}.$$

Contrary to the Sierpinski carpet case, the separation constant ε_0 for the Sierpinski gasket cannot be evaluated through the first order subsets $G_i,\ i = 1, 2, 3$. This is why we considered the minimum of the distance between any two disjoint subsets of the second order, that is,

$$\varepsilon_0 = \{\min\{d(G_{i_1 i_2}, G_{j_1 j_2})\} : G_{i_1 i_2}, G_{j_1 j_2} \text{ are disjoint},\ i_n = 1, 2, 3\} = \frac{\sqrt{3}}{8},$$

where d is the usual Euclidean distance. Thus, one can see that separation condition is valid.

The similarity map acting on the Sierpinski gasket, G, can be defined by $\varphi(G_{i_1 i_2 i_3 \ldots}) = G_{i_2 i_3 \ldots}$. Consequently, the triple (G, d, φ) is a self-similar space and φ is chaotic in the sense of Poincaré, Li–Yorke, and Devaney.

The same idea can be extended to the fractals associated with Pascal's triangles. It is well known that Pascal's triangle in mod 2 creates the classical Sierpinski gasket. Different fractals associated with Pascal's triangles in different moduli can be considered as abstract similar sets and it can also be proved that the similarity map defined on these sets possesses chaos.

12.4.4 Koch Curve and Chaos

Let us consider the Koch curve, K, constructed from an initial unit line segment. To identify an abstract self-similar set corresponding to the Koch curve, we start

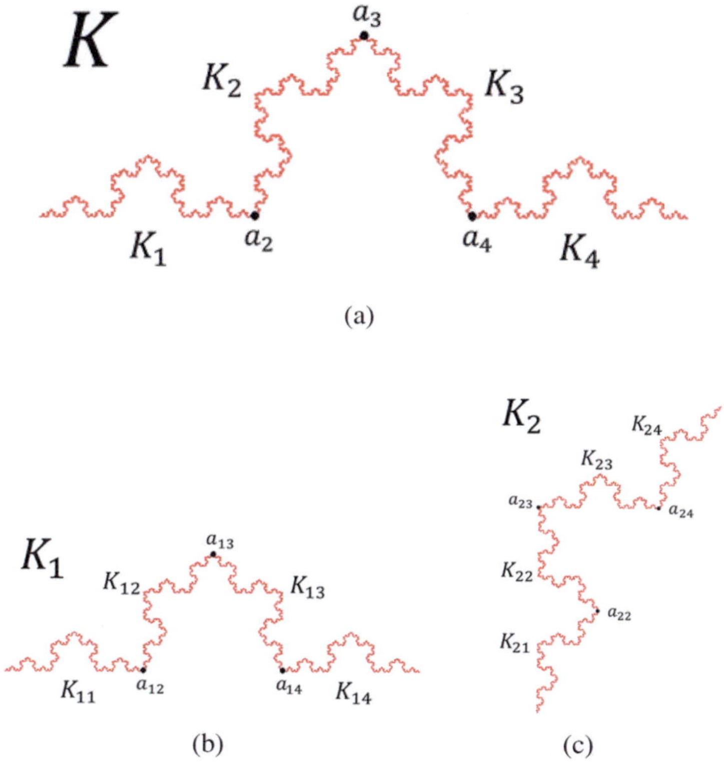

Fig. 12.5 The construction of abstract self-similar set corresponding to the Koch curve

by dividing K into four equal parts (subsets) and denoting them as K_1, K_2, K_3, and K_4 as shown in Fig. 12.5a. Since the Koch curve is a connected set, each two adjacent subsets share a single point as a common boundary. Let us denote the end points of each subset $K_i,\ i = 2, 3$ by a_i and a_{i+1}. Figure 12.5a illustrates these points and for clarification the points are shown by black thick dots. It is seen in the figure that K_1 and K_2 share the point a_2, K_2 and K_3 share the point a_3, and K_3 and K_4 share the point a_4. In the second step, each subset $K_i,\ i = 1, 2, 3, 4$ is again subdivided into four subsets $K_{ij},\ j = 1, 2, 3, 4$. Figure 12.5b and c illustrate the second step for the subsets K_1 and K_2, respectively. Again here we see that each two adjacent subsets share a single boundary point. We continue in the same way such that at each step, n, every set $K_{i_1 i_2 \ldots i_{n-1}},\ i_k = 1, 2, 3, 4$ is divided into four subsets $K_{i_1 i_2 \ldots i_{n-1} i_n},\ i_n = 1, 2, 3, 4$ and denote the end points of each $K_{i_1 i_2 \ldots i_n},\ i_n = 2, 3,$ by $a_{i_1 i_2 \ldots i_n}$ and $a_{i_1 i_2 \ldots i_n + 1}$.

As in the previous cases, to determine the abstract self-similar set, we need to consider the following boundary agreement: For each adjacent subsets $K_{i_1 i_2 \ldots i_{n-1} j}$ and $K_{i_1 i_2 \ldots i_{n-1} j+1}$, the common boundary point $a_{i_1 i_2 \ldots i_{n-1} j+1}$ belongs to $K_{i_1 i_2 \ldots i_{n-1} j+1}$. This condition means that the common boundary point a_2 shown in

Fig. 12.5a, for instance, belongs to K_2 not to K_1 and the common boundary point a_{23} shown in Fig. 12.5c belongs to K_{23} not to K_{22}.

By applying the boundary agreement to all subsets at each step, we have fully described the disjoint subsets of the proposed abstract self-similar set for the Koch curve. From the construction of the Koch curve and by using the usual Euclidean distance, one can deduce that the distance between the end points of each subset $K_{i_1 i_2 \ldots i_n}$ is $\frac{1}{3^n}$ which clearly represents the diameter of the subset. Therefore, the diameter condition holds. A point in K can be represent by $K_{i_1 i_2 \ldots i_n \ldots}$, so that,

$$K = \left\{K_{i_1 i_2 \ldots i_n \ldots} : i_k = 1, 2, 3, 4, \ k \in \mathbb{N}\right\}.$$

The separation condition is also valid with degree 1, since for any K_i, $i = 1, 2, 3, 4$ one can find K_j, $j = 1, 2, 3, 4$, $j \neq i$ such that they are separated from each other by a distance of not less than ε_0. The separation constant, ε_0, can be defined by

$$\varepsilon_0 = \min\{d(K_1, K_3), d(K_1, K_4), d(K_2, K_4)\} = \frac{\sqrt{7}}{9}.$$

The similarity map for the abstract fractals of Koch curve is given by $\varphi(K_{i_1 i_2 i_3 \ldots}) = K_{i_2 i_3 \ldots}$, and thus, we have shown that the triple (K, d, φ) defines a chaotic self-similar space.

12.4.5 Chaos for Cantor Set

A perfect example of chaos in fractals is the Cantor set. As we previously mentioned, the chaoticity in the Cantor set is determined by finding a topological conjugacy with the symbolic dynamics. To show that the Cantor set is not an exception to our approach for chaos, we shall establish an abstract self-similar set corresponding to the Cantor set. Let us consider the middle third Cantor set, C, initiated from a unit line segment. The first step consists of dividing C into two subsets and denoted them by C_1 and C_2 (see Fig. 12.6a). In the second step each of C_1 and C_2 is subdivided

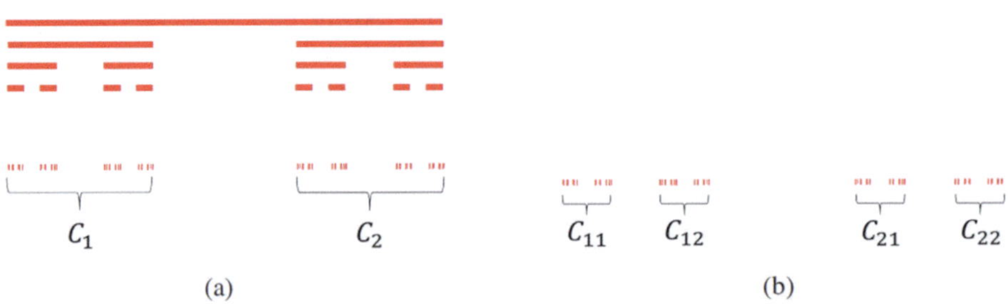

Fig. 12.6 The 1st and the 2nd order subsets of the abstract self-similar set for the Cantor set

into two subsets as shown in Fig. 12.6b. These subsets are denoted by C_{11}, C_{12}, C_{21} and C_{22}. In every next step, we repeat the same procedure for each subsets resulting from the preceding step. We denote the resultant subsets at each nth step by $C_{i_1 i_2 \ldots i_n}$, $i_k = 1, 2$.

In the Cantor set case, we do not need any boundary agreement since all subsets are disjoint. Considering the usual Euclidean distance, the diameter of a subset, $C_{i_1 i_2 \ldots i_n}$, is $1/3^n$. Therefore, the diameter condition holds. The points in C are represented by $C_{i_1 i_2 \ldots i_n \ldots}$. Hence the abstract self-similar set is defined by

$$C = \left\{ C_{i_1 i_2 \ldots i_n \ldots} : i_k = 1, 2, \ k \in \mathbb{N} \right\}.$$

From the construction, the separation condition is clearly valid and the constant ε_0 is defined by the distance between the subsets C_1 and C_2, so that $\varepsilon_0 = 1/3$.

The similarity map is defined by $\varphi(C_{i_1 i_2 i_3 \ldots}) = C_{i_2 i_3 \ldots}$, and the triple (C, d, φ) defines a self-similar space. Theorems 12.1, 12.2 and 12.3 are also applicable for this case.

In connection with the above examples of chaos, we remark that the Sierpinski carpet and Koch curve indicate that a non-continuous map can have a domain which is a connected set while for continuous maps, the domains of chaos are usually disconnected. Examples of disconnected chaotic domains are the Cantor set for the logistic map [13], the modified Sierpinski triangle with exceptions in [10], the set associated with Smale horseshoe map [25], and the Poincaré section of the Lorenz attractor [20].

12.5 Domain-Structured Chaos

In this section, we describe a dynamical determination of abstract self-similar set by utilizing the roles of the domain and the map simultaneously. In other words, a map is used to describe the structure of $\mathscr{F}$ and the relationships between its subsets. The set constructed by this way, we call it Dynamical Abstract Similarity Set (DASS). We start by considering the triple (X, d, φ), where X denotes the underlying set, d is a metric, and $\varphi : X \to X$ is a map.

Fix subsets F and F_0 of X such that F is compact, F_0 is bounded and $\varphi(F) \cap F_0$ is a nonempty set. Let m be a fixed natural number. Denote by $F^{(1)}$ the preimage of the set $\varphi(F) \cap F_0$ under the function φ in F and assume that there exist disjoint nonempty subsets $F_i \subset F$, $i = 1, 2, \ldots m$, such that $\cup_{i=1}^{m} F_i = F^{(1)}$.

Denote by $F^{(2)}$ the preimage of the set $\varphi(F) \cap F_0$ under φ^2 in $F^{(1)}$ and assume that there exist disjoint nonempty subsets $F_{ij} \subset F_i$, $j = 1, 2, \ldots m$, such that $\cup_{j=1}^{m} F_{ij} = F^{(2)}$.

Once more, denote by $F^{(3)}$ the preimage of the set $\varphi(F) \cap F_0$ under φ^3 in $F^{(2)}$ and assume that there exist disjoint nonempty subsets $F_{ijk} \subset F_{ij}$, $k = 1, 2, \ldots m$, such that $\cup_{k=1}^{m} F_{ijk} = F^{(3)}$.

In general, if the sets $F^{(n-1)}$ are determined, we denote by $F^{(n)}$ the preimage of the set $\varphi(F) \cap F_0$ under φ^n in $F^{(n-1)}$ and assume that there exist disjoint nonempty subsets $F_{i_1 i_2 \ldots i_n} \subset F_{i_1 i_2 \ldots i_{n-1}}$, $i_n = 1, 2, \ldots m$, such that $\cup_{i_n=1}^{m} F_{i_1 i_2 \ldots i_n} = F^{(n)}$.

We continue in this procedure, and assume that the following condition is satisfied:

$$\max_{i_k=1,2,\ldots,m} \operatorname{diam}(F_{i_1 i_2 \ldots i_n}) \to 0 \text{ as } n \to \infty. \tag{12.5.9}$$

Let us construct a sequence, p_n, of points in F such that $p_0 \in F$, $p_1 \in F_{i_1}$, $p_2 \in F_{i_1 i_2}, \ldots, p_n \in F_{i_1 i_2 \ldots i_n}$, $n = 1, 2, \ldots$. It is clear that,

$$F \supset F_{i_1} \supset F_{i_1 i_2} \supset \ldots \supset F_{i_1 i_2 \ldots i_n} \supset F_{i_1 i_2 \ldots i_n i_{n+1}} \ldots, \; i_k = 1, 2, \ldots, m, \; k = 1, 2, \ldots.$$

That is, the sets form a nested sequence. Therefore, due to the compactness of F and condition (12.5.9), there exists a unique limit point for the sequence p_n. According to the indexes of the nested subsets, we denote the point as $\mathscr{F}_{i_1 i_2 \ldots i_n \ldots} \in F$. Conversely, it can be verified that each point $p = \mathscr{F}_{i_1 i_2 \ldots i_n \ldots}$ admits a corresponding p_n. Based on this, one can justify that the representation of each such point is a unique one. The collection of all such points constitutes the set $\mathscr{F}$, i.e.,

$$\mathscr{F} = \left\{\mathscr{F}_{i_1 i_2 \ldots i_n \ldots} : i_k = 1, 2, \ldots, m\right\},$$

and for fixed indexes $i_1, i_2, \ldots, i_n$ the subsets of $\mathscr{F}$ can be represented by

$$\mathscr{F}_{i_1 i_2 \ldots i_n} = \bigcup_{j_k=1,2,\ldots,m} \mathscr{F}_{i_1 i_2 \ldots i_n j_1 j_2 \ldots}.$$

Since $\mathscr{F}_{i_1 i_2 \ldots i_n} \subset F_{i_1 i_2 \ldots i_n}$, the condition (12.5.9) implies that the diameter condition (12.2.1) is valid for the set $\mathscr{F}$. Thus, the set $\mathscr{F}$ is a DASS. Moreover, from the above construction, we see that the map φ satisfies the relations (12.2.4) and (12.2.5). Therefore, φ is a similarity map and triple $(\mathscr{F}, d, \varphi)$ is a self-similar space.

Now, let us formulate the following condition: For arbitrary $i_1 i_2 \ldots i_n$ one can find $j_1 j_2 \ldots j_n$ and a positive number ε such that

$$d\left(F_{i_1 i_2 \ldots i_n}, F_{j_1 j_2 \ldots j_n}\right) \geq \varepsilon. \tag{12.5.10}$$

From the construction, it is clear that if the condition (12.5.10) holds for the sets $F_{i_1 i_2 \ldots i_n}$, then the separation condition (12.3.7) is valid for the set $\mathscr{F}$ with a separation constant $\varepsilon_0 \geq \varepsilon$. If this is the case, then in view of Theorem 12.1, 12.2 and 12.3, the similarity map φ is chaotic in the sense of Poincaré, Li–Yorke, and Devaney.

The approach described above is not only an alternative way of the abstract similarity construction but it can be an essential part of the subject. For instance, it gives us a method for indexing. That is, the similarity map can be simultaneously used to number the subsets of each order depending on the numeration of their images. The following examples illustrate the idea of DASS, indexing, and chaos.

12.6 Examples

Example 12.3 ([4]) Let $X = \mathbb{R}^n$ and $F_0 = F = [0, 1]^n$ is the n-dimensional unit cube. Consider the n-dimensional logistic map $\varphi = (\varphi_1, \varphi_2, \ldots \varphi_n) : \mathbb{R}^n \to \mathbb{R}^n$ defined by

$$
\begin{aligned}
x_{k+1}^1 &= \varphi_1(x_k^1) = r_1 x_k^1(1 - x_k^1),\\
x_{k+1}^2 &= \varphi_2(x_k^2) = r_2 x_k^2(1 - x_k^2),\\
&\vdots\\
x_{k+1}^n &= \varphi_n(x_k^n) = r_n x_k^n(1 - x_k^n),
\end{aligned}
\tag{12.6.11}
$$

where $r_i > 4,\ i = 1, 2, \ldots, n$ are parameters. Proceeding upon the properties of the logistic map, ETA is applied for the map (12.6.11). We iterate the points in F under the map (12.6.11) such that in each iteration, we keep only the points whose images do not escape the domain F_0. The resulting points from the first iteration belong to the subsets $F_i,\ i = 1, 2, \ldots, 2^n$. In the second iteration, the non-escaped points belong to 2^{2n} subsets and each subset is indexed as $F_{ij},\ j = 1, 2, \ldots, 2^n$ such that $F_{ij} \subset F_i$ and $\varphi(F_{ij}) = F_j$. Similarly, a subset resulting at the kth iteration is indexed as $F_{i_1 i_2 \ldots i_k}$ such that $F_{i_1 i_2 \ldots i_k} \subset F_{i_1 i_2 \ldots i_{k-1}}$ and $\varphi(F_{i_1 i_2 \ldots i_k}) = F_{i_2 i_3 \ldots i_k}$.

Based on the algorithm, it is clear that the condition (12.5.9) holds. Thus, we describe the points $\mathscr{F}_{i_1 i_2 i_3 \ldots} = \lim_{k\to\infty} F_{i_1 i_2 \ldots i_k}$, and then the DASS, the self-similar set $\mathscr{F}$ corresponding to the above algorithm, is defined as the collection of the points $\mathscr{F}_{i_1 i_2 i_3 \ldots}$.

For $r_i,\ i = 1, 2, \ldots, n$ larger than 4, the separation condition is guaranteed to be valid for the set $\mathscr{F}$, and therefore, Theorem 12.1, 12.2 and 12.3 imply that the similarity map φ is chaotic in the sense of Poincaré, Li–Yorke, and Devaney.

For numerical simulation, let us consider the 2-dimensional system

$$
\begin{aligned}
x_{n+1} &= \varphi_1(x_n) = r_1 x_n(1 - x_n),\\
y_{n+1} &= \varphi_2(y_n) = r_2 y_n(1 - y_n),
\end{aligned}
\tag{12.6.12}
$$

with $r_1 = 4.2$ and $r_2 = 4.3$. We fix $F_0 = F = [0, 1] \times [0, 1]$ and apply ETA for (12.6.12). The first iteration will generate the sets $F_i,\ i = 1, 2, 3, 4$. In the second iteration, we get the sets $F_{ij},\ j = 1, 2, 3, 4$, and so on. Figure 12.7 shows the subsets constructed in the first three iterations. The DASS corresponding to the

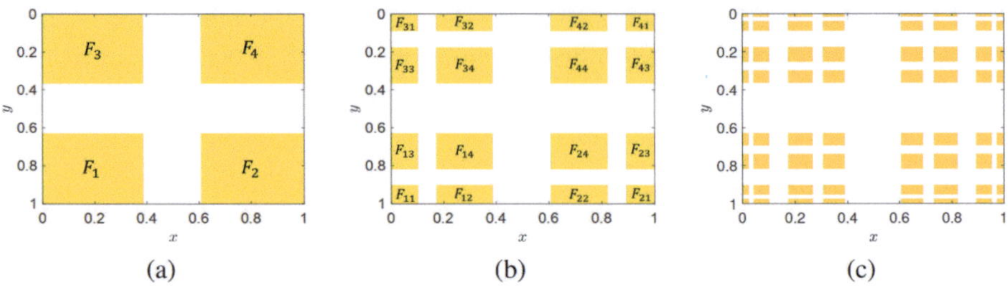

Fig. 12.7 The construction of abstract self-similar set using the map (12.6.12)

system (12.6.12) is the set resulting from an infinite iteration of this procedure. The set is a sort of Cantor dust which is the Cartesian product of two Cantor sets [17].

Example 12.4 Let $F_0 = F$ denote the initial set $[0, 1] \times [0, 1]$ and consider the 2-dimensional perturbed logistic map $\varphi = (\varphi_1, \varphi_2) : \mathbb{R}^2 \to \mathbb{R}^2$ defined by

$$\begin{aligned} x_{n+1} &= \varphi_1(x_n, y_n; \mu_1) = r_1 x_n(1 - x_n) + \mu_1 y_n, \\ y_{n+1} &= \varphi_2(x_n, y_n; \mu_2) = r_1 y_n(1 - y_n) + \mu_2 x_n, \end{aligned} \tag{12.6.13}$$

where r_1, r_2, μ_1, and μ_2 are parameters. The last term in the left-hand side of the both equations in system (12.6.13) can be considered as a perturbation of a unimodal map. It is known that, if a unimodal map is regular [15, 25], then it is structurally stable, and therefore, any small perturbation does not affect the topological properties of the map [9]. Such an inference can be extended for high-dimensional unimodal maps. Numerical simulations can provide an adequate verification of the unimodal properties of the perturbed map.

Similar to Example 12.3, we apply ETA to the map (12.6.13). The points that do not escape F_0 in the first iteration belong to the subsets $F_i,\ i = 1, 2, 3, 4$. In the second iteration, the resulting points belong to the subsets indexed by $F_{ij},\ j = 1, 2, 3, 4$ such that $F_{ij} \subset F_i$ and $\varphi(F_{ij}) = F_j$. Similarly, a subset resulting at the nth iteration is indexed as $F_{i_1 i_2 \ldots i_n}$ such that $F_{i_1 i_2 \ldots i_n} \subset F_{i_1 i_2 \ldots i_{n-1}}$ and $\varphi(F_{i_1 i_2 \ldots i_n}) = F_{i_2 i_3 \ldots i_n}$. Figure 12.8 shows the subsets constructed in the first three iterations with the parameters $r_1 = 4.2$, $r_2 = 4.5$, $\mu_1 = 0.03$, and $\mu_2 = -0.05$.

Depending on the choice of the coefficients r_i and relying on the smallness of the coefficients μ_i, we have that the diameter and separation conditions for abstract similarity and chaos are fulfilled. Moreover, the simulation results confirm that both conditions hold. Therefore, we could say that the similarity map (12.6.13) is chaotic on the self-similar set $\mathscr{F}$. Figure 12.9 depicts the trajectories of some points that approximately belong to the set $\mathscr{F}$. The irregular behavior of the trajectories reveals the presence of chaos in (12.6.13).

Example 12.5 ([4]) Consider the space $X = \mathbb{R}^n$ and let F be a compact set in X such that it contains an open neighborhood of the n-dimensional unit cube and

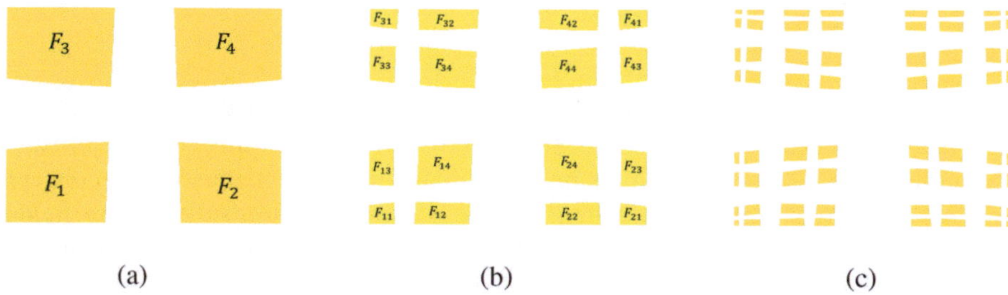

Fig. 12.8 The first three iterations of DASS construction using the map (12.6.13)

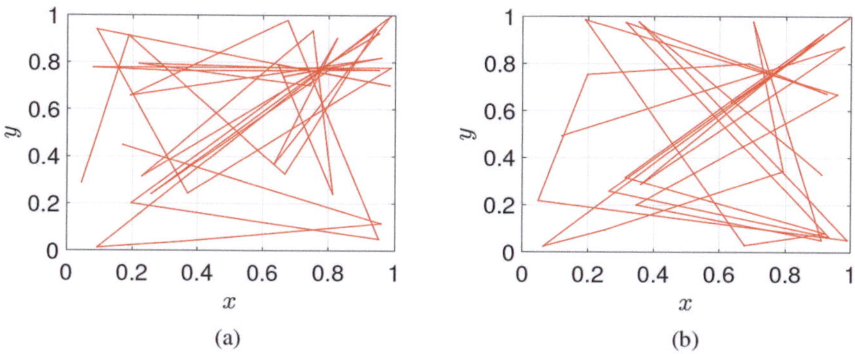

Fig. 12.9 The two trajectories of the system (12.6.13). (**a**) The trajectory which starts at the point (0.044608921784357, 0.287657531506301). (**b**) The trajectory which starts at the point (0.910182036407281, 0.329865973194639)

the set F_0 to be sufficiently near to the cube. Consider the n-dimensional perturbed logistic map $\varphi = (\varphi_1, \varphi_2, \dots \varphi_n) : \mathbb{R}^n \to \mathbb{R}^n$ which is defined by

$$\begin{aligned} x_{k+1}^1 &= \varphi_1(x_k^1, x_k^2, \dots, x_k^n; \mu_1) = r_1 x_k^1(1 - x_k^1) + \mu_1 \chi_1(x_k^1, x_k^2, \dots, x_k^n), \\ x_{k+1}^2 &= \varphi_2(x_k^1, x_k^2, \dots, x_k^n; \mu_2) = r_2 x_k^2(1 - x_k^2) + \mu_2 \chi_2(x_k^1, x_k^2, \dots, x_k^n), \\ &\vdots \\ x_{k+1}^n &= \varphi_n(x_k^1, x_k^2, \dots, x_k^n; \mu_n) = r_n x_k^n(1 - x_k^n) + \mu_n \chi_n(x_k^1, x_k^2, \dots, x_k^n), \end{aligned} \tag{12.6.14}$$

where μ_i, $i = 1, 2, \dots, n$ are parameters and $\chi = (\chi_1, \chi_2, \dots, \chi_n; \mu_i)$ is a continuous function. Due to the continuity of χ, if $r_i > 4$ for each $i = 1, 2, \dots, n$ and μ_i, $i = 1, 2, \dots, n$ are sufficiently small in absolute value, then one can show that a DASS can be constructed using (12.6.14), and thus, chaotic behavior in the sense of Poincaré, Li–Yorke, and Devaney for the n-dimensional perturbed logistic map (12.6.14) would be expected to appear.

12.7 Notes

The abstraction of the self-similarity concept is now accomplished. Furthermore, we have shown that the set of symbolic strings satisfies the definition of abstract self-similarity. This example illustrates how the abstraction of a mathematical concept can be significant not only to extract its essence but also to explore more fields where it can be manifested. In addition to equipping the self-similar set with a metric, the similarity map is introduced to define abstract self-similar space. The map is proven to be chaotic in the sense of Poincaré, Li–Yorke, and Devaney. The building of chaos usually begins with a map defined over its domain and then saying about a chaotic attractor that appears as a part of the domain. In our research, we start by describing a chaotic set, and only then introduce a similarity map which admits chaotic dynamics. We utilize infinite sequences to index the points of the domain. The action of the map is not just a shifting in the string space as much as a transforming of the domain points.

Self-similarity is widely spread in nature, but it is usually associated with fractal geometry [19, 21]. Proceeding from this point, we have shown that the Sierpinski fractals, Koch curve, and Cantor set can be associated with abstract self-similarity, and consequently possess chaos. This covers already known fractals constructed through self-similarity and possibly other fractals that generated by FJI such as Julia and Mandelbrot sets. The suggested abstract similarity definition can be elaborated through fractal sets defined by fractal dimension, chaotic dynamics development, topological spaces, physics, chemistry, and neural network theories development [11, 14, 22, 24].

The examples of chaotic fractals [4], domain-structured chaos [3], and multidimensional chaotic cubes [5] demonstrate that the concept of the abstract self-similarity can be useful to prove that chaos is a generic property for the world. At least our results show that chaotic dynamics is more extended in the reality than we can imagine.

References

1. T. Addabbo, A. Fort, S. Rocchi, V. Vignoli, Digitized chaos for pseudo-random number generation in cryptography, in *Chaos-Based Cryptography. Studies in Computational Intelligence*, vol. 354, ed. by L. Kocarev, S. Lian (Springer, Berlin, Heidelberg, 2011)
2. M. Akhmet, E.M. Alejaily, Abstract Similarity, Fractals and Chaos. ArXiv e-prints, arXiv:1905.02198, 2019 (submitted)
3. M. Akhmet, E.M. Alejaily, Domain-structured chaos in a Hopfield neural network. Int. J. Bifurc. Chaos, 2019 (in press)
4. M. Akhmet, E.M. Alejaily, Abstract fractals. Discontinuity Nonlinearity Complexity (in press) (arXiv: 1908.04273)
5. M. Akhmet, E.M. Alejaily, Chaos on the Multi-Dimensional Cube. ArXiv e-prints, arXiv:1908.11194, 2019 (submitted)

6. M.U. Akhmet, M.O. Fen, Replication of chaos. Commun. Nonlinear Sci. Numer. Simul. **18**, 2626–2666 (2013)
7. M. Akhmet, M.O. Fen, Unpredictable points and chaos. Commun. Nonlinear Sci. Numer. Simul. **40**, 1–5 (2016)
8. M. Akhmet, M.O. Fen, Poincaré chaos and unpredictable functions. Commun. Nonlinear Sci. Numer. Simul. **48**, 85–94 (2017)
9. A. Avila, C.G. Moreira, Bifurcations of unimodal maps, in *Dynamical systems, Part II. Pubbl. Cent. Ric. Mat. Ennio Giorgi, Scuola Norm. Sup.*, 2003, pp. 1–22
10. M.F. Barnsley, *Fractals Everywhere* (Academic Press, London, 1988)
11. G. Boeing, Visual analysis of nonlinear dynamical systems: Chaos, fractals, self-similarity and the limits of prediction. Systems **4**, 1–18 (2016)
12. G. Chen, Y. Huang, *Chaotic Maps: Dynamics, Fractals and Rapid Fluctuations, Synthesis Lectures on Mathematics and Statistics* (Morgan and Claypool Publishers, Texas, 2011)
13. R.L. Devaney, *An Introduction to Chaotic Dynamical Systems* (Addison-Wesley, USA, 1989)
14. M. Hata, Topological aspects of self-similar sets and singular functions, in *Fractal Geometry and Analysis*, ed. by J. Belair, S. Dubuc (Kluwer Academic Publishers, Dordrecht, 1991), pp. 255–276
15. B.R. Hunt, V.Y. Kaloshin, Prevalence, in *Handbook of Dynamical Systems*, vol. 3, ed. by H. Broer, F. Takens, B. Hasselblatt (Elsevier Science, Amsterdam, 2010), pp. 43–87
16. P.E.T. Jorgensen, *Analysis and Probability: Wavelets, Signals, Fractals, Graduate Texts in Mathematics*, vol. 234 (Springer, New York, 2006)
17. G.C. Layek, *An Introduction to Dynamical Systems and Chaos* (Springer, India, 2015)
18. R. Li, X. Zhou, A note on chaos in product maps. Turk. J. Math. **37**, 665–675 (2013)
19. B.B. Mandelbrot, *The Fractal Geometry of Nature* (Freeman, New York, 1983)
20. C. Masoller, A.C. Schifino, R.L. Sicardi, Characterization of strange attractors of Lorenz model of general circulation of the atmosphere. Chaos Solitons Fractals **6**, 357–366 (1995)
21. H-O. Peitgen, H. Jürgens, D. Saupe, *Chaos and Fractals* (Springer, New York, 2004)
22. S. Sato, K. Gohara, Fractal transition in continuous recurrent neural networks. Int. J. Bifurc. Chaos. **11**, 421–434 (2001)
23. S. Wiggins, *Global Bifurcation and Chaos: Analytical Methods* (Springer, New York, Berlin, 1988)
24. G.M. Zaslavsky, M. Edelman, B.A. Niyazov, Self-similarity, renormalization, and phase space nonuniformity of Hamiltonian chaotic dynamics. Chaos **7**, 159–181 (1997)
25. E. Zeraoulia, J.C. Sprott, *Robust Chaos and Its Applications* (World Scientific, Singapore, 2012)

Index

M. Akhmet et al., *Dynamics with Chaos and Fractals*, Nonlinear Systems and Complexity 29, https://doi.org/10.1007/978-3-030-35854-9

MIX
Papier aus verantwortungsvollen Quellen
Paper from responsible sources
FSC® C105338

If you have any concerns about our products,
you can contact us on
ProductSafety@springernature.com

In case Publisher is established outside the EU,
the EU authorized representative is:
Springer Nature Customer Service Center GmbH
Europaplatz 3, 69115 Heidelberg, Germany

Printed by Libri Plureos GmbH
in Hamburg, Germany